Climate Change and Rice

Springer
Berlin
Heidelberg
New York
Barcelona
Budapest
Hong Kong
London
Milan
Paris
Santa Clara
Singapore
Tokyo

S. Peng K.T. Ingram
H.-U. Neue L.H. Ziska (Eds.)

Climate Change and Rice

With a Foreword by Klaus J. Lampe

With 102 Figures and 67 Tables

Springer

IRRI
INTERNATIONAL RICE RESEARCH INSTITUTE

Dr. Shaobing Peng
Dr. Keith T. Ingram
Dr. Heinz-Ulrich Neue
Dr. Lewis H. Ziska

International Rice Research Institute (IRRI)
P.O. Box 933
1099 Manila
Philippines

ISBN-13:978-3-642-85195-7 e-ISBN-13:978-3-642-85193-3
DOI: 10.1007/978-3-642-85193-3

Library of Congress Cataloging-in-Publication Data. Climate change and rice/[edited by] S. Peng . . . [et al.]: with a foreword by Klaus Lampe. p. cm. Papers from a symposium held Mar. 1994. Includes bibliographical references (p.) and index.ISBN-13:978-3-642-85195-7 1. Rice–Climatic factors – Congresses. 2. Climatic changes – Congresses. 3. Crops and climate – Congresses. I. Peng, S. (Shaobing), 1962– . SB191.R5C635 1995 584'.93–dc20 95–12491

Softcover reprint of the hardcover 1st edition 1995

Cover design: Springer-Verlag, Design & Production

Typesetting: Thomson Press (India) Ltd., New Delhi

SPIN: 10493742 31/3137/SPS – 5 4 3 2 1 0 – Printed on acid-free paper

Foreword

Among basic human needs, food is foremost. Finding sufficient, affordable food is still the major concern of one in every five persons on earth. Rice is the basic staple for more than half of the world – and will continue to be well into the future. For this reason, the International Rice Research Institute (IRRI) has set the goal of improving the well-being of present and future generations of rice farmers and consumers, particularly those with low incomes.

As we look toward the 21st century, several important issues cloud food security. World population is growing by almost 100 million people per year, and it is not likely to plateau before the end of the next century. This means that food demand will at least double, and may quadruple. The technologies of the Green Revolution have provided an unprecedented leap in annual rice production, enough to feed 600 million more people, but have had unforeseen social and environmental costs. Loss of genetic diversity and misuse of pesticides endanger the sustainability of the Green Revolution. Soil erosion, rural to urban population migration, and increased competition for land and water between urban and agricultural sectors further threaten the sustainability of rice production. Finally, human activities may be changing the global climate. Any change in global climate is a potential challenge to the food security of future generations. The world cannot afford to be unprepared.

As yet, there is no definitive answer to the causes and effects of the thinning of the stratospheric ozone layer, or to the possibility of global warming. There is strong evidence that levels of atmospheric carbon dioxide and other greenhouse gases are on the rise. Rates of global climate change and their interactions with agricultural production, however, are less clear. The best current estimates are that rice cropping contributes 15–20% of the total methane emissions, but there are no reliable estimates of nitrous oxide emissions. It is possible that the rice fields of Asia, particularly the flooded fields, are the modern global sinks for carbon and nitrogen. An understanding of the carbon and nitrogen balance is fundamental, not only to the global environment, but also to the productivity and sustainability of intensively cropped irrigated rice systems.

Although we cannot predict future climates, we can prepare for probable changes. In September 1990, IRRI and the United States Environmental Protection Agency began a cooperative program to examine two aspects of global climate change: the effects of likely changes in global climate on the rice plant and its production, and the effects of flooded rice cropping on gases that may contribute to global warming. This book summarizes our current understanding of these issues.

IRRI is grateful for the financial support, through cooperative agreement 817426 with the US Environmental Protection Agency, for a symposium on climate change and rice held in March 1994.

KLAUS J. LAMPE
Director General
International Rice Research Institute

Preface

Climate change is not a new phenomenon. Change has been a consistent feature of global climate. For the past 10 000 years, however, the earth has experienced the longest period of consistently warm temperatures since the beginning of life. This warm period almost exactly matches the period over which modern agriculture has evolved.

The reason that current predictions of climate change are compelling is that, for the first time, climate may be changing as a direct result of human activity. People have released chlorofluorocarbons into the atmosphere – leading to the degradation of stratospheric ozone and increasing biologically harmful ultraviolet-B radiation. Through mining and combustion of fossil fuels, deforestation, maintenance of livestock herds, and even through rice cultivation, enormous quantities of carbon dioxide, methane, and other greenhouse gases have been released to the atmosphere. Global circulation models predict that these greenhouse gases will increase global average temperatures by 1–8 °C over the next century. The outputs from these models are only predictions. We do not yet have any firm evidence to prove that global warming will occur.

In contrast to predictions of mass disruptions to crop production and their resulting adverse social and environmental impacts, some components of global change may benefit agriculture. Carbon dioxide is a substrate for photosynthesis. As atmospheric carbon dioxide levels increase, photosynthesis rates and crop productivity may increase. Similarly, global warming may increase the duration of the cropping season in temperature and higher latitudes.

We are, in essence, conducting a real-time, unreplicated, global experiment. In this experiment, humans play the roles of both experimenter and subject. As experimenters we affect several independent variables, including atmospheric greenhouse gas concentrations and stratospheric ozone thickness. As subjects, we respond to the impacts of climate change on food production and human health. Unfortunately, we do not have precise control over the independent variables. Other biological and physical systems also affect the concentrations of atmospheric gases, both greenhouse gases and those that catalyze the destruction of the stratospheric ozone layer.

Rice, especially the flooded rice ecosystem, plays a major role in this global experiment. Effects of global change on rice productivity will influence global food security. As a source of atmospheric methane and nitrous oxides, and as a potential sink for carbon, rice cultivation is a system by which human activity may affect climate change.

Recently, there has been an increasing amount of research to improve our understanding of the interactions of climate change with rice cultivation, and on options to mitigate adverse effects while optimizing benefits of climate change for rice production. This book summarizes some of that research.

The book hasfive sections. First, we introduce the issues of global change and their relation to global food security and environmental sustainability. The next three sections address emissions of greenhouse gases from rice fields, and major areas of climate change that impact on rice, namely, ultraviolet-B radiation effects and interactive effects of carbon dioxide and temperature. The last section discusses prediction of effects of climate change using simulation models.

Under even the most pessimistic future scenario for ozone depletion, field research suggests that increasing ultraviolet-B flux will probably not directly affect rice productivity. There are still concerns that ultraviolet-B radiation will affect rice pest and disease organisms, and thereby indirectly affect rice productivity. Ultraviolet-B radiation may also cause genetic mutations, or may affect secondary metabolites such as flavonoids that could change rice-grain quality. Although these important issues remain unresolved, the current attitude toward ultraviolet-B radiation effects on rice is one of guarded optimism.

The growth response of rice to carbon dioxide is not linear. Growth increases only with carbon dioxide concentrations up to 500 $\mu l/l$. Fortunately, rice germplasm varies in growth response to carbon dioxide and this may provide a way to improve growth and productivity responses to carbon dioxide levels above 500 $\mu l/l$. The reason for this lack of response is not clear. Sink limitations, acclimation of Rubisco, plant nutrition, developmental responses, and carbon allocation processes must be investigated for their possible involvement in this phenomenon.

The situation is less sanguine for temperature and its interactions with carbon dioxide. Higher temperatures, especially in tropical areas that are already near or above the optimum temperature range for rice, will reduce growth and yields. Higher temperatures reduce productivity because of the combined effects of reduced vegetative growth and spikelet sterility or abortion caused by heat injury during panicle emergence. Global warming, if it occurs, is likely to negate any potential benefits from increased carbon dioxide. A few experiments suggest that temperature and carbon dioxide have interactive effects that will mean that higher temperatures with high carbon dioxide concentrations will be more detrimental than high temperatures alone.

Flooded ricefields emit significant amounts of methane. The current estimate of global emissions of methane from rice is 20–100 Tg/year. This range is far too broad. Further comparisons and complementary applications of micrometeorological and closed-chamber measurements, and observations in nonirrigated rice ecosystems can help refine estimates of methane emissions that are scaled up from the field to the global level.

Tactics to reduce methane emissions from rice fields include changes in rice variety and land use, reduced disturbance of soils, and improved timing of irrigation and fertilizer amendments. Unfortunately, many options to mitigate methane emissions enhance nitrous oxide emissions or reduce productivity. When adjusted for residence time in the atmosphere and radiative trapping efficiency, nitrous oxides may be more important than methane. Any strategy to reduce methane emissions must be judged with respect to costs, especially in light of anticipated increases in food demand.

The book editors wish to acknowledge the valuable technical editing assistance of Michael Graham and L. Reginald MacIntyre.

THE EDITORS

Contents

Contributors

You will find the addresses at the beginning of the respective contributions.

Issues of Global Climate Change

Global Change: Problems and Uncertainties

R.G. PRINN

The current evidence for growth in the concentration of carbon dioxide and certain other atmospheric gases, the simultaneous evidence for an increase in the average temperature of the earth, and climate model results that predict a significant global warming over the next century have made global warming perhaps the most talked about manifestation of worldwide human influence on the environment.

The two major greenhouse substances, water vapour, and water and ice clouds, collectively are responsible for over 90% of the greenhouse effect that keeps the earth about 33 °C warmer than it would be in the absence of this effect. These substances have short lifetimes and, therefore, abundances that adjust fairly quickly to the longer-term climate forcing represented by changes in the so-called long-lived greenhouse gases. The latter gases include carbon dioxide, methane, nitrous oxide, chlorofluorocarbons, and ozone. These gases, although responsible for only a small percentage of the direct greenhouse warming, have a great indirect influence. The increase in temperatures that results from their presence leads to increases in water vapour and clouds, which, in net, accelerate the warming. These water-related feedback processes are very important but not quantitatively well understood and lead to much of the uncertainty in current climate models.

Past and Present Atmospheric Changes

The detailed glacial record in Greenland and Antarctica indicates clearly that the climate of the earth has changed remarkably in the past and that concurrent changes in greenhouse gases certainly accelerated, and may even have helped instigate, the climate changes. Concern today is for the long-lived greenhouse gases because their levels are currently rising at rates that, if continued into the future, are predicted to cause significant global warming (Houghton et al. 1992; Prinn 1994).

Center for Global Change Science, Massachusetts Institute of Technology, Building 54-1312, Cambridge, MA 02139, USA

S. Peng et al. (Eds.) Climate Change and Rice

Carbon dioxide, increasing at 0.3–0.4% per year, is added to the air through the combustion of fossil fuels and deforestation, and removed through net uptake by land ecosystems and the oceans. Methane, rising on the average over the past 20 years at 0.8% per year, is produced, in poorly understood proportions, by the gas and coal industries, rice fields, cattle, and naturally in all wetlands. It is destroyed by chemical reactions that are driven by ultraviolet sunlight and involve the very reactive hydroxyl radical (OH). Methane was rising at 1.2% per year in the late 1970s but is now increasing at less than 0.3% per year. Either global methane emissions are decreasing or the methane destruction rate is increasing (or both).

Nitrous oxide, increasing at 0.2–0.3% per year, has, like methane, a wide range of poorly understood natural and anthropogenic regional sources. It is removed in the ozone layer, as are the purely industrial chlorofluorocarbons (CFCs). The CFCs were increasing at 5% per year in the 1980s but they are now increasing at less than 3% per year as a result of the Montreal Protocol. It is important to note that the potency of both nitrous oxide and the CFCs as greenhouse gases is offset significantly by the stratospheric ozone that they destroy.

Ozone, a key environmental chemical and protective ultraviolet shield in the stratosphere, is also an important greenhouse gas. The common urban air pollutants nitric oxide, nitrogen dioxide, hydrocarbons heavier than methane, and carbon monoxide, although not directly important as greenhouse gases, play a key role through their influence on the tropospheric ozone, hydroxyl radical, and methane concentrations. Gaseous sulfur compounds, both natural and from coal burning, are efficiently oxidized to sulfate particles that can be very important contributors to reflection of sunlight and can somewhat offset the influence of the greenhouse gases.

Climate Change Predictions and Uncertainty

Predictions of increases in global average temperature with rising greenhouse gases are a common conclusion from climate models. However, these predictions are not useful in answering a very critical societal question: what is the impact on humanity of climate change? The answer to this question instead requires predictions of important regional climate variables such as those associated with water availability for agriculture and natural ecosystems. The current leading climate models lead, unfortunately, to significantly different predictions for regional climate change over the globe. For example, for the important agricultural state of Indiana in the United States, two leading models predict increased winter dryness; whereas another two leading models predict increased wetness as we go into the next century (Stone 1992).

Current models of global climate are uncertain tools for climate prediction because they do not presently include in a fully realistic way some of the fundamental recognized processes that determine climate (Prinn and Hartley 1992). These processes include cloud formation and convection, oceanic circulation and its interaction with the atmosphere, land hydrology and vegetation coupling, and the chemical and biological processes involved in production and removal of greenhouse substances and reflective aerosols. It is not surprising that the ability of these models to predict regional climate change is presently quite poor.

Although most current climate models predict a steady warming because of increases in greenhouse gases over the next century, the real climate system may not respond so evenly. The thermohaline circulation of the ocean serves in the current climate both to maintain higher-than-normal temperatures at high northern latitudes and to move carbon dioxide from the surface into the deep ocean (the ocean sink for carbon). The circulation is delicate and could be halted by as simple a climate change as increased rainfall over the North Atlantic. Such a halt could lead to a significant cooling in northern Europe and to a deceleration of the oceanic uptake of carbon dioxide. Volcanoes add another unpredictable element to climate. Two recent volcanic eruptions, El Chichon and Pinatubo, produced such large amounts of atmospheric particles reflective to sunlight that they cooled the earth by 0.2–0.4 °C over a period of 1–2 years after the eruptions.

It is expected that advances in oceanic, atmospheric, and biospheric observations, computer capabilities, and theory over the next decade will improve significantly the ability to predict climate change, but improvements are not likely to come much quicker than that. Policy decisions will need to be made in the face of uncertain (yet hopefully improving) climate science. A major policy question is whether to take action now to curb emissions of greenhouse gases from our energy and agricultural industries or to wait perhaps a decade or so until the climate predictions become more accurate or until the observations show clear evidence for warming. This policy question is complicated by at least three facts: most of the greenhouse gases of interest have atmospheric lifetimes of decades to centuries, the time scales inherent in the evolution of energy technology are at least decadal, and the impacts of climate change are predicted to fall not on the current generation but two or three generations into the future.

Anthropogenic Processes and Global Change

Two anthropogenic activities, combustion and land use, play major roles as sources of climatically important gases (or their precursors). Combustion of fossil, biomass, and detrital fuels are well recognized sources of CO_2, SO_2, NO, NO_2, and hydrocarbons. Assessing the current and predicting the future emissions of these gases from these sources is a major challenge. The inventory of current emissions is already underway. Land use and land-use change are dominating factors in the CH_4 budget (through rice and cattle production and land fills) and the N_2O budget (through fertilization).

One specific example of the effects of both land use and combustion is provided by biomass burning. Because of this burning, the influence of the short-lived pollutants from combustion is not seen just in the industrialized northern hemispheric midlatitudes. Tropospheric column ozone maps constructed by subtracting the stratospheric from the total atmospheric ozone columns as measured from satellites gave the first indication of very high ozone levels in the September to November period in the tropical central Atlantic. The only serious candidate for the phenomenon is air advected off Africa and South America and containing copious ozone and/or its precursors (e.g., NO_x and CO). Recent research by a large team of scientists indicates that the precursors are produced by biomass burning associated with land clearing and nomadic agriculture. Studies of the distribution of fires over the globe show clearly a high frequency of fires in the tropics. Also, the fires are quite seasonal. They occur in the dry season following the wet growth season. It is now clear that biomass burning, in addition to being very important to the land system itself, actually drives a significant portion of tropical atmospheric chemistry. The gases from the fires are the very ingredients needed to pump up ozone and OH levels. Current research on biomass burning tends to support a picture of seasonal burning that pumps chemicals into the air in many tropical areas. These chemicals appear to increase the oxidizing capability of the atmosphere (Andreae et al. 1994).

Obtaining Answers

The need to understand better the complex natural and anthropogenic processes that lead to global change has led to the development of an internationally integrated plan for the scientific study of the global environment. The goal is to determine the processes that govern its present state and to predict how this state might change because of human and natural forcing. The international plan is embodied in the sister programs of the World Climate Research Program (WCRP), the International Geosphere-Biosphere Program (IGBP), and the Human Dimensions of Global Enviornmental Change Program (HDP).

The major goal of IGBP is to describe and understand the interactive physical, chemical, and biological processes that regulate the total earth system, the unique environment that it provides for life, the changes that are occurring in this system, and the manner in which these changes are influenced by human actions. Examples of major IGBP projects are the International Global Atmospheric Chemistry (IGAC) Project, the Joint Global Ocean Flux Study (JGOFS), and the Global Change and Terrestrial Ecosystems (GCTE) Project.

The major goal of WCRP is to determine to what extent climate can be predicted and the extent of human influence on climate. Examples of major WCRP projects are the Tropical Oceans and Global Atmosphere (TOGA) project, the World Ocean Circulation Experiment (WOCE), and the Global Energy and Water Cycle Experiment (GEWEX).

The objectives of HDP include understanding human interactions with the earth system, studying social change affecting the global environment, identifying mitigation, prevention and adaption strategies, and analyzing policy options in response to global change.

Two IGBP projects have a focus on rice agriculture. IGAC in its Rice Cultivation and Trace Gas Exchange (RICE) Activity, led by H.-U. Neue and R. Sass, seeks to determine if rice cultivation is an important cause for changes in the atmospheric concentration of methane and other trace gases, to understand the causes of the regional and temporal variability in methane emission from rice cultivation, and to identify management practices that can stabilize or reduce emission of trace gases as rice production is increased in the future to meet the demands of an expanding world population. This task is underway, but there is much work to be done (Neue and Sass 1994).

GCTE, in its Effects of Global Change on Key Agronomic Species Activity led by J. Goudriaan, seeks to determine how anticipated global changes in rainfall and temperature patterns together with increases in atmospheric CO_2 and UV-B will affect the production of crops, including rice, throughout the world (Steffen et al. 1992).

Food Security in an Uncertain Climate

P. Usher

At the dawn of history, simple precepts were evolved to regiment society and adapt to the changing seasons. People lived by the weather and organized their agricultural activities round it; but their ability to predict events was extremely limited. So limited, in fact, that they made gods of natural events, and created mythologies for whatever natural disasters befell them or their crops.

Today, we rely on science, not myth, to understand droughts and loss of soil fertility. Moreover, in recent years we have developed to a much greater degree the capacity to predict events. It is still a science in its infancy, but it has already told us of dangers that loom for humanity.

Over the last 20 years, more people have starved to death or died of hunger-related diseases than ever before. In that time, land degradation has expanded by some 120 million ha. Each year the world loses an estimated 26 billion tonnes of topsoil, about the same amount as covers the entire area devoted to wheat in Australia. Today, over two billion people depend directly on agriculture for their livelihood – in developing countries it is the driving force for economic and social development. Millions of hectares of agricultural land become unproductive each year because of deforestation and soil erosion. As a result, 950 million adults and children are undernourished – 50,000 die each day, and each minute 20 children die because of hunger.

The overall impact of global warming on food production is still uncertain, but it is clear that climate change poses a threat not only to global food security, but to the stability of societies across the entire globe.

When farmers began harvesting the first crops about 8000 B.C., the world population was about 4 million. Today, every 10 days, that same number of people is born. If this trend continues, the world will have to grow as much food in the first two decades of the next century as was produced over the past 10000 years.

The pressure on land to produce more food will increase, as will the demand for energy. Traditional farming has had to change to satisfy this demand. There is a reliance on new super crops, high-yielding, water-and energy-efficient, single varieties; but there is a risk – a shrinking biodiversity base upon which the added

Contribution on behalf of Ms Elizabeth Dowdeswell, Executive Director, United Nations Environment Programme, P.O. Box 30552, Nairobi, Kenya.

S. Peng et al. (Eds.) Climate Change and Rice

effects of changing climate is unknown. In Sri Lanka, where farmers grew some 2000 traditional varieties as recently as 30 years ago, only five varieties are grown today. In India, which once had 30000 varieties of rice, three-quarters of the total production comes from fewer than 10 varieties. We are losing wild strains of rice everywhere. Without IRRI, which plays a crucial role in protecting biodiversity through its gene bank of over 80000 varieties of rice from all over the world, the situation would be critical.

The World Watch Institute in its most recent *State of the World* report has identified a 3-month doubling of rice prices and a 20-year low in rice stocks as indications that the biological limits of the planet may have been reached. Whether or not this reflects a global trend or merely reflects a third year of aberrant climate associated with El Niño is incidental to the fact that the limit of long-term agricultural sustainability is being reached, and care and caution are needed to avoid overstepping that boundary.

Global warming must be factored into the future food equation or agriculture will suffer. One consequence of global warming could be that climate zones may shift over hundreds of kilometers of latitude, and over hundreds of meters in altitude. This shift might allow farmers to grow wheat in parts of the United States, Canada, and Europe that are currently too cold for cultivation. In some areas, the rise in carbon dioxide levels could lead to increased fertilization and to increase in crop yields of between 10 and 80%.

However, the negative consequences outweigh the positive ones. In this greenhouse planet of the 21st century, broad swathes of fertile land in semiarid, subtropical, and mediterranean climates could be transformed into barren dustbowls. A 200-km shift southward of the Sahara would leave Sahelian countries, which have been in the grip of drought and famine for the past two decades, permanently unable to feed 30% of their populations.

Because of disruptions in rainfall patterns around the world, winters in temperate zones could be even wetter. Meanwhile the dry subtropics could become drier still. Many areas of land will simply disappear. By the middle of the next century, without expensive protection, the oceans may swallow up large tracts of low-lying arable land in the fertile deltas of China, Indonesia, Burma, Nigeria, and South America, and create millions of environmental refugees.

These predictions do not need to be 100% accurate. They need merely to be sufficient for us to change our attitudes. They need merely to illustrate the potential cost of inaction. Ultimately, science tells us that climate change is a global concern that affects all of us. It is a concern that must be tackled by the world community cooperatively – developed and developing nations must act in concert.

There is already plenty of good news. The international community has begun to act. In 1988, WMO and UNEP jointly established the Intergovernmental Panel on Climate Change, and assigned it the task of making regular assessments of the various aspects of the climate-change issue. One direct result of the heightened concern was the signature of the United Nations Framework Convention on Climate Change by 166 governments, and the subsequent deposit of the 50th instrument of ratification of the treaty allowing it to enter into force on 21 March 1994. The Convention defines objectives that provide a focus for everyone in the business, whether it be monitoring, assessing, doing research, policymaking, or negotiating. It was a process in which UNEP played a vital part – a part recognized subsequently by the document called Agenda 21, adopted by the Rio de Janeiro Summit as a blueprint for sustainable development.

In fact, Agenda 21 marks a milestone in efforts to tackle the interconnected environmental and development issues epitomized by climate change and agricultlure. It calls for stronger ties between UNEP and associated agencies, especially UNDP, and recognizes that UNEP is particularly well positioned to bring about the convergence of sound science, informed social consensus, and coherent, and timely policies. Acting as a catalyst, UNEP is able to create bridges between the various sectors of the environment-development nexus, connections between global warming and agriculture, global warming and biodiversity, and global warming and the potential changes in our oceans.

However, there is still much more to be done, especially at the national level. This applies most of all to the developed countries. In Agenda 21, the governments of the world acknowledged that developed countries, having put more pressure on the world environment, bear some responsibility for putting it right. Many have already agreed to roll back the emissions of carbon dioxide and other greenhouse gases by the end of the decade. They need to live up to those agreements. If the battle against climate change and its aftermath must be fought today in the North, it will be fought in coming years in the global South, in the developing countries. There is need, on the one hand, to stem the emergence of vast new greenhouse gas-producing constituencies, and on the other to provide effective mechanisms to ensure continued food security. In this, the developed countries of the North must bear the responsibility and the costs by making available environmentally effective technology and by absorbing the incremental costs involved.

Fundamentally, this means building the capacity of national governments in those countries so that they can take advantage of new productive capabilities. They must have an effective infrastructure, including national climate programs, that allows the interaction between the physical, natural, and political science communities to jointly influence government action and national agricultural policy. There is a need for effective environmental education and training of all levels of the community. This spirit of cooperation has been captured in a traditional Asian saying *you cannot pick up a grain of rice with one finger alone*.

We must act now and set concrete plans in motion. Already, we know well enough the types of weather-induced disruption to the food supply that could arise for future generations. The key word is preparedness. Many of the problems of Africa could disappear if there were effective early warning systems for droughts and effective national drought preparedness mechanisms. Current response to drought is donor relief often after thousands are dead and millions of others are too malnourished to do more than survive.

We must not take previous bounties for granted; instead we must be prepared for an uncertain future. Scientific discovery and its application can guide us toward effective policies for global food security.

Emission of Greenhouse Gases

Measuring Fluxes of Greenhouse Gases Between Rice Fields and the Atmosphere

O.T. DENMEAD

The theoretical and operational aspects of chamber and micrometeorological methods for measuring fluxes of the greenhouse gases CO_2, CH_4, and N_2O between water, land, or vegetative surfaces and the atmosphere are reviewed in this chapter, with an emphasis on techniques suitable for the smaller, but variably sized fields used in rice production. More extensive general reviews of trace-gas flux measurement are given by Denmead and Raupach (1993), Hutchinson and Livingston (1993), Lenschow (1993), and Denmead (1994).

Exchange of Greenhouse Gases in Rice Cultivation

Although the gains and losses of atmospheric CO_2 by cropping systems are generally considered to be nearly in balance, there is some evidence that rice soils might sequester carbon from the atmosphere over many years of cultivation (Cassman et al. 1994). It is doubtful if this could be confirmed by direct measurements of the flux of CO_2 between the atmosphere and rice fields, but there are other reasons for making such measurements. One is to explore the links between CO_2 uptake by rice plants, photosynthesis, and the production of carbonaceous root exudates from which methane is generated. Another might be to establish baseline values of CO_2 uptake by rice crops against which to evalute the effects of future climate change.

Both CH_4 and N_2O are produced in soils under anaerobic conditions by numerous microbial processes. Flooded rice fields, which constitute about 9% of all cultivated land, are thought to be significant sources of trace gases, especially CH_4 (Rennenberg et al. 1992). Recent estimates place the annual emission of CH_4 from rice cultivation at between 20 and 100 Tg CH_4, or 4–20% of the total global sources (Watson et al. 1992). Much of the uncertainty in these estimates exists because most of our information comes from a limited number of chamber investigations in which the variation between locations, soils, and cropping systems is large. This chapter describes complementary micrometeorological

CSIRO Centre for Environmental Mechanics, GPO Box 821, Canberra, ACT 2601, Australia

S. Peng et al. (Eds.) Climate Change and Rice

approaches that do not disturb the natural environment or the emission process as chambers can do, and that integrate over larger areas. As the number of published reports of CH_4 production from rice cultivation has increased, estimates of the size of the rice source have decreased. However, the level of uncertainty has remained at about two; the true production could be twice as much or half as much. By removing small-scale variability, micrometeorological flux measurements should improve our estimate of source size and reduce the uncertainty factor.

Data for N_2O emissions from rice fields are even fewer than for CH_4. Most of the N_2O production is believed to result from denitrification of nitrate derived from N fertilizers. Estimates of the production of N_2O-N from rice fields range from 0.01 to 1% of the fertilizer N applied (Rennenberg et al. 1992; Bronson and Singh 1994). Emissions of N_2O from rice fields seem to be more important during fallow periods, no matter how brief (Bronson and Singh 1994). There are insufficient data to estimate the global production of N_2O in rice cultivation. The information that is available for CH_4 comes from chamber measurements, and there is a need for methods that can provide direct measurements of N_2O fluxes from areas larger than the 1 m^2 typical of chambers.

Chamber Methods

Chambers have been preferred for CH_4 and N_2O flux measurement because, until recently, they were the only feasible way to detect the small fluxes involved, which are typically less than 1 μmol m^{-2} s^{-1} for both gases (Table 1). New developments in instumentation make the more demanding micrometeorological approaches possible, although it may be some years before their use becomes routine. Tables 1 and 2, adapted from Denmead (1993), set out some operating requirements for chambers and conventional micrometeorological methods. For chambers, the gas sensors can have relatively slow response (Table 2) and can be of much lower precision (Table 1). For CH_4 and N_2O, and depending on chamber design, the precision required to estimate fluxes is approximately 100 times less than for conventional micrometeorological approaches. Put another way, for comparable sensor precision, chambers can detect fluxes 100 times smaller.

As well as high sensitivity, chambers offer portability, versatility, and relative simplicity. Those used for measurement of CH_4 and N_2O fluxes typically enclose land areas < 1 m^2. Therefore, they permit process studies and experiments with many treatments in numbers that could not be contemplated with micrometeorological approaches because of the large land areas that the latter require. For the above reasons, chambers and micrometeorological methods should be seen as being complementary rather than alternatives.

Micrometeorological methods integrate over larger space scales, but chambers can provide the fine detail on production processes and the point to point variation needed for prediction on a regional scale. Another important consideration in this context is the length of time over which flux measurements are

Table 1. Sensor precision required to measure fluxes to within 10%

Gas	Surface	Typical flux ($kg\ m^{-2}\ s^{-1}$)	Method	Required precision (ppb)	Reference
CO_2	Forest	10^{-6}	Open chamber	150	Denmead et al. (1993)
	Forest	10^{-6}	Micrometeorology[a]	30	Denmead et al. (1993)
	Rice	10^{-6}	Micrometeorology	300	Denmead (this chapter)
CH_4	Rice field	10^{-9}	Closed chamber	200	Seiler et al. (1984)
	Rice field	6×10^{-9}	Micrometeorology	6	Kanemasu et al. (1994)
	Rice field	10^{-9}	Micrometeorology	2	Denmead (this chapter)
N_2O	Soil	2×10^{-10}	Closed chamber	80	Matthias et al. (1980)
	Grassland	3×10^{-11}	Open chamber	3	Denmead et al. (1979)
	Pasture	4×10^{-11}	Micrometeorology	0.01	Galle et al. (1994)

[a] All micrometeorological methods have much the same requirements for precision.

Table 2. Characteristics of various methods for measuring trace-gas fluxes

Method[a]	Footprint	Maximum instrument response time(s)
Closed chamber	1 m	500
Open chamber	1 m	100
Mass balance	20 m	100
Aerodynamic and energy balance	200 m–1 km	100
Eddy accumulation	200 m–1 km	100
Eddy correlation	200 m–1 km	0.1

[a] The sequence of methods follows the order of increasing footprint size and decreasing instrument response time.

required. Many measurements are made for only short times during various phases of the growing season. If the requirement is to track gas fluxes through the whole of the growing season, special precautions need to be taken and systems become more elaborate. It is difficult to make definite pronouncements, but both chamber and micrometeorological measurements can be made over long periods of time.

Closed and Open Systems

The operating principle for chambers is to restrict the volume of air with which gas exchange occurs, thereby magnifying changes in the concentration of the emitted (or absorbed) gas in the headspace. In closed chamber systems there is no replacement of air in the headspace and the gas concentration changes continuously. The rate of concentration change in the headspace is used to calculate the gas flux to or from the soil-water-plant system:

$$F = (V/A)\,d\rho_g/dt, \tag{1}$$

where F is the flux density of the gas, V is the volume of the headspace, A is the area of land enclosed, ρ_g is the gas density within the chamber, and t is time. In open systems, a constant flow of air through the headspace is maintained and the gas concentration attains a steady difference from the background concentration in the ambient air. The flux is calculated from:

$$F = v(\rho_g - \rho_b) / A, \quad (2)$$

where v is the gas flow rate and ρ_b is the background gas concentration. In practice, v is adjusted to keep $(\rho_g - \rho_b)$ small. Closed systems are more common because a larger concentration change occurs and they are mechanically simpler. However, open systems are much to be preferred for reactive gases like NO, which is easily oxidized to NO_2 in a closed vessel. As well, open chambers must be used for measurements of CO_2 exchange because of the rapid depletion of atmospheric CO_2 in closed chambers through photosynthesis. Growing plants can reduce atmospheric CO_2 concentrations to ineffective levels within the 15–60 min enclosure time typically employed in closed chambers.

Problems common to all chambers, closed or open, such as the possibility of pressure-induced soil-gas flows, the effects of increasing gas concentrations in the headspace on the emission rate, site disturbances, point to point variability, and changes to the microclimate, have received much attention (Denmead and Raupach 1993; Hutchinson and Livingston 1993). However, one microclimatic problem that has relevance to CH_4 and N_2O production should be mentioned: the effects of temperature. Like most biological processes, the production of both gases is strongly temperature-dependent. A Q_{10} (relative change in activity for a 10 °C change in temperature) of 2.8 has been calculated for N_2O production (Denmead et al. 1979) and Q_{10} values of 2.5–3.5 have been reported for CH_4 production (Conrad 1989). Large diurnal variations in soil emissions of these gases are to be expected in the field, and maintaining soil temperatures inside the chamber close to those outside is important. Soil temperatures are determined by energy exchange by the plants and the water and the soil making up the chamber ecosystem, and these exchanges are likely to be different from those outside. Both open and closed chambers affect air flow, radiant energy receipt, and energy transfer to the atmosphere. A solution is to determine the temperature response of gas production and then adjust chamber production rates to match the outside soil temperature.

Some particular problems arise in the use of chambers for measuring CH_4 production in rice cultivation. We might expect that because root exudates are substrates for CH_4 production in the rhizosphere, the rate of CH_4 production might depend partly on CO_2 assimilation rate. Realistic measurements of CH_4 production in chambers require preservation at all times of those environmental conditions that influence stomatal functioning and CO_2 exchange such as ventilation, atmospheric CO_2 concentrations, and light intensity. Another problem is that some 70% of the CH_4 generated in the rhizosphere remains there and is gradually oxidized (Holzapfel-Pschorn et al. 1986). Disturbances to the soil during chamber installation can release this trapped CH_4 and produce artifact results. It is desirable that in rice fields, chambers be installed permanently with tops that can open, or that permanent footings be in place for installing chambers from time to time.

Soil disturbances do occur during rice cultivation, for example, during episodes of weeding, and CH_4 releases then constitute a real and perhaps important part of the CH_4 emission of the crop (Kanemasu et al. 1994). These releases are difficult to quantify with chambers or conventional micrometerorological methods, but could be measured very precisely with mass balance methods.

Multichambers

To combat the notoriously large point to point variability of soil-gas emissions and to overcome the problems of CO_2 depletion and soil disturbance, multichamber installations have been developed (Schütz and Seiler 1989; Kanemasu et al. 1994). Both systems are developed for rice and employ 16 chambers that each enclose a land area of 1 m^2, and have lids that are closed periodically under computer control. When the lids close, air from the chamber is pumped to a self-calibrating gas chromatograph (GC). The GC output is stored by the computer and analyzed. The system allows automatic determination of the CH_4 emission rates at each of the 16 field plots eight times per day and remains in place for the whole growing season.

Micrometeorological Methods

Footprints

Fluxes measured in the surface layer are a result of contributions from many ground-level sources upwind. The influence of individual sources can be estimated from footprint analyses that use theories of atmospheric dispersion to predict particle trajectories (Leclerc and Thurtell 1990). A general conclusion for crop surfaces like rice is that in the near-neutral conditions of atmospheric stability that often pertain over rice by day, more than 85% of the flux seen by an instrument mounted at height z comes from within a distance of 100 z upwind. Alternatively, for a uniform crop, a fetch to height ratio of 100 to 1 is required to ensure that micrometeorological measurements of the flux at some height above the crop are mostly representative of the surface. Leclerc and Thurtell (1990) calculated that required fetch to height ratios for unstable conditions will be less than 100 to 1, but ratios for stable conditions (nighttime) approach 200 to 1. For a fetch of 100 m, these ratios would give a maximum working height of less than 1 m. Many eddy correlation instruments, however, can be operated successfully only at heights of 2 m or more above the surface because of limitations to their frequency response. Likewise, the small size of atmospheric gradients often requires that concentration differences be measured over a height interval of 2–3 m. It is evident that experimental areas suitable for conventional micrometeorological approaches need to be hundreds of meters in extent.

Footprint analyses also indicate that surface emissions at a distance x from the measuring point will have a significant effect on concentrations measured downwind (at $x = 0$) up to a height Z of about 0.1 x. The actual magnitude of Z depends on fetch, surface roughness, and atmospheric stability. This result is the basis for small-plot, mass-balance techniques. The aim is to measure gas concentration profiles right through the boundary layer developed over the plot, rather than just in the much smaller, equilibrium layer close to the surface. Unlike conventional micrometeorological methods (eddy correlation, eddy accumulation, and gradient diffusion), this technique is most successful when the fetch is small and, therefore, Z is also conveniently small. In typical applications, $x < 40$ m.

Eddy Correlation

This is the preferred micrometeorological method for measuring trace-gas fluxes because it is a direct, basic approach that requires no simplifying assumptions. Simultaneous measurements of the vertical wind speed w and the gas concentration ρ_g are multiplied to give instantaneous vertical fluxes that are then averaged over a sampling period long enough to encompass all the effective transporting eddy sizes. Required instrumental frequency responses for various techniques are shown in Table 2, from which it is evident that eddy correlation instruments must not only be of high precision, but also of rapid response. For land-based systems, the required maximum frequency response is typically about 10 Hz; whereas, the required averaging time is > 20 min. Such high-frequency responses are quite attainable with sonic anemometry and water vapor and CO_2 sensors, but it is only in very recent years that appropriate high-frequency sensors have become available for CH_4 and N_2O.

The mean flux density of a trace gas F is given by:

$$F = \overline{w\rho_g}, \tag{3}$$

where the overbar denotes a time average over an appropriate measuring period. Both w and ρ_g can be represented as sums of means, $\overline{w}$ and $\overline{\rho}_g$, and fluctuations about those means, w' and ρ'_g. Therefore:

$$F = \overline{W}\overline{\rho}_g + \overline{W'\rho'_g} \tag{4}$$

The second term on the right-hand side of Eq. (4) is the one usually measured by eddy correlation systems operated in situ. The first term is virtually impossible to measure directly because the mean vertical wind speed $\overline{w}$ is usually < 1 mm s^{-1} (Webb et al. 1980). However, the mean background concentration $\overline{\rho}_g$ can be relatively large and for all three of the greenhouse gases considered here, the product $\overline{w}\overline{\rho}_g$ can be of the same magnitude as the second term in Eq. (4) and may even have the opposite sign. The fact that $\overline{w}$ is not usually zero is due to the effects on air buoyancy of simultaneous fluxes of sensible heat H and water vapor E. The magnitude of $\overline{w}$ can be calculated from knowledge of those fluxes. Webb et al. (1980) show that Eq. (4) can be written as:

$$F = \overline{w'\rho_g'} + \left(\overline{\rho}_g / \overline{\rho}_a\right) \{\mu / (1+\mu\sigma)\} E + \left(\overline{\rho}_g / \overline{\rho}\right) H / c_p \overline{T}. \qquad (5)$$

In Eq. (5), $\mu = m_a/m_v$, m_a is the molecular weight of dry air and m_v that of water vapor; $\sigma = \overline{\rho}_v/\overline{\rho}_a$, $\overline{\rho}_v$ and $\overline{\rho}_a$ are the mean densities of water vapor and dry air, respectively; $\overline{\rho}$ is the mean density of air (i.e., $\overline{\rho} = \overline{\rho}_a + \overline{\rho}_v$), c_p is the specific heat of air at constant pressure, and $\overline{T}$ (K) is air temperature.

The first term on the right hand side of Eq. (5) is the conventional eddy flux, calculated on the assumption that $\overline{w} = 0$. The other two terms are the corrections for the density effects because of water vapor and heat transfer. Denmead (1994) gives the relative magnitudes of the correction terms in Eq. (5) for some published investigations of fluxes of greenhouse gases. The importance of the corrections is highly variable. In one instance, investigations of CH_4 flux over a ricefield, the corrections were negligible. Those associated with a typical negative heat flux of -50 W/m^2 very nearly balanced those arising from a typical latent heat flux λE of 250 W/m^2 (λ is the latent heat of evaporation). In an example of N_2O flux from grassland, corrections for a positive H of 100 W/m^2 reinforced those for a λE of 300 W/m^2 to produce an overall correction more than ten times the true flux. Attractive as the eddy correlation approach is in principle, it would seem to be a quite inappropriate technique in such a case. Gradient diffusion approaches that can avoid the corrections would be more reliable.

Eddy Accumulation

Eddy correlation requires fast-response anemometers and sensors. As Table 2 indicates, the required sampling frequency is usually >> 1 Hz. An alternative that retains the need for a fast anemometer, but does not require fast gas measurement, is eddy accumulation. Fast-response sampling valves are substituted for a fast-response sensor. The most promising version of this method uses a relaxed eddy-accumulation scheme, based on conditional sampling (Businger and Oncley 1990; Baker et al. 1992c). A fast-response anemometer measures the vertical velocity and controls a simple valving system by which air is sampled at a constant rate and is diverted into updraft or downdraft bins depending on the direction of the vertical wind. The flux is given by:

$$F = \overline{\rho}_a \, b \, \sigma_w \left(\overline{s}_u - \overline{s}_d\right), \qquad (6)$$

where b is a coefficient, σ_w is the standard deviation of the vertical wind speed, and $\bar{s}_u$ and $\bar{s}_d$ are mixing ratios of the gas in the up and down bins. Both probability theory and experiment indicate that b is a constant, independent of stability, and has a value close to 0.6.

This relatively simple measurement and calculation scheme has many attractions for trace-gas measurement. First, it does not require a fast-response sensor although the precision required will be much the same as for eddy-correlation or gradient-diffusion methods. The sensor need not necessarily be in the field; therefore, high-resolution laboratory-based instruments can be used. Second, the samples can be preconditioned before measurement to avoid the corrections for density effects. Third, like eddy correlation, the system provides a direct point measurement and therefore does not have the potential problem associated with gradient methods (air samples from different levels have different footprints). The system does impose rigorous demands on the switching circuitry, plumbing, and flow control. Precautions must be taken to remove any offsets in the vertical velocity, not by software after the event, but in real time. So far, it has been used successfully for CO_2 measurement, but not for measurements of CH_4 or N_2O.

Gradient Diffusion

Vertical transport of gases in the lower atmosphere may also be described by the relationship:[1]

$$F = -\bar{\rho}_a K_g \partial\bar{s} / \partial z, \tag{7a}$$

where K_g is the eddy diffusivity for the gas and s $(= \rho_g/\rho_a)$ is the mixing ratio of the gas with respect to dry air (Webb et al. 1980). For the evaporation of water Eq. (7a) becomes:

$$F = -\bar{\rho}_a K_g \partial\bar{r} / \partial z, \tag{7b}$$

where r is the mean mixing ratio of water vapor, defined formally as ρ_v/ρ_a and calculated as $0.622e/(p-e)$, e is vapor pressure, and p total pressure. The counterpart for sensible heat transfer is:

$$H = -\bar{\rho} c_p K_h \partial\bar{\theta} / \partial z, \tag{7c}$$

[1] Wind speed in the air layers above plant communities appears to approach zero not at the ground, but at some height above it. That height is called the zero-plane displacement (d) and is a substantial fraction of the height of the community. When dealing with plant communities, it is usual in micrometeorology to reckon height as height above the zero plane, i.e., as actual height above the ground (z) less the zero-plane displacement. The term $z - d$ should be substituted for z whenever height-dependent relationships are applied to plant communities.

where $\bar{\theta}$ is the mean potential temperature ($\bar{\theta} = \bar{T} + \Gamma z$, where Γ is the adiabatic lapse rate). Unlike molecular diffusion, which results from the random motion of molecules, eddy diffusion results from the movements of parcels of air from one level to another. Consequently, eddy diffusivities are usually several orders of magnitude greater than their molecular counterparts. Their actual magnitudes are determined by wind speed, height above the surface, the aerodynamic roughness of the surface, and the vertical temperature gradient.

The more usual statement of the gradient diffusion equation is:

$$F = -K_g \, \partial \bar{\rho}_g / \partial z, \tag{8}$$

If ρ_g were measured in situ, i.e., on samples of moist air and in the presence of the mean temperature gradient, the flux calculated from Eq. (8) would correspond to the term $\overline{w'\rho'_g}$ in Eq. (5), and the same corrections for E and H would need to be added to obtain the true flux. Equation (7a), however, specifies the gas concentration in terms of its mixing ratio with respect to dry air (i.e., as ρ_g/ρ_a). The latter is essentially what is measured when samples of air from different heights are predried before their gas concentrations are measured at a common temperature and pressure. If gas concentrations are measured in this way, no corrections for the density effects associated with the simultaneous transfers of water vapor and heat are needed (Webb et al. 1980), although other corrections may be needed to account for the effects of temperature stratification on the determination of K_g when aerodynamic methods are used.

Equations (7a) and (8) provide means for calculating the flux density of gases from measurements of their vertical concentration profiles. The calculations require knowledge of K_g, but it is not possible to calculate K_g a priori or to measure it directly. The usual procedure in applying Eq (7a) and (8) is to infer K_g in situ from some other tracer entity whose flux is known and whose concentration gradient is measured simultaneously. Momentum and energy are the tracer fluxes used most commonly, and these lead to so-called aerodynamic and energy-balance approaches.

Aerodynamic Methods

If momentum is used as the tracer, the flux density of the gas of interest is evaluated from measurements of horizontal wind speed u and gas concentration at two or more heights. Theoretical developments lead to specific formulations for different stability conditions. In neutral conditions and assuming measurements at just two heights z_1 and z_2:

$$F = \frac{\bar{\rho}_a k^2 (\bar{u}_2 - \bar{u}_1)(\bar{s}_1 - \bar{s}_2)}{\{\ln(z_2/z_1)\}^2}, \tag{9}$$

where k is the von Karman constant (= 0.41). Under stable conditions:

$$F = \frac{\bar{\rho}_a k^2 (\bar{u}_2 - \bar{u}_1)(\bar{s}_1 - \bar{s}_2)}{\{\ln(z_2/z_1) + 5(z_2 - z_1)/L\}^2}, \tag{10}$$

whereas, under unstable conditions:

$$F = \frac{\bar{\rho}_a k^2 (\bar{u}_2 - \bar{u}_1)(\bar{s}_1 - \bar{s}_2)}{\{\ln(z_2/z_1) - [\psi_1(z_2) - \psi_1(z_1)]\}\{\ln(z_2/z_1) - [\psi_2(z_2) - \psi_2(z_1)]\}}, \tag{11}$$

where the Monin-Obukhov stability length L is a parameter that can be calculated from measurement of wind speed, temperature, and humidity and specifies the relative influences of wind shear and buoyancy on transport in the atmosphere, and the ψ values are functions of L (Paulson 1970). Equations (9), (10), and (11) are written in finite difference form for two measuring heights. If measurements at more than two heights are available, vertical gradients of $\bar{u}$ and $\bar{s}$ can be calculated (i.e., $\partial\bar{u}/\partial z$ and $\partial\bar{s}/\partial z$) and used to calculate gas fluxes from an analogous set of equations that employ corresponding unintegrated formulations for the stability corrections (Denmead 1994).

Energy-Balance and Alternative Methods

Because neutral conditions occur very seldom in the field, the stability corrections contained in Eq. (10) and (11) will usually be required if aerodynamic approaches are employed. Alternative approaches for calculating gas fluxes, which obviate the need to apply stability corrections, can be made by invoking similarity in K for heat and the gases (i.e., by assuming that $K_h = K_g$). A common procedure is to use the energy balance: the energy gained by natural surfaces from solar radiation is balanced by the loss of energy to the atmosphere through the transfer of sensible heat and water vapor and by a change in energy storage. For a rice field:

$$R = H + \lambda E + W + G, \tag{12}$$

where R is the net radiation at the surface (the incoming short- and long-wave radiation less what is reflected and reradiated), W is the change in heat stored in the floodwater, and G is the change in heat stored in the soil underneath.

The Bowen ratio β is the ratio of the fluxes of sensible and latent heat:

$$\beta = H/\lambda E. \tag{13}$$

From Eq. (7b) and (7c):

$$\beta = \left(\frac{\bar{\rho}}{\bar{\rho}_a}\right)\left(\frac{c_p}{\lambda}\right)\frac{\partial\bar{\theta}/\partial z}{\partial\bar{r}/\partial z}, \tag{14}$$

which in finite difference form becomes:

$$\beta = \left(\frac{\bar{\rho}}{\bar{\rho}_a}\right)\left(\frac{c_p}{\lambda}\right)\left(\frac{\bar{\theta}_1 - \bar{\theta}_2}{\bar{r}_1 - \bar{r}_2}\right). \tag{15}$$

From Eq. (12) and (13)

$$H = \frac{\beta(R - W - G)}{(1 + \beta)} \tag{16}$$

and

$$\lambda E = \frac{R - W - G}{(1 + \beta)}, \tag{17}$$

and from Eq. (7a), (7b) and (7c)

$$F = \left(\frac{\bar{\rho}_a}{\bar{\rho}}\right)\left(\frac{H}{c_p}\right)\left(\frac{\bar{s}_1 - \bar{s}_2}{\bar{\theta}_1 - \bar{\theta}_2}\right) \tag{18}$$

or

$$F = E\,(\bar{s}_1 - \bar{s}_2)/(\bar{r}_1 - \bar{r}_2). \tag{19}$$

Of course, Eq. (18) or Eq. (19) could be invoked if H or E were measured directly by eddy correlation along with the gradients of θ or r.

The prime measurements required to calculate F are therefore R, W, and G, and the temperatures, humidities, and gas concentrations at a minimum of two heights above the surface. The energy terms R and G can be measured with commercial net radiometers and soil heat-flux plates, but W must be calculated from the rate of change of the temperature of the floodwater, T_w:

$$W = \rho_w c_w \int_0^D (\partial T_w / \partial t)\, dz. \tag{20}$$

where ρ_w is the water density, c_w is the specific heat of water, D is floodwater depth, and t is time. In some situations, W can be so large that the accuracy of its measurement determines the success of the energy-balance approach. Figure 1 shows the magnitudes of the energy terms R, G, and W for an Australian rice field (at Griffith, NSW) with a floodwater depth of 11 cm, which is admittedly larger than the typical depth for Asian rice fields. In this case, however, W accounts for about half the daytime energy retention by the rice crop, namely R, and is the dominant energy term by night when release of the heat stored during the day provides the energy for evaporation.

The employment of alternative methods is advisable at any time, but particularly when nighttime fluxes are being investigated. Calm periods may invalidate the aerodynamic approach, and difficulties in measuring R and G, which are small, and W, which is large, can make an approach via the energy balance difficult. The problem is acute for rice crops for which nighttime fluxes of at least two of the greenhouse gases, CH_4 and CO_2, can be large. Figure 2 presents a comparison of fluxes of those gases for the Australian rice field (Fig. 1) as measured by aerodynamic and energy-balance methods, i.e., by Eqs. (9)–(11) and Eqs. (12)–(19). In this case, the crop was 0.65 m tall, and wind speeds, temperatures, humidities, and gas concentrations were measured at heights of 1.25 and 2.25 m above the soil surface. Similar comparisons are presented by Kanemasu

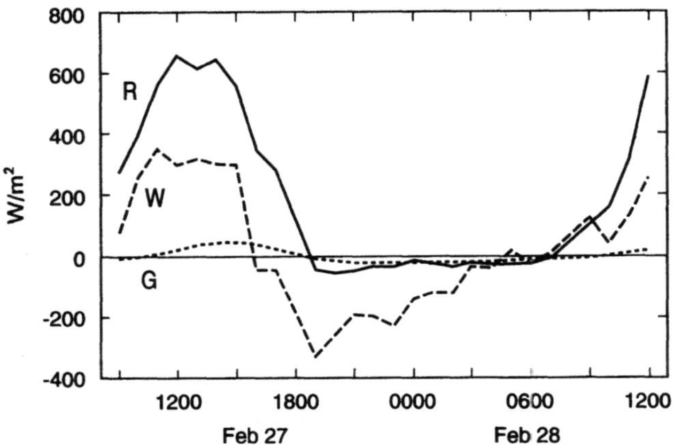

Fig. 1. Relative magnitudes of net radiation *R*, the change in heat stored in the floodwater *W*, and soil heat flux *G* for a rice crop with a water depth of 11 cm. *W* accounts for half the net energy retention by day and dominates energy exchange by night

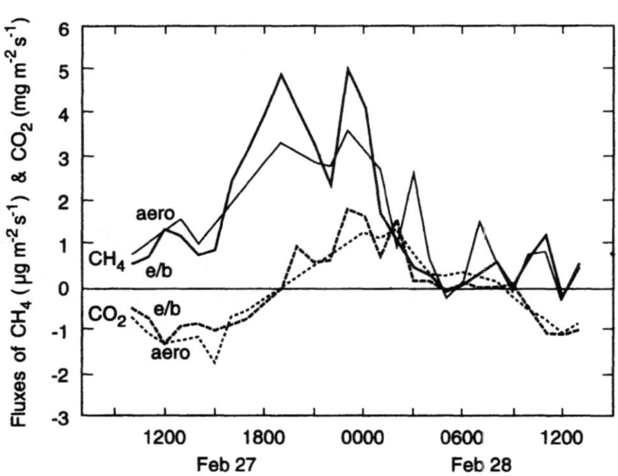

Fig. 2. Fluxes of CH_4 and CO_2 over a rice field as measured by aerodynamic (*aero*) and energy balance (*e/b*) techniques

et al. (1994). In both cases, the agreement between alternative micrometeorological procedures during both day and night gives much confidence in the calculated fluxes.

Mass Balance Methods

The large point to point variability in soil-gas emissions and the need to measure over large areas and long time scales have been discussed. Alternative conventional micrometeorological methods, however, require uniform land areas hundreds of meters in extent, which precludes their use in experimental situations where several land treatments are to be investigated. There is scope for micrometeorological methods that can be used to measure gas exchange from smaller land areas, areas larger than the usual chamber size (1 m^2), but with dimensions of tens rather than hundreds of meters. Methods based on mass balance have been developed in recent years to fill this gap (Beauchamp et al. 1978; Wilson et al. 1982; Denmead 1983).

In general terms, mass-balance methods equate the horizontal gas flux across the downwind edge of a treated area with the vertical flux from the treated surface upwind. The horizontal flux at any height is the product of wind speed and gas concentration. The total horizontal (and surface) flux is obtained by integrating that product over the depth Z of the layer affected by the emission. As discussed earlier, Z is in the order of one-tenth of the fetch X. The conservation of mass gives:

$$F = \frac{1}{X}\int_0^Z \overline{u(\rho_g - \rho_b)}\,dz, \qquad (21)$$

where ρ_b is the upwind, background concentration.

One restriction on this approach is that the effective gas concentration is the concentration in excess of background. Not only must the upwind concentration profile be measured as well as the downwind concentration profile, but calculation of the flux using Eq. (21) also requires subtraction of experimentally determined data, which can be an error-prone procedure. The technique is suited best to investigations of trace-gas fluxes in situations where ρ_b is small and experimental treatments generate fluxes that are large compared with normal emissions. A further practical difficulty is that the fetch X must be known precisely to calculate F. If the experimental area is the usual rectilinear plot used by agronomists, X will vary with wind direction and will need to be determined frequently. This complication can be overcome by working with a circular plot and measuring u and ρ_g at its center. Regardless of compass direction, the wind will always blow toward the center and X will always be equal to the plot radius. Circular plots with radii between 3.5 and 36 m have been employed.

Other aspects of this approach are discussed by Denmead and Raupach (1993). These include the possibility of inferring the surface flux from measurements of the horizontal flux at just one, stability-independent height above the surface and the need to account for horizontal gradient diffusion and convection. The latter usually requires a correction of about 15% (Wilson and Shum 1991).

Apart from its eminent suitability for small plots, this method has some distinct advantages over other micrometeorological methods. First, it has a simple theoretical basis: it requires no special form for the wind profile and no corrections for thermal stratification. Second, the instrumentation is relatively simple. It needs neither to be of fast response, nor to have the high precision required for gradient measurements. So far the circular plot technique has been used mainly to measure ammonia fluxes, but could be employed equally well in rice research for measuring N_2O production from plots receiving various N treatments. For CH_4 production, however, large background emissions seem likely to limit its potential uses. In that case, another variant of the mass-balance approach, which requires more detailed measurements, can be employed. Developed for measuring CH_4 production by grazing animals, it requires measurements of gas concentration to be made on the boundaries of the test area rather than within it (Denmead 1994).

In the animal application described by Denmead (1994), cattle were grazed in a small field (22 × 22 m). Sampling tubes were mounted at four levels along the length of each boundary. The top level was 3.5 m. Air was drawn into the sampling tubes equally along their length through capillary tubes inserted at 1-m intervals, and pumped through separate air lines to a mobile laboratory for CH_4 analysis. Coupled with measurements of wind speed and direction, the concentration measurements permitted calculation of the horizontal flux of CH_4 across each boundary. The boundaries are numbered 1 to 4, and u is resolved into components U normal to boundaries 2 and 4 and V normal to boundaries 1 and 3. The total flux across each boundary is obtained by integrating the horizontal fluxes over the boundary length X and height Z. The difference between the fluxes of gas into and out of the field is the total rate of production or consumption and the average surface flux density is given by:

$$F = (X/A)\int_0^Z \left[\overline{U_z}\left(\left\langle \overline{\rho}_{4,z}\right\rangle - \left\langle \overline{\rho}_{2,z}\right\rangle\right) + \overline{V_z}\left(\left\langle \overline{\rho}_{3,z}\right\rangle - \left\langle \overline{\rho}_{1,z}\right\rangle\right)\right] dz, \qquad (22)$$

where A is the area of the experimental field, the overbars denote time averages, and the angular brackets denote spatial averages. For a plot size similar to that used in the cattle experiment, the precision required to measure CH_4 fluxes from rice fields by this mass-balance method would be about the same as given in Table 1 for other micrometeorological approaches (i.e., about 10 ppb to measure the flux to within 10%). The method would be ideal for quantifying CH_4 emissions during cultivation operations (e.g., weeding) when the sediments are disturbed and very large CH_4 emissions occur (Kanemasu et al. 1994).

Gas Sensors

Sensors adequate for measuring CO_2 fluxes by any of the methods described here have been readily available for some time. For CH_4 and N_2O, it can be asserted that whereas 10 years ago the availability of sensors with the required fast response and/or precision severely limited the application of any of the micrometeorological methods, new instrumentation now promises to make such measurements routine. Table 2 shows that for eddy correlation, instruments need to be able to respond to changing atmospheric concentrations at frequencies of 10 Hz; whereas Table 1 shows that for the flux densities of CH_4 and N_2O common in agriculture, resolutions in the parts-per-billion range or better are required. Both requirements can now be met. Particularly noteworthy is the advent of tunable diode laser systems that have fast response and high precision for both CH_4 and N_2O measurement (Fan et al. 1992; Ritter et al. 1992; Hargreaves et al. 1993), flame ionization detection systems for CH_4 (Fan et al. 1992), and Fourier-transform infrared spectroscopy that, although not of fast response, permits simultaneous, high-precision measurements of CH_4, N_2O, H_2O and CO_2 in the one scan (Galle et al. 1993). Long-path, nondispersive infrared gas analyzers for N_2O measurement are also becoming available (Smith et al. 1993). In summary, appropriate instrumentation now exists for all the micrometeorological methods described. Much of it is presently expensive and requires specialist operators. It is reasonable to expect, however, that in the next 5 years, this new suite of gas sensors will be refined, commercialized, and used routinely.

Dynamics of Methane in Rice Fields: Emissions to the Atmosphere in Japan and Thailand

M. Kimura[1] and K. Minami[2]

Emission rates (fluxes) of CH_4 from rice fields were measured during the cultivation period at four different sites in the central region of Japan (Table 1; Yagi and Minami 1990a, 1991). Successive applications of organic matter (e.g., rice straw and its compost) have been made at three of the sites for more than 10 years. The experimental fields in these sites consist of non-nitrogen, mineral, compost, and rice-straw plots. Organic matter was applied at a rate of 12 t/ha to the compost plots and at 6–9 t/ha to the rice-straw plots in addition to the mineral fertilizers. A closed-chamber method was used for the measurements of the CH_4 emission flux from the rice fields (Minami and Yagi 1988).

Methane Emissions from Japanese Rice Fields

Seasonal variations in the CH_4 fluxes from the rice fields in Ryugasaki, the daily mean temperature of the soil and air, soil Eh, and agricultural practices are shown in Fig. 1. A strong seasonal variation in CH_4 fluxes from rice fields was observed during the cultivation period. The CH_4 emissions began to increase as soil Eh decreased for nearly a month after submergence. The fluxes then increased as the rice plants grew, although considerable depressions were observed during the period of midsummer drainage and shortly after the supplementary application of mineral fertilizers (top dressing). The fluxes rapidly dropped after the rice field was intermittently drained at the end of August.

The CH_4 fluxes from a rice-straw plot were much greater than from the other plots throughout the period of submergence. The application of compost prepared from rice straw did not appreciably enhance the CH_4 flux. The differences in the magnitude and variation of the CH_4 fluxes between the mineral and the non nitrogen plots were negligible.

[1] School of Agricultural Sciences, Nagoya University, Furocho, Chikusaku, Nagoya, 464-01 Japan
[2] National Institute of Agro-Environmental Sciences, Kannondai, Tsukuba, 305 Japan

S. Peng et al. (Eds.) Climate Change and Rice

Table 1. Soil properties of the rice fields

Site	Ryugasaki	Kawachi	Mito	Tsukuba
Soil	Gley soil	Peat soil	Humic Andosol	Light-colored Andosol
Texture	SCL[a]	CL	L	L
Percolation rate (mm/day)	10	10	30	30
pH	5.9	5.6	6.5	5.9
Total C (%)	1.4	3.4	6	2.2
Total N (%)	0.13	0.26	0.4	0.21
C/N	10	13	15	10
RMC[b] (μg C/g per 28 days)	78.9	232.9	38.3	19.4

[a] SCL sandy clay loam, CL clay loam, and L loam.
[b] RMC readily mineralizable carbon.

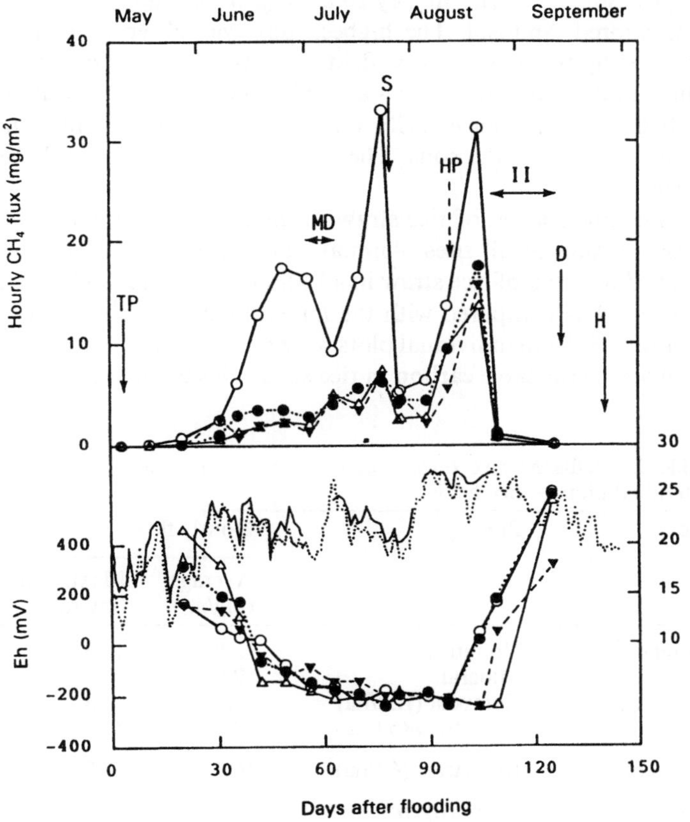

Fig. 1. Seasonal variations of the CH_4 fluxes from rice fields in Ryugasaki (1988), along with the daily mean temperature of soil and air, soil Eh, and the agricultural practices. *TP* Transplanting; *MD* midsummer drainage; *S* supplementary application of mineral fetilizer; *HP* heading period; *II* intermittent irrigation; *D* drainage; *H* harvest. Straw plot (○), compost plot (●), mineral plot (◊), and non-nitrogen plot (▼)

In Mito, the trends in seasonal variations of CH_4 fluxes and the effects of organic matter application were similar to those observed in Ryugasaki. In addition, the CH_4 flux from the rice field increased as a greater amount of rice straw was applied. The fluxes in Mito were appreciably smaller than in Ryugasaki. This was probably because the soil Eh was relatively higher in the rice field in Mito. In Kawachi, the pattern of seasonal variation in CH_4 flux was very similar to that observed in Ryugasaki. The fluxes were, however, much greater. The CH_4 fluxes observed in Tsukuba remained considerably low throughout the cultivation period.

From the integration of the observed seasonal variations in CH_4 fluxes, annual emission rates of CH_4 from the individual plots were calculated (Table 2). Table 2 also provides the average and maximum values of the observed fluxes. Emission rates of CH_4 varied according to soil type and the organic matter applied. Annual emission rates of CH_4 from the rice-straw plots were 44.8 g/m^2 in the peat soil, 27.0 g/m^2 in gley soil, 9.8 g/m^2 in humic Andosol, and 1.1 g/m^2 in light-colored Andosol. The highest flux was observed in the peat rice field, followed by the gley-soil rice field. Because the management of irrigation and other cultivation practices in Kawachi and Ryugasaki were almost the same, it is likely that the difference in CH_4 fluxes was caused by the difference in the organic matter content of the soils. The CH_4 fluxes from Andosols were significantly lower.

The application of rice straw to the rice fields significantly increased CH_4 emission rates at all sites. Annual emission rates of CH_4 from the plots that received 6–9 t/ha of rice straw in addition to the mineral fertilizer increased 1.8- to 3.5-fold as compared with the mineral fertilizer plots. The annual emission rates of CH_4 from individual plots were positively correlated with the contents of readily mineralizable carbon in rice soils collected before submergence.

Table 2. Methane flux during cultivation period and annual emission rates from individual plots of rice fields

Site	Plot	Hourly flux		Annual emission rate (g/m^2)
		Average (mg/m^2)	Maximum (mg/m^2)	
Ryugasaki	No nitrogen	2.8	15.3	8.0
	Mineral	2.9	13.3	8.2
	Compost (12 t/ha)	3.8	17.3	10.5
	Rice straw (6 t/ha)	9.6	32.6	27.0
Kawachi	Rice straw (6 t/ha)	16.3	67.2	44.8
Mito	No nitrogen	1.4	5.1	4.1
	Mineral	1.2	5.7	3.6
	Compost (12 t/ha)	1.9	5.7	5.9
	Rice straw (6 t/ha)	3.2	10.2	9.8
	Rice straw (9 t/ha)	4.1	8.9	12.6
Tsukuba	Mineral	0.2	0.5	0.6
	Rice straw (6 t/ha)	0.4	0.7	1.1

Seasonal Variation in Methane Emissions From Thai Rice Fields

Field measurements were performed in the Rice Experimental Stations at Suphan Buri, Khlong Luang, and Chai Nat in 1991 (Yagi et al. 1994a) and Ratchaburi, Pathumthani (Rangsit), and Surin in 1992 (Jermsawatdipong et al. 1994); (Fig. 2). Soil classifications and properties are listed in Table 3. The soils at Khlong Luang and Pathumthani are acid sulfate soils mixed with brackish alluvium. Suphan Buri, Chai Nat, and Ratchaburi soils were formed from recent river alluvium. The soil of Surin is a low-humic gley soil with a low content of free iron. The fields were irrigated before the rice plants were transplanted and the surface water was maintained continuously until the rice plants matured.

Indica rice plants were cultivated according to the conventional methods of the areas. In the experiments at Suphan Buri, Khlong Luang, and Chai Nat in 1991, ammonium phosphate (50–90 kg N/ha) was applied as a basal fertilizer and urea (30–40 kg N/ha) was applied as a top dressing. In the experiments conducted at Ratchaburi, Pathumthani, and Surin in 1992, some plots received both organic materials (2.0–3.1 t/ha) and mineral fertilizers; other plots received only mineral fertilizer or no fertilizer (i.e., *Sesbania rostrata* at Ratchaburi, rice straw at

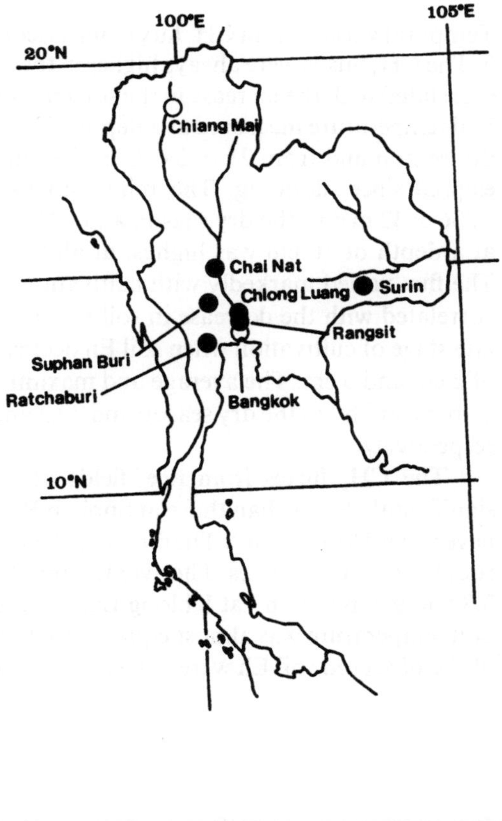

Fig. 2. Map of Thailand showing location of study sites (●)

Table 3. Classification and properties of the rice soils

	Khlong Luang (ST)[a]	Suphan Buri (VT)	Chai Nat (AT)	Ratchburi (FT)	Pathumthani (TT)	Surin (AP)
Texture	C[b]	C	CL	SiC	HC	LFS
pH	4.3	4.9	5.2	7.1	4.0	4.7
Total C (%)	1.2	1.3	0.9	1.0	2.4	0.1
Total N (%)	0.10	0.10	0.08	0.07	0.18	0.02
Available N[c]	69	84	48	–	–	–
Free Fe_2O_3 (%)	2.4	1.6	2.3	1.1	1.8	0.11
ER Mn (μg/g)[d]	17	27	145	387	91	29

[a] ST Sulfic Tropaquepts, VT Vertic Tropaquepts, AT Aeric Tropaquepts, FT Fulvic Tropaquepts, TT Thionic Tropaquepts, and AP Anthraquic Paleaqualts.
[b] C clay, CL clay loam, SiC silty clay, HC heavy clay, and LFS loamy fine sand.
[c] Amount of nitrogen mineralized during 28 days under anaerobic incubation at 30 °C.
[d] Easily reducible manganese.

Pathumthani, and rice-straw compost at Surin). A closed chamber method was used to measure the CH_4 emission flux from the rice fields (Minami and Yagi 1988; Kimura et al. 1991b).

Methane Emissions from Thai Rice Fields

Temporal variations in CH_4 flux from a rice field at Suphan Buri are shown in Fig. 3. The CH_4 fluxes were always higher in the afternoon than in the morning, which correlated with the increase in the soil temperature of the surface layer of the soil. Soil temperature measured at a depth of 2 cm ranged from 28 to 33 °C during the dry season and from 21 to 34 °C in the rainy season. Soil Eh decreased as time elapsed since flooding. The minimum values of the observed soil Eh were –236 ± 32 mV in the dry season and –241 ± 14 mV in the rainy season. The Eh at a depth of 10 cm was highest at all times during the measurement periods. The flux varied markedly with cultivation stage. The variation in CH_4 flux was correlated with the decrease in soil Eh. The highest fluxes were observed at the late stage of cultivation when soil Eh decreased to lower than –200 mV at depths of 2 cm and 5 cm. The average and maximum values of the fluxes were 19.5 and 46.6 mg m^{-2} h^{-1} in the dry season and 32.2 and 55.9 mg m^{-2} h^{-1} in the rainy season, respectively.

The CH_4 fluxes from rice fields at Khlong Luang and Chai Nat were significantly lower than those at Suphan Buri. At Khlong Luang, the fluxes were never over 15 mg m^{-2} h^{-1}. The fluxes at Chai Nat were much lower and maintained roughly constant values. The average and the maximum values of the fluxes were 3.8 and 9.4 mg m^{-2} h^{-1} at Khlong Luang and 1.6 and 3.0 mg m^{-2} h^{-1} at Chai Nat. Soil temperature was almost equal to that at Suphan Buri. The minimum values of the observed soil Eh were –189 ± 21 mV and –171 ± 13 mV at Khlong Luang

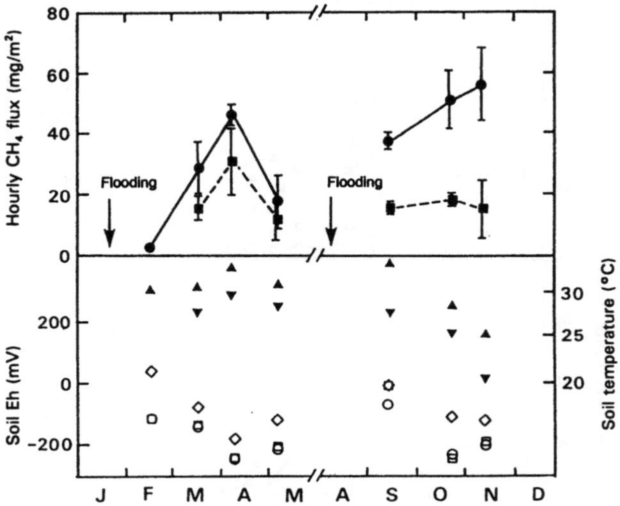

Fig. 3. Temporal variations in CH_4 flux, soil temperature, and soil Eh measured in the rice field at Suphan Buri in the dry and rainy seasons. The CH_4 flux in the morning (■) and in the afternoon (●), soil temperature at a depth of 2 cm in the morning (▼), and in the afternoon (▲), soil Eh at depth of 2 (O), 5 (□), and 10 cm (◊), *Bars* indicate the deviation between duplicate measurements

and Chai Nat, respectively. Soil Eh at Khlong Luang and Chai Nat did not reach values as low as –200 mV at a depth of 2–10 cm, although reduction occurred in both soils with time elapsed since flooding.

Figure 4A shows the CH_4 emission rates from the rice field at Ratchaburi throughout the growth period in 1992. Methane emissions from the no fertilizer (NF) plot were detected from the first measurement, although the emission rates remained low throughout the growth period. The CH_4 emission rates from the chemical fertilizer (CF) plot were 1.5–2.5 times as high as those from the NF -plot throughout the growth period. The CH_4 emission rates from the organic material (OM) plot were very high (51–69 mg m^{-2} h^{-1}) in the early growth period. They decreased during September and remained at about 29 mg m^{-2} h^{-1} until harvest.

The CH_4 emission rates from the rice field with acid sulfate soil at Pathumthani (Fig. 4B) were very low throughout the growth period in the NF and CF plots. Methane was emitted at a rate of 5–23 mg m^{-2} h^{-1} from the OM plot. The highest emission rate was recorded on August 21, when the first measurement was conducted. The CH_4 emission rates decreased during the tillering stage and then remained at 5–8 mg m^{-2} h^{-1} after the end of September.

Methane emission rates from the NF- and CF plots were higher than those from the OM plot in the Surin rice field (Fig. 4C), where rice-straw compost was not incorporated into soil but scattered on the soil surface. The maximum emission rates from the NF- and CF plots were recorded in the early growth

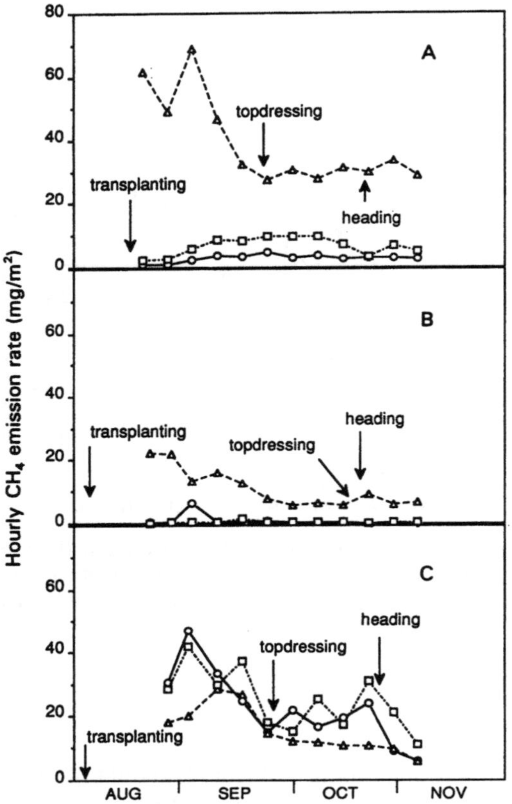

Fig. 4 A–C Seasonal variations in CH_4 emission rates. NF plot without fertilizer application (O), CF plot with chemical fertilizer application (□), and OM plot with organic matter application (Δ). **A** Ratchaburi. **B** Pathumthani. **C** Surin

period (1 September), as was observed in the OM plots in the Pathumthani and Ratchaburi rice fields. Total emissions of CH_4 from rice fields during the cultivation period were 3.6–77.0 g/m^2 at Suphan Buri, Khlong Luang, and Chai Nat in 1991, and 1.3–78.6 g/m^2 at Ratchaburi, Pathumthani, and Surin in 1992.

Characteristics of Methane Emissions in Japanese and Thai Rice Fields

Four major categories of factors control CH_4 emissions from rice fields (Yagi et al. 1994b). These are summarized in Fig. 5.

Soil Properties

A number of physical, chemical, and biochemical factors affect CH_4 production in rice soils. Soils containing higher amounts of readily decomposable organic

Fig. 5. Factors controlling CH_4 emissions from rice fields

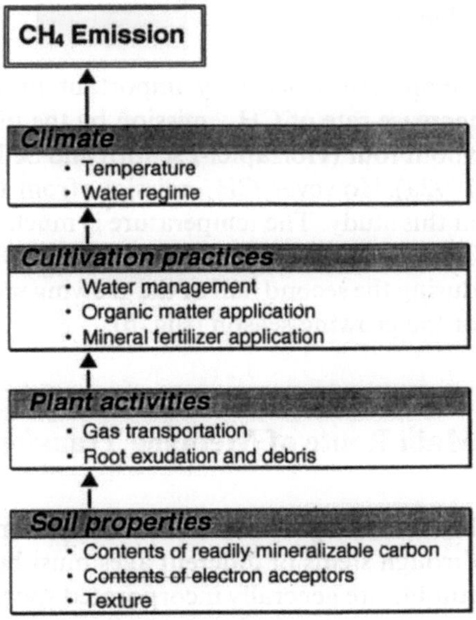

matter and lower amounts of electron acceptors generally produce higher amounts of CH_4.

The freshwater alluvial soils at Suphan Buri, Chai Nat, and Ratchaburi in Thailand, and at Ryugasaki in Japan are rice soils typical of the tropical and temperate regions. Acid sulfate soils (Khlong Luang and Pathumthani) and low humic gley soils (Surin) are grown only to rice in the tropics. The Andosols at Mito and Tsukuba are specific rice soils in Japan.

The CH_4 emissions from the rice fields of freshwater alluvial soils were moderate, whereas emission from the low-humic gley soil (Surin) was very high because of the very low content of free iron. Emissions from the acid sulfate soils (Khlong Luang and Pathumthani) were very low because of the strong acidity. The CH_4 emissions from the two Andosol fields in Japan (Mito and Tsukuba) were significantly lower than from the other Japanese sites.

Cultivation Practices

Some common cultivation practices for rice have substantial effects on CH_4 emissions from rice fields. Cultivation practices vary according to water availability, rice cultivar, soil fertility, and mechanical resources available.

Water management of rice fields is critically important to CH_4 emissions as an anaerobic environment develops in the soil because the oxygen supply from the atmosphere is limited by flooding. Short-term drainage at the tillering stage (midsummer drainage) is commonly performed in Japan. During this period, an increase in soil Eh and a decrease in CH_4 flux were observed.

Climate

Temperature has very important influence on CH_4 emissions. The Q_{10} (the increase rate of CH_4 emission by the increase of 10 °C) has been reported to be about four (Holzapfel-Pschorn and Seiler 1986; Schütz et al. 1989b; Miura et al. 1992a). However, CH_4 emissions from Thai ricefields were not found to be higher in this study. The temperature is much higher at the beginning of rice growth in Thailand than in Japan. Therefore, the main emission stage was found in Japan during the second half of the growing season, and in Thailand during the first half of the growing season (Fig. 6).

Main Route of Methane Transfer Through Rice Plants

Because rice plants consist of many stems of varying ages, the CH_4 emission rates through stems of different ages must be considered. In addition, rice straw and stubble are generally incorporated sporadically in rice fields. Therefore, it is also important to consider where rice straw and stubble are incorporated into the soil. Pot experiments were conducted to compare CH_4 emissions from rice stems of different ages, from soils in which rice straw was incorporated near the soil surface and deep in the plow layer, and from stems of a rice plant with roots exposed to soil with and without incorporated rice straw (Miura et al. 1992a; Watanabe et al. 1994a). The treatments were Treatment 1 – 3 kg soil with 12 g rice straw, Treatment 2 – 1.5 kg soil with 12 g rice straw (upper layer) + 1.5 kg soil without RS (lower layer), Treatment 3 – 1.5 kg soil without rice straw (upper layer) + 1.5 kg soil with 12 g RS (lower layer), Treatment 4 – 1.5 kg soil with 12 g rice straw (one side) + 1.5 kg soil without rice straw (the other side), Treatment 5 – 3 kg soil without rice straw, and Treatment 6 – 3 kg soil with 18 g rice straw. In Treatment 4, soils were separated vertically by a plastic plate that had a V-shape cut on the top. A seeding was transplanted over the plastic plate to develop roots from tillers into both sites equally.

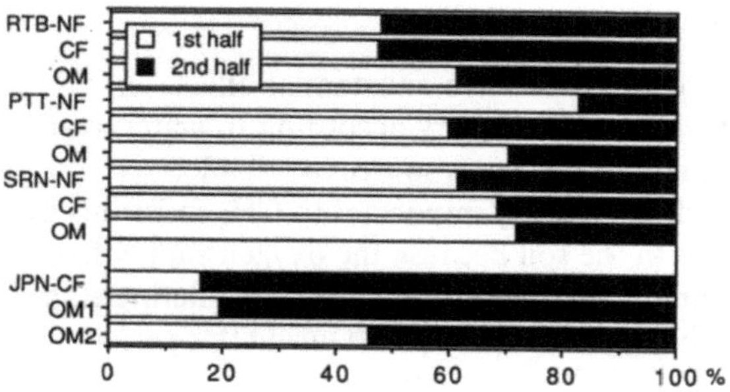

Fig. 6. Percentage distribution of the total amount of CH_4 emitted in the first and the second halves of the growth period

Methane emission rates were measured by the closed-chamber method (Kimura et al. 1991b). For Treatment 4, an acrylic pipe with two compartments was used to measure rates of CH_4 emissions from each side of the pot. To measure CH_4 emission rates from younger and older stems, CH_4 emission rates from six pots of Treatment 6 were measured first. One half of the stems of each tiller were then cut from the outer stems and from the inner stems below the water surface (two pots each). The CH_4 emission rates were then measured again. As the control, CH_4 emission rates from the two pots that remained without any stems being cut were also measured. This measurement was conducted three times (18 July, 8 September, and 13 October).

Effect of Site of Incorporation of Rice Straw

The CH_4 emission rates from Treatment 3, in which rice straw was incorporated in the lower site, were larger than those from Treatments 1 and 2 in the first 49 days after transplanting, although they contained the same amounts of rice straw. The CH_4 emission rates from Treatment 2, in which rice straw was applied in the upper site, were similar to those from Treatment 1, in which rice straw was applied uniformly in the whole soil. The effect of incorporation site was not observed after 27 July. Therefore, the site of rice-straw incorporation influenced CH_4 emissions from rice fields, especially at the early growth stage.

Methane Emission Rates Among Stems

The CH_4 emission rates of Treatment 4 differed markedly according to whether the roots grew in the soil with or without rice straw (Fig. 7). The CH_4 emission rates of tillers with roots growing in straw-amended soil were higher than those

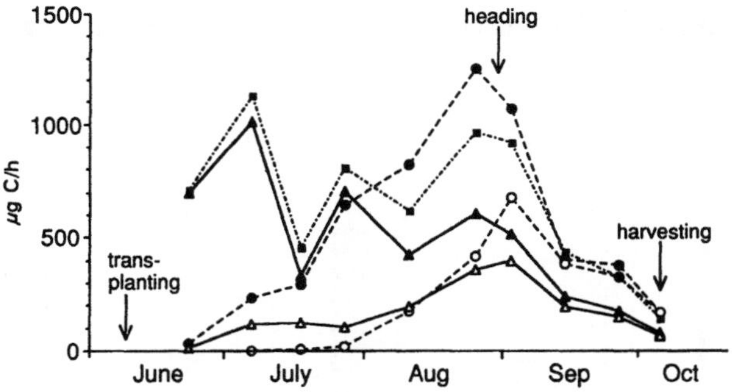

Fig. 7. Effects of incorporation sites of rice straw on CH_4 emission rates. Plot 1 (●); Plot 4 (+RS side) (▲); Plot 4 (–RS side) (Δ); Plot 4 (total) (■); Plot 5 (O). Details of treatments are given in text

from roots in unamended soil. The site of rice-straw incorporation influenced CH_4 emission rates from the stems of a rice plant.

Variations in Methane Emission Rates Between Younger and Older Stems

At the tillering stage (18 July), cutting the older stems decreased the CH_4 emission rates more than cutting the younger stems (Fig. 8). The contribution of older and younger stems to the total CH_4 emission rate from a whole plant was about 60 and 40 %, respectively. Differences in the contribution between older and younger stems became wider at the milky stage (8 September). These results suggest that the older stems are the main transfer routes of CH_4 to the atmosphere in the early and middle growth stages. At harvest (13 October), there was no significant difference in CH_4 emission rates between younger and older stems.

Methane Flux Distribution to the Atmosphere and the Subsoil

The flux of CH_4 to the atmosphere is not the sole direction of CH_4 produced in the rice field. Laboratory experiments have shown that a portion of the CH_4 produced in the plow layer is carried to the subsoil by percolation (Kimura et al. 1992b). A pot experiment was conducted to elucidate the effects of percolation rate and of rice-straw application on CH_4 fluxes both to the atmosphere and the subsoil (Murase et al. 1993).

During rice cultivation, pots were percolated every 2 days with 0, 200, and 600 ml (corresponding to 0, 5, and 15 mm/day of percolation). Leachate was collected every 6 days, and the CH_4 concentration in the leachate was determined by gas chromatography.

There was no difference in CH_4 amounts emitted to the atmosphere among planted RS (rice straw) treatments with different percolation rates. However, CH_4 concentrations in the leachate from the planted RS treatment with 15 mm day percolation was higher than that with 5 mm/day percolation (Table 4). The total amount of CH_4 leached was more than three times higher in the former (251 mg C)

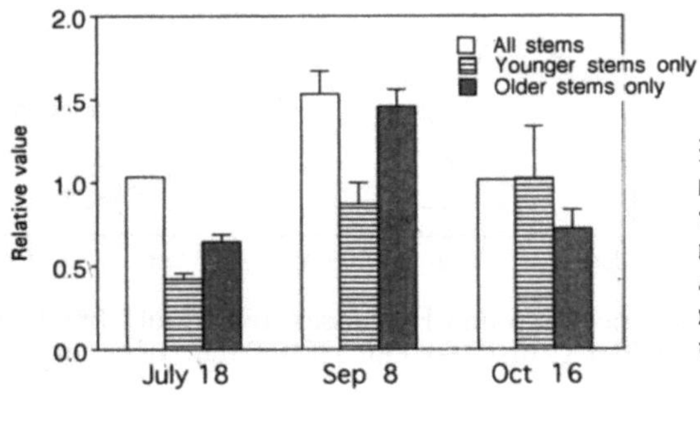

'ig. 8. Contribution of younger ınd older stems to CH_4 emission 'ates from a rice plant (relative valıes of CH_4 emission rates of the econd measurement against those)f the first measurement). *Bars* represent half ranges of standard error

Table 4. Total CH_4 fluxes to subsoil (mg C/pot)

	To atmosphere			To subsoil		
	Early stage	Late stage	Total	Early stage	Late stage	Total
CF treatments (5 mm/day)						
Planted	267	1327	1594	2.4	36.1	38.5
Nonplanted	1.6	5.4	7	3.7	73.1	76.8
RS-treatments						
0 mm/day						
Planted	1124	1638	2761	0	0	0
Nonplanted	40.4	8.8	49.2	0	0	0
5 mm/day						
Planted	1080	1441	2521	29.7	41.3	71
Nonplanted	20.6	25.9	46.5	59.1	89.3	148
15 mm/day						
Planted	1211	1457	2668	99.6	152	251

CF: Chemical fertilizer, RS: rice straw.

than in the later (71 mg C). This suggests that CH_4 production in a submerged soil is accelerated by a higher percolation rate.

Cumulative CH_4 fluxes were divided into early and late stages at the end of July, when rice plants reached the panicle initiation stage (Table 4). The physiological properties of the rice plant induce a reductive environment in the rhizosphere about this stage (Kimura et al. 1982). In planted RS treatment, nearly equal amounts of CH_4 were released to the atmosphere in both stages, but in the planted CF (chemical fertilizer) treatment, the late stage was dominant (83% of the total CH_4 emissions). The amount of CH_4 emitted to the atmosphere from the planted CF treatment in the late stage was almost the same as from planted RS treatments at the late stage. The amount of CH_4 leached to the subsoil was biased more to the late stage: 94–95% of the total in CF treatments and 58–60% in RS treatments, irrespective of the presence of plants or percolation rate. In addition, the amounts of CH_4 leached to the subsoil in the late stage were similar in CF and RS treatments with a percolation rate of 5 mm/day, irrespective of plant growth. The total CH_4 flux to the subsoil accounted for 2–9% of the sums of emitted plus leached CH_4 in planted and 76–92% in nonplanted treatments, respectively.

Methane in Soil and Its Emission Rates (Apparent Turnover Time)

The relationships between the amounts of CH_4 in the soil and the rates of CH_4 emission were studied. Rice plants were grown in the submerged soil in pots, and the CH_4 emission rates were measured throughout the growth period (Watanabe et al. 1994b). Figure 9 shows the CH_4 emission rates and the amounts of CH_4 in the soil (CH_4 in bubbles plus CH_4 in soil solution). The CH_4 content in the soil with RS reached a peak on 2 July and decreased until 6 August. The CH_4 content then increased again until harvest (Fig. 9A). On the contrary, the CH_4 content in the soil without RS was very low until 6 August (less than 0.2 mg/pot, Fig. 9B).

The apparent turnover time of the CH_4 in the soil was calculated from the ratios of CH_4 in soil to CH_4 emission rates. The apparent turnover times for the treatment with RS were 14 days on 2 July, 2.0–3.2 days during 16 July to 24 August, and 6.3–12 days in September and later. In comparison, the apparent turnover times were less than 1 day for the treatment without RS until September (except on 6 August, 2.1 days), and increased from 2.9 to 13 days from 10 September to 9 October. Therefore, the CH_4 in the soil was replaced by new CH_4 within 2–3 days during the late tillering stage and the flowering stage for the treatment with RS. For the treatment without RS, the CH_4 emitted to the atmosphere until September was replaced as soon as CH_4 was produced in the soil.

Methane Oxidation in the Plow Layer and Subsoil

A model experiment using soil columns was conducted to clarify CH_4 production and decomposition (anaerobic oxidation) in the plow layer of rice soils (Murase and Kimura 1994). The plow layer soil with and without rice straw was packed in a glass column consisting to two soil layers. The column was submerged and the leachate was collected periodically. The CH_4 content in the leachate from the upper soil layer with rice straw was decreased by percolation through the lower soil layer without rice straw (Fig. 10). This indicates that the CH_4 produced in the upper soil layer was oxidized in the lower anaerobic layer.

The rate of CH_4 consumption in the subsoil was also observed. Methane produced in the plow-layer soil column was leached into the subsoil column. Some portion of the leached CH_4 was consumed by percolation through the

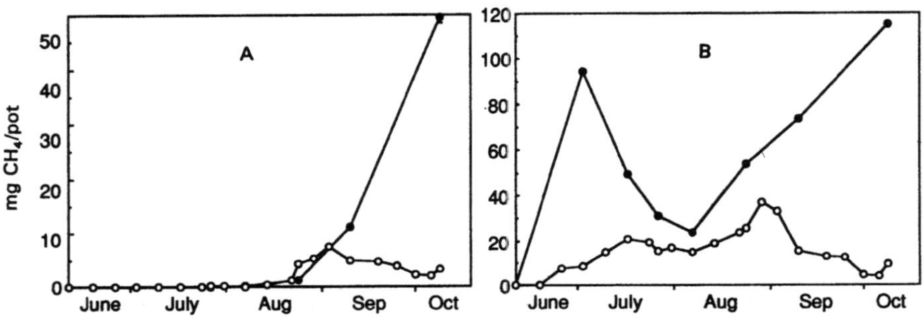

Fig. 9A,B. Methane emission rates per day (O) and amounts of CH_4 in soil (●). **A** treatment without RS application. **B** Treatment with RS application

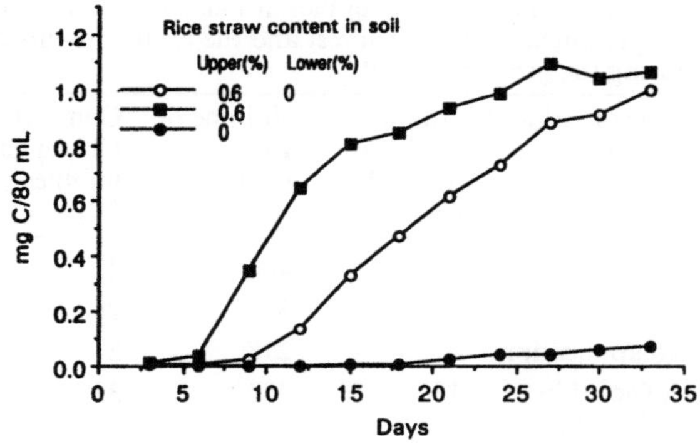

Fig. 10. Time course of CH_4 content in leachate from plow layer soils. One (●, ■) or two (O) layer(s) of soil received with rice straw at the concentrations shown in the figure (w/w %)

subsoil (Miura et al. 1992b). A pot experiment with rice plants showed that about 70% of the CH_4 leached from the plow-layer soil was consumed in the subsoil during the cultivation period (Murase and Kimura 1995).

Fate of Methane After Drainage

Pot experiments were conducted to elucidate the fate of CH_4 in rice fields at harvest. The amounts of CH_4 retained in the soil at harvest were measured. Then, the amounts of CH_4 released to the atmosphare and to the subsoil following drainage, and the amount of CH_4 that remained in soil after drainage were recorded (Watanabe et al. 1994a).

Amounts of Methane in the Soil at Harvest

Table 5 shows the volume of gas bubbles, the concentration of CH_4 in gas bubbles, and the amounts of CH_4 in the soil, as well as the total amounts of CH_4 emitted to the atmosphere during the whole growth period (Watanabe et al. 1993, 1994b). The amounts of CH_4 retained in the soil were markedly different, depending on the amounts of rice straw applied to the soil, although CH_4 emission rates were similar at harvest. The amount of CH_4 was smaller in the treatment with 18 g of rice straw that decomposed in the field (before collection in February) than in the treatment with 18 g of fresh rice straw (Murase et al. 1993). The amounts of CH_4 retained in the soil at harvest corresponded to 3.7–6.2% of the CH_4 emitted to the atmosphere during the whole growth period.

Fate of Methane

The amount of CH_4 drained, and the amount of CH_4 remaining in soil after 4 days of drainage, were much smaller than the amounts of CH_4 released to the atmos-

Table 5. Volume of gas bubbles in soil, concentration of CH_4 in gas bubbles, amounts of CH_4 retained in soil at harvest and the total amounts of CH_4 emitted to the atmosphere during the growth period[a]

Type of straw applied	Amount of rice straw applied (g)	Volume of gas (ml)	Conc. of CH_4 in gas bubbles (%)	CH_4 retained in soil (mg C)	Total CH_4 emission during growth period (mg C)[b]
Fresh	18	460	32	78.9	1790
Fresh	12	365	28.3	55.3	1480
No application	–	218	27.5	32.1	516
Collected in February	18	383	33	67.7	1520

[a] Values are means of duplicates.
[b] Watanabe et al. (1993).

Table 6. Fate of CH_4 after drainage at harvest

	Retained in soil before drainage (mg C)	After drainage			
		Released to the atmosphere (mg C)	Drained (mg C)	Remained in soil (mg C)	Total (mg C)
1[a]	59.5	75.3	0.29	0.33	75.9
2	95.1	79.0	1.30	0.17	80.5

[a] Treatment 1 – pot treated with RS collected in April, Treatment 2 – pot treated with RS with lipid fraction removed.

phere after drainage (Table 6). The amounts of CH_4 released to the atmosphere equaled 95–99% of the total CH_4 recovered. The percentage of CH_4 released to the atmosphere to CH_4 retained in the soil at the onset of drainage was 83–126%. Therefore, release to the atmosphere is the main fate of CH_4 after drainage, and oxidation of CH_4 in the soil after drainage may not be important even if it does take place. Emission to the atmosphere during the growth period is the main route of CH_4 in rice-planted sites, whereas leaching to the subsoil and flux to the atmosphere are the dominant fates of CH_4 after drainage in nonplanted sites.

Fate of Methane in the Rice Field Ecosystem

Methane produced at the reduced soil layer in rice fields in not only transferred to the atmosphere and to the subsoil (Murase et al. 1993), but also oxidized at various sites in the rice field. The oxidation of CH_4 is known to occur aerobically at the oxidized layer in rice fields (Holzapfel-Pschorn et al. 1985), the rice rhizosphere (Holzapfel-Pschorn et al. 1985, 1986), and probably also anaerobi-

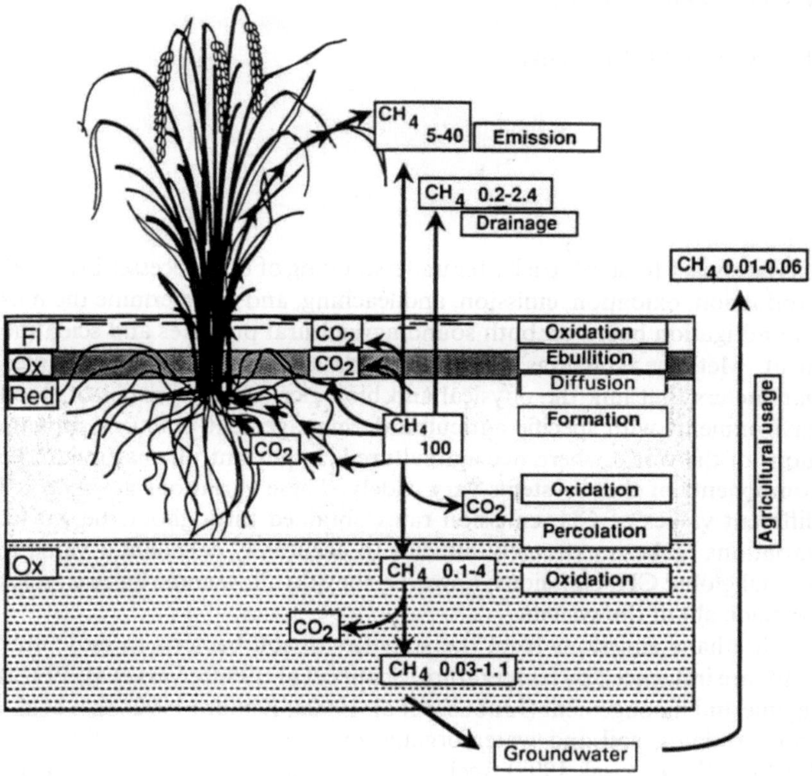

Fig. 11. The fate of CH_4 produced in rice fields

cally at the reduced plow layer and the subsoil (Miura et al. 1992a,b; Murase and Kimura 1994). Methane is also sometimes retained in the plow layer (Denier van der Gon et al. 1992; Watanabe and Kimura 1995) and then undergoes various fates.

Figure 11 summarizes the fates of CH_4 in the rice field ecosystem with the probable distribution amounts of CH_4 (the amount of CH_4 produced in the plow layer is assigned a value of 100). The release of CH_4 to the atmosphere from groundwater used in agriculture was calculated according to Watanabe et al. 1994c. Other values in Fig. 11 were compiled from various field and pot experiments (Holzapfel-Pschorn et al. 1985, 1986; Inubushi et al. 1989; Miura et al. 1992b; Murase et al. 1993; Watanabe and Kimura 1995).

Methane Emissions from Texas Rice Fields: a Five-Year Study

R.L. Sass and F.M. Fisher

It is necessary to develop a better understanding of the processes involved in CH_4 production, oxidation, emission, and leaching, and to determine the most effective mitigation based on both sound agricultural practices and scientific judgement. Methane emissions from rice fields are governed by a complex set of parameters that link the physical and biological characteristics of flooded soil environments with specific agricultural management practices. In various regions of the world where rice agriculture is important, the natural and cultural components of these systems vary widely. These variations are reflected in the different values of CH_4 emission rates obtained throughout the world. These variations make it extremely difficult to accurately calculate a value for total annual global CH_4 emissions; however, they do allow some basic conclusions to be made about the system.

Methane emissions from the activities of soil bacteria in agricultural wetlands are influenced by temperature (Schütz et al. 1989b; Sass et al. 1992a), water regime and management (Inubushi et al. 1990a; Kimura 1992; Sass et al. 1992a), root exudates, soil and water organic residues (Neue et al. 1990; Sass et al. 1991a,b; Wang et al. 1992) (including fertilizers and other chemical additives), plant physiology (Nouchi et al. 1990) and soil physical, chemical, and biological properties (Yagi and Minami 1990b; Van der Gon et al. 1992). It is the net result of opposing bacterial processes, production in anaerobic microenvironments, and consumption and oxidation in aerobic microenvironments. It is also the net result of two opposing plant-biophysical processes, transport of oxygen (O_2) to the rhizosphere and transport of CH_4 out of the soil. Methane transport takes place through the aerenchyma, a specialized gas-conduit system common to wetland plants. These processes may well show independent responses to the same environmental forcing function. Therefore, emission measurements must be obtained in conjunction with and correlated with, measurements of these processes.

Department of Ecology and Evolutionary Biology, Rice University, Houston, TX 77251, USA

S. Peng et al. (Eds.) Climate Change and Rice

Site Description

Field measurements were performed in rice fields at the Texas A&M University Agricultural Research and Extenstion Center near Beaumont, Texas (94°30′W, 29°57′N) and its stations near Bay City and Eagle Lake, Texas. Rice is the main crop in this area, which has an annual growing season of approximately 275 days and only 15 days with temperatures below 0 °C. Annual rainfall averages 1340 mm, of which about 50% (122 mm/month) occurs during the rice-growing season in April through September. Soybeans are generally rotated with rice. Native vegetation is coastal prairie (Kuchler 1964; Correll and Johnston 1970).

The fields selected for this study represented the soil types typical of the Texas coastal prairie (Crout et al. 1965). Most studies were done on one of three clay soils: Beaumont clay, an Entic Pelludert, Lake Charles clay, a Typic Pelludert that is slightly less acid and stronger in structure, and Bernard-Morey, a fine clayey silty loam thermic Vertic Ochraqualf. All three soils have poor internal and surface drainage with percolation rates of less than 0.5 mm/day after inital saturation (Brown et al. 1978). In addition, some comparison studies were done on a Katy-Crowley soil association (Alfisols), a fine sandy loam, at the Eagle Lake Station west of Houston, Texas. These four soils are representative of the majority of the rice-growing areas of the Texas Gulf Coast.

Factors Effecting Methane Emissions

To date, full seasonal field production and emission data sets have been collected using 36 different experimental conditions. In addition, five different soil types are represented in our work in fields located along the Texas coast of the Gulf of Mexico. Investigations have been conducted to establish the magnitude of the emission of CH_4 and the interaction of CH_4 production, consumption, and emission processes in the field. We have studied the effects on CH_4 emission of agricultural practices such as nitrogen fertilization, organic amendments, flood water regime, and field preparation, as well as environmental factors such as temperature, plant growth, weather, climate, and soil properties.

Diel variations in CH_4 emissions were largely due to temperature variation, and observed daily cycles in CH_4 emission levels can be explained by the daily cycle in soil temperature (Sass et al. 1991a). Figure 1 shows a typical plot of both CH_4 emissions and soil temperature (10 cm depth). The response of CH_4 emissions to soil temperature change is very rapid and consequently no phase difference is observed between the temporal course of the soil temperature and that of CH_4 emissions. Figure 2 shows the Arrhenius plot using data collected over a 24-h period in two fields, one with and one without straw additions. Figure 3 compares experimental diel emission data with values calculated from the Arrhenius parameters.

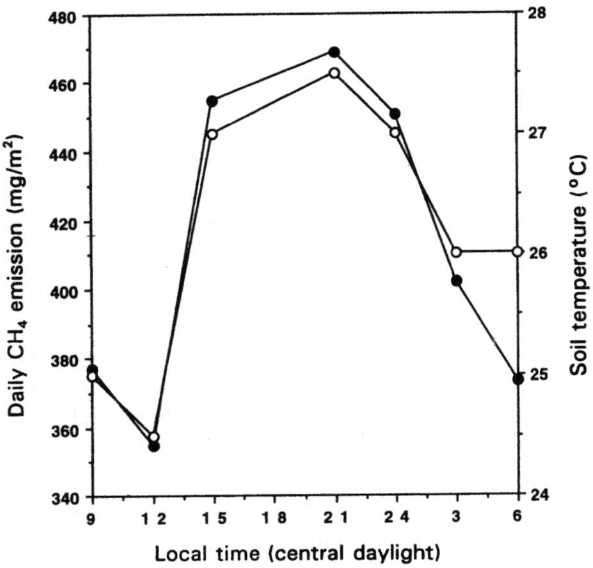

Fig. 1. Diel relationship between soil temperature (O) and CH_4 emissions (●) with time. Data collected in 1990

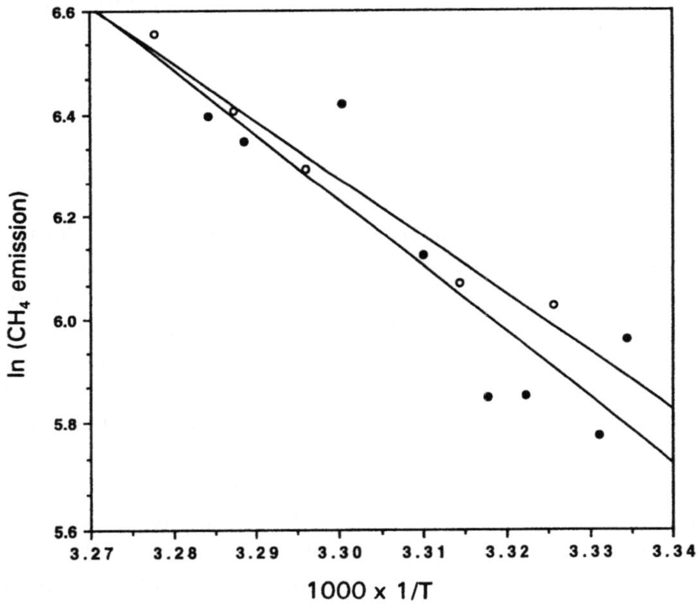

Fig. 2. Arrhenius graph of the diel dependence of the rate of CH_4 emission with temperature. Data are from two fields, one with straw incorporation (O) and one without (●). For ● ln NS_{ave} $y = 47.915 - 12.632\ x$ ($r^2 = 0.790$); for O ln S_{ave} $y = 43.196 - 11.189\ x$ ($r^2 = 0.976$). Data collected in 1990

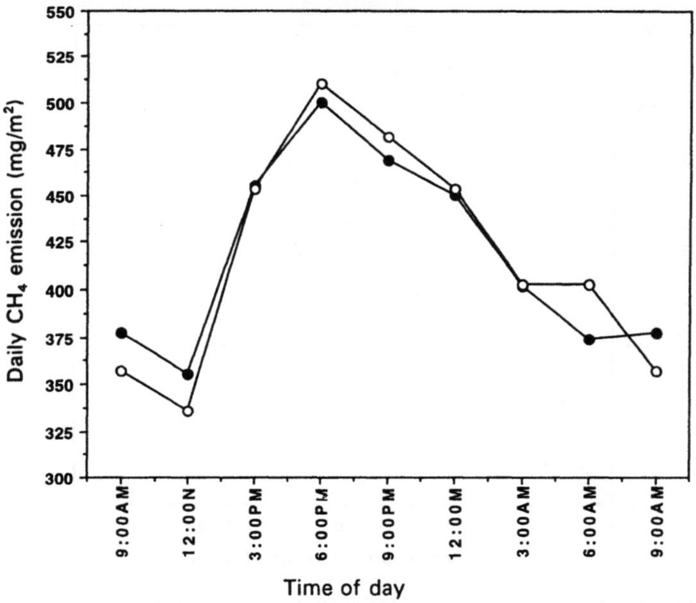

Fig. 3. Diel relationship of observed (●) CH_4 emissions and CH_4 emissions calculated (O) from Arrhenius parameters. Data are plotted against the time of day on 31 July 1990

The temperature dependence of CH_4 production observed in anaerobic soil incubations shows the same Arrhenius dependence (same activation energy). This indicates that the limiting step that determines the rate of CH_4 emission is the same as that for CH_4 production (Sass et al. 1991b). We conclude from these experiments that the majority of the CH_4 emissions to the atmosphere were through the rice plant and took place rapidly after being produced. Furthermore, in these dense clays with low porosity and percolation rates, only minor buildup of CH_4 in the pore water has been observed (< 400 μm) and CH_4 emission by ebullition or bulk diffusion has been observed to be minimal.

Although daily variations in CH_4 emissions were strongly temperature-dependent, seasonal variations in CH_4 production and emission followed plant development with no apparent temperature dependence (Sass et al. 1990, 1991a,b, 1992a). From negligible values at permanent flood, CH_4 emissions generally rose during the vegetative phase. Emissions peaked at panicle differentiation during a period of rapid root development, probably due to increased root exudation from the rapidly growing root tips. Emissions were relatively constant during the reproductive stage, and decreased during late grain filling. During the period from permanent flood to past the end of the reproductive stage (65–75 days), CH_4 emissions correlated with above-ground biomass (Fig. 4). Prior to the end of the flooded season, an emission peak was generally observed. This late-season increase in emissions was attributed to an increase in carbon substrate in the soil because of leaf and root senescence. Methane emissions increase while

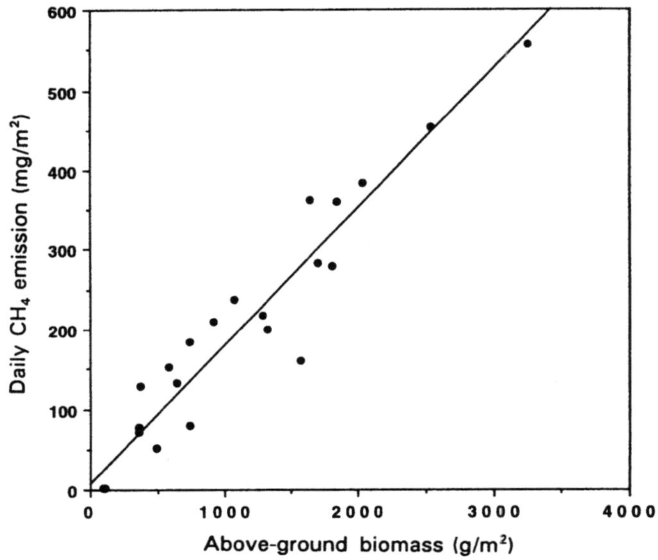

Fig. 4. Methane emissions correlate with above-ground biomass in the Lake Charles soil. *Line* is the least squares best fit regression through all data. ($y = 6.7021 + 0.17279$ x, $r^2 = 0.919$). Data collected in 1989

live biomass decreases. The addition of readily degradable carbon (e.g., rice grass or straw) before planting increased early-season CH_4 emissions. In a soil of given clay content, CH_4 production is correlated with the local root biomass (Sass et al. 1990). In the early season, production is concentrated near the base of the rice plants. As the season progresses and the root system extends deeper and laterally further from the base of the plant, CH_4 production in these regions increases along with root biomass.

By varying planting date during the same season, three different fields, both with and without incorporated straw, were subject to different climate variables, including integrated solar radiation (Sass et al. 1991b). Seasonal emission rates of CH_4 and the yield of rice grain from individual fields were positively correlated with accumulated solar radiation for both straw-incorporated and control plots (Fig.5). Linear regression analyses of these data showed that: a 1% increase in accumulative solar radiation is accompanied by a 1.1% increase in CH_4 emission and a 1% increase in the yield of rice grain. In the presence of incorporated straw, a 1% increase in solar radiation is accompanied by a 1.7% increase in CH_4 emission and a 2.2% increase in the yield of rice grain. However, straw incorporation resulted in an overall decrease in grain yield and an overall increase in CH_4 emission. We hypothesize that solar radiation and, therefore, photosynthetic activity of the rice plant correlates with CH_4 production and grain yield through partitioning of nonstructural carbohydrates to the root system and grain panicle. If photosynthates are available to form root exudates, then the amount of plant-

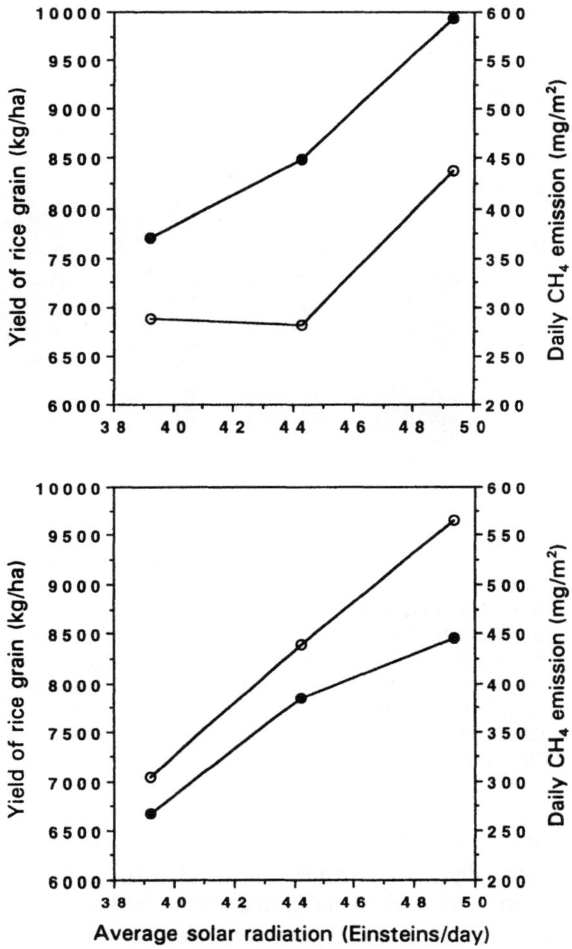

Fig. 5. Grain yield (●) and seasonal CH_4 emissions (O) plotted against the average solar radiation during the 42-day critical growing period (panicle initiation 21 days) for three different planting dates. The *top graph* is for fields without incorporated straw; the *bottom graph* is for fields with incorporated straw. Note that the relative amounts of CH4 and grain yield are reserved in the two cases. Data collected in 1991

derived substrate available for methanogenesis is directly associated with solar radiation. If straw incorporation affects root respiration in such a way as to cause additional root carbohydrate fermentation or loss, then the partitioning of photosynthates may be altered from grain formation to increased root exudation and subsequent CH_4 production and emission. Figure 6 illustrates this hypothesis with data from this experiment in a Bernard-Morey soil and another experiment in Beaumont and Lake Charles soils (Sass et al. 1991a). Rough rice grain, dried to 12% moisture, contains 65.2% carbohydrate. Therefore, 1 g of rice grain is equivalent to 0.65 g of carbohydrate. Emitted CH_4 is ultimately derived from plant carbohydrate by the overall reaction:

$$C_6H_{12}O_6 \rightarrow 3\,CH_4 + 3\,CO_2\,. \qquad (1)$$

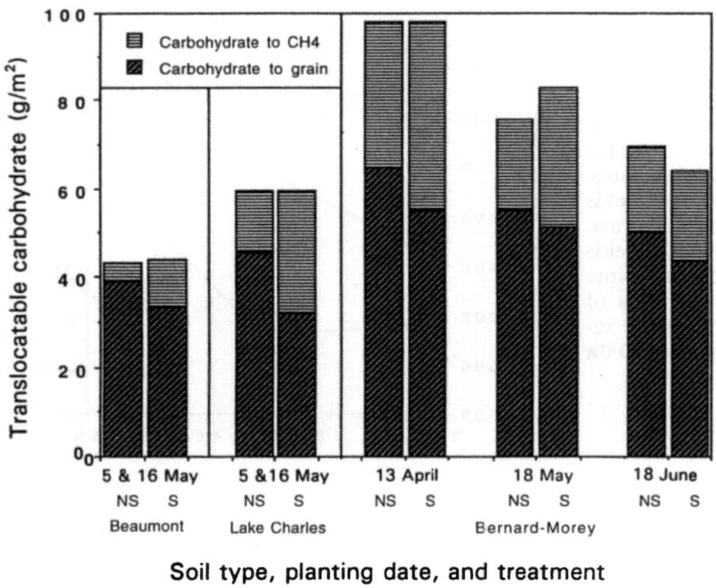

Fig. 6. Calculated amount of translocatable carbohydrate allocated to CH_4 emission and grain yield for fields with (*S*) and without (*NS*) incorporated straw. Three different soils (Beaumont, Lake Charles, and Bernard-Morey) are represented with data collected over a 3-year period. *Dates* refer to planting date. With incorporated straw, some carbon that would have gone into grain production is redirected to the root system and ultimately into increased CH_4 emissions with subsequent reductions in grain yeild. Data collected in 1989, 1990, and 1991

Incubation data (Sass et al. 1990) show that 58% of the CH_4 produced is oxidized bacterially prior to emission. Therefore, by combining the compositional information on the yield of rice grain with the reaction stoichiometry for the formation and oxidation of CH_4, we can calculate that 1 g of emitted CH_4 is equivalent to 8.93 g of plant carbohydrate.

Emission data obtained between 1989 and 1991 were collected from fields composed of three different soils (Beaumont, Lake Charles, and Bernard-Morey). Averages of the seasonal values of CH_4 emissions obtained from each soil (Fig. 7, open squares) show a strong correlation (straight line) when plotted with the percentage of sand in the soil. To test this correlation, we have measured both emitted CH_4 and percentage of sand in ten experimental sites established along a transect through a field containing a soil sand-clay-silt gradient ranging from 15 to 35% sand (Sass et al. 1994). Duplicate plots were established at each site (A and B). The CH_4 emissions were positively correlated with sand content and negatively correlated with clay content. The dependence on sand was stronger ($r^2 = 0.6$). The slope of the regression line through these 20 points was the same as that of the line through the average values from the original three soils. In the same experiment, the yield of rice grain was not significantly influenced by the sand-silt-clay ratio.

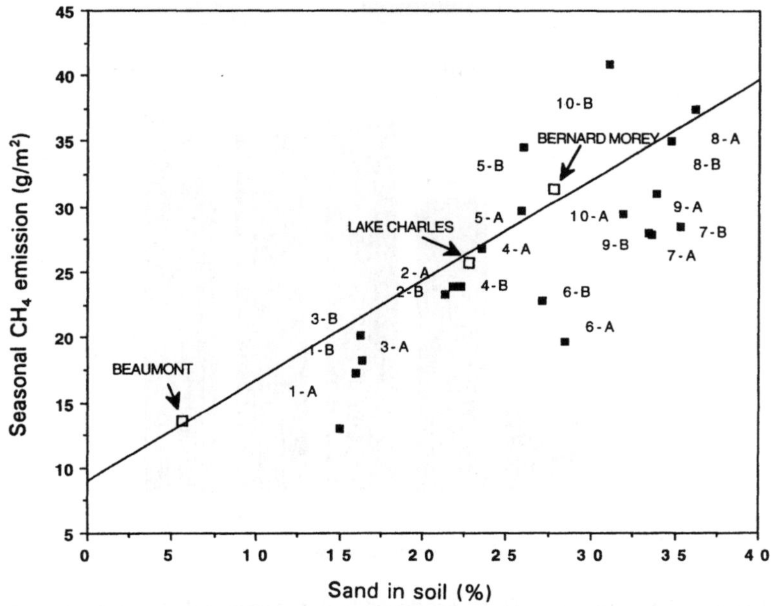

Fig. 7. Effect of sand content of the soil on seasonal CH_4 emissions. The *line* is the best fit linear regression line through average CH_4 emission values (*Open squares*) for all data obtained in the three soils during the past 5 years. Other data (*closed squares*) are from a duplicate (*A* and *B*) ten plot (*1–10*) transect through a soil-sand gradient. Data collected in 1992

Four water management schedules were investigated: normal permanent flood (46 days postplanting to harvest drain), normal permanent flood with a midseason drainage aeration (6 days immediately following panicle differentiation), normal flood with multiple drainage aeration of 2–3 days each, and late flood (76 days postplanting) (Sass et al. 1992a). The CH_4 emission rates varied markedly with water regime. Periodic drainage of irrigated ricefields significantly decreases CH_4 emissions. A single midseason drain reduced CH_4 emissions by approximately 50% compared with a normal water-management schedule (4.86 g/m^2 compared with 9.27 g/m^2). A short period of drainage (2 days) approximately every 3 weeks during the growing season can reduce seasonal CH_4 emissions from irrigated ricefields to an insignificant amount (<1 g/m^2). Methane emissions can be reduced to near zero by field drainage, and CH_4 production and oxidation values remain high. In the normally treated field, CH_4 oxidation increases as the season progresses and may account for as much as 81% of the CH_4 produced (Fig. 8). In the fields with late flooding and with a midseason drainage, CH_4 oxidation was as high as 94% of the CH_4 produced. Periodic short periods of water drainage do not appear to reduce the yield of rice grain. However, delaying initial flooding for too long may result in a delayed but intensified pattern of CH_4 emissions and a significant loss of yield of rice grain (Sass et al. 1992a).

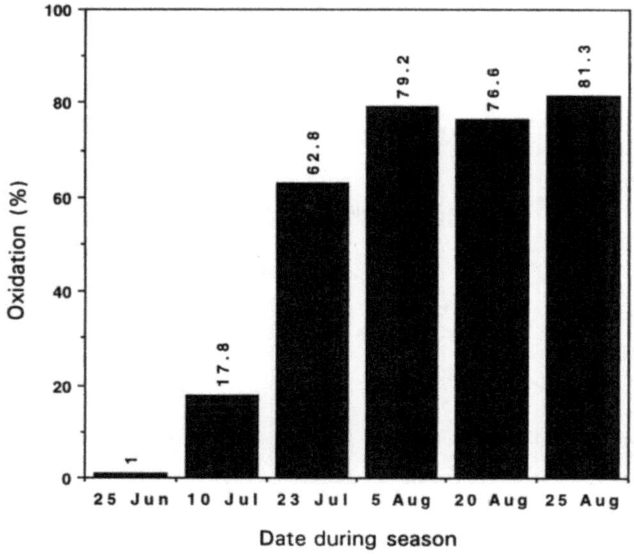

Fig. 8. The percentage of CH_4 produced that was oxidized at six different times during the rice-growing season of 1991. Methane emission data were obtained in the field, whereas CH_4 production data were obtained from anaerobic incubation of field soils. Methane oxidation is assumed to be equal to CH_4 emission minus CH_4 production. The fraction oxidized equals (production minus emission)/production

Straw incorporation influences CH_4 emission in two ways. Depending on the amount of straw added, CH_4 emisions can be increased only during the 2–3 weeks following permanent flooding or can be increased throughout the flooded season (Sass et al. 1991a). When straw incorporation causes an increase in CH_4 emissions over the whole season, yields of rice grain decrease proportionately. Over a 3-year period, the degree of seasonal CH_4 emissions from a specific field was lowest when the field remained fallow for an extended period before planting, intermediate in following years when only the roots and low stubble from the previous year were tilled into the soil, and highest when additional straw was added before planting. The increase in CH_4 emissions with additional straw amendments depended on the method of incorporation. The lowest increase occurred when the straw was tilled into the field before the winter season. This treatment gave the maximum time for aerobic decomposition before rice planting. Higher increases in CH_4 emissions were observed when the straw was tilled into the field immediately before planting and when the rice stubble from the previous year was not tilled. The highest increase in emissions was observed when the applied straw was partially burned (Sass et al., unpubl.).

Preliminary studies in 1993 indicated that the choice of rice cultivar has a substantial effect on the amount of CH_4 emitted to the atmosphere during the growing season (Sass et al., unpubl.). The wide variation of plant traits and related CH_4 emission rates among cultivars provides an opportunity to explain the wide variability in observed CH_4 emission from individual regions of the world. It also helps explain the uncertainties in estimated source strength and presents an opportunity to stabilize or reduce global CH_4 emissions while maintaining grain yields through the characterization and development of appropriate rice cultivars. We have surveyed ten rice cultivars appropriate to temperate and subtropical irrigated rice fields. The seasonal CH_4 emission rates from these cultivars varied from 17.95 to 41.05 or by a factor of 2.3 (Fig. 9).

Conclusion

Methane emissions from rice fields are governed by a complex set of parameters that link the physical and biological characterisitics of flooded soil environments with specific agricultural management practices. In various regions of the world where rice agricultrue is important, the natural and cultural components of these systems vary widely. These variations are reflected in the different values of CH_4 emission rates recorded in the results of these experimetns (Table 1) and those of other investigators in other parts of the world. Although these variations make it extremely difficult to accurately calculate a total value for annual global CH_4

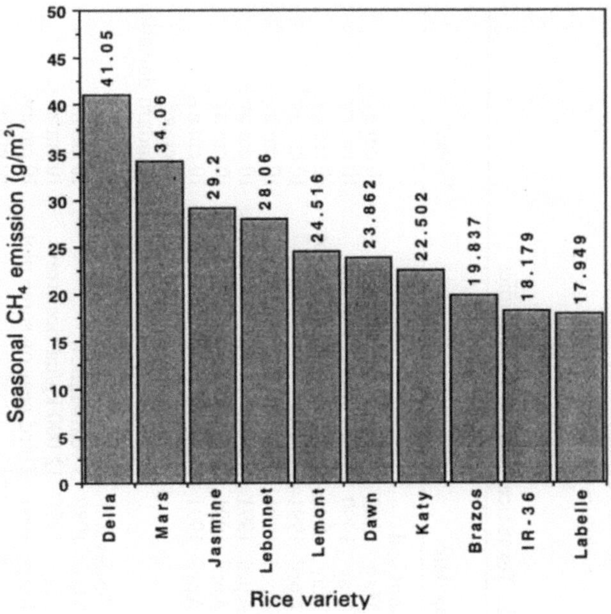

Fig. 9. Seasonal CH_4 emission values from ten different rice cultivars, all adapted to grow in the Gulf Coast region of Texas. Data collected in 1993

Table 1. Texas rice field methane studies

Planting date	Treatment	Water	Solar Rad (Ei/m^2)	Daily CH_4 (mg/m^2)	Seasonal CH_4 (g/m^2)	Rice yield (kg/ha)
Beaumont soil						
5 May 1989	3 years fallow	10 cm av	46.08	59.9	4.5	6000
16 May 1990	8000 kg/ha rice straw	10 cm av	43.99	134.2	11.4	5160
7 May 1991	Third year on same plot	10 cm av	35.39	182.5	15.33	6985
3 April 1991	Fallow	10 cm av	44.05	152.4	13.11	8320
3 April 1991	Late till (late March)	10 cm av	44.05	183.9	15.81	8025
3 April 1991	No till	10 cm av	44.05	191.95	16.51	8532
3 April 1991	Burned straw	10 cm av	44.05	209.39	18.01	8595
3 April 1991	Early till (September)	10 cm av	44.05	165.31	14.22	8742
		Average		159.94	13.61	7545
		Standard deviation		46.77	4.21	1349
Lake Charles soil						
5 May 1989	3 years fallow	10 cm av	46.08	209.7	15.9	7000
16 May 1990	12 000 kg/ha rice straw	10 cm av	43.99	364.6	31	4900
7 May 1991	Third year on same plot	10 cm av	35.39	358.69	30.13	6256
		Average		311	25.68	6052
		Standard deviation		87.78	8.48	1065
Bernard-Morey soil						
13 April 1990	4 years fallow	10 cm av	49.3	438.5	37.27	9942
13 April 1990	6000 kg/ha grass straw	10 cm av	49.3	563.9	47.93	8461
18 May 1990	4 years fallow	10 cm av	44.24	282.1	22.85	8496
18 May 1990	6000 kg/ha grass straw	10 cm av	44.24	438.9	35.55	7843
18 June 1990	4 years fallow	10 cm av	39.2	289.1	21.97	7698
18 June 1990	6000 kg/ha grass straw	10 cm av	39.2	304.5	23.15	6680
		Average		386.17	31.45	8187
		Standard deviation		113.16	10.54	1084

Lake Charles to Bernard-Morey						
23 April 1992	37.9% clay/19.08% sand	10 cm av	42.86	162.415	15.11	5294
23 April 1992	37.1% clay/18.4% sand	10 cm av	42.86	253.5	23.58	5373
23 April 1992	34.8% clay/19.6% sand	10 cm av	42.86	206.35	19.19	6055
23 April 1992	33.2% clay/19.8% sand	10 cm av	42.86	272.789	25.37	5296
23 April 1992	21.6% clay/42.5% sand	10 cm av	42.86	345.386	32.12	
23 April 1992	25.7% clay/29.8% sand	10 cm av	42.86	228.253	21.23	6909
23 April 1992	23.1% clay/30.5% sand	10 cm av	42.86	303.647	28.24	5919
23 April 1992	24.0% clay/31.1% sand	10 cm av	42.86	390.061	36.27	5433
23 April 1992	25.6% clay/29.7% sand	10 cm av	42.86	318.104	29.58	5894
23 April 1992	27.0% clay/28.2% sand	10 cm av	42.86	378.739	35.22	6510
		Average		285.92	26.59	5854
		Standard deviation		74.89	6.96	573
Bernard-Morey soil-special water treatment						
2 May 1991	2nd year planting	Late flood	36.63	151.29	14.98	6396
2 May 1991	2nd year planting	Evap. area	36.63	230.7	20.07	7499
2 May 1991	2nd year planting	Normal flood	36.63	106.54	9.27	8253
2 May 1991	2nd year planting	Midseason	36.63	55.89	4.86	8199
2 May 1991	2nd year planting	Multiple-aeration	36.63	13.21	1.15	8131
		Average		111.53	10.07	6499
		Standard deviation		84.51	7.61	2154
Lake Charles soil-rice variety study						
7 June 1993	Della cultivar	10 cm		647.74	41.05	
7 June 1993	Mars cultivar	10 cm		536.73	34.06	
7 June 1993	Jasmine cultivar	10 cm		401.27	29.20	
7 June 1993	Lemont cultivar	10 cm		403.89	24.52	
7 June 1993	Lebonnet cultivar	10 cm		434.55	28.06	
7 June 1993	Brazos cultivar	10 cm		316.15	19.84	
7 June 1993	Dawn cultivar	10 cm		365.93	23.86	
7 June 1993	Katy cultivar	10 cm		355.54	22.50	

Table 1. (*Contd.*)

Planting date	Treatment	Water	Solar Rad (Ei/m²)	Daily CH_4 (mg/m²)	Seasonal CH_4 (g/m²)	Rice yield (kg/ha)
7 June 1993	Labelle cultivar	10 cm		272.98	17.95	
7 June 1993	IR 36 cultivar	10 cm		285.10	18.18	
		Average		401.99	25.92	
		Standard deviation		115.83	7.37	
Eagle Lake station						
1991	14.1% clay 54.88% sand			250.5	16.032	
1991	14.1% clay 54.88% sand			172.43	11.036	
		Average		211.47	13.53	
		Standard deviation		55.20	3.53	
Bay City site						
1991	43.18% clay 21.62% sand			15.68	1.11	
1991	43.18% clay 21.62% sand			16.33	1.16	
		Average		16.00	1.14	
		Standard deviation		0.33	0.02	
All types of soil and all treatments		Number of observations		46	46	31
		Average		270.03	21.29	7058
		Standard deviation		142.88	10.75	1337

emissions, they do allow one to establish some basic conclusions about the system.

The production of CH_4 depends on the presence of a readily metabolizable source of organic carbon. The primary source is the rice plant itself and depends on the extent of photosynthetic activity of the plant and the extent of the partitioning of the resultant carbohydrate to the rhizosphere. The higher the level of photosynthetic activity in the rice plant, the higher the production of CH_4 in the soil. If the root system of the plant is stressed by high reducing conditions in the soil, more translocatable carbohydrate is transported to the root environment, eventually resulting in an increase in CH_4 emissions and a concomitant decrease in the yield of rice grain. The presence of additional decomposable organic matter will also lead to an increase in CH_4 production and emission.

A primary mitigation switch of the production and emission of CH_4 is the presence of oxygen in the soil-root-water environment. The removal of oxygen from this unique ecosystem is normally accomplished by its consumption by soil bacteria and takes place over a period of about 3 weeks after the field is flooded. The presence of the flood water impedes the diffusion of oxygen from the atmosphere into the soil and keeps it anoxic. By periodically removing the flood water through drainage, the soil may be reoxidized and the emission of CH_4 curtailed. If this procedure is repeated every 3 weeks throughout the rice-growing season, CH_4 emissions can be almost completely prevented. This procedure can only be used effectively on irrigated fields and only if the supply of water is not limiting.

Another candidate for mitigation is the use or development of new rice cultivars. Our data show that CH_4 emissions varied by as much as a factor of 2.3 among a group of different cultivars. Of all atmospheric CH_4 sources, irrigated rice cultivation accounts for approximately 10% of the annual total (50 Tg). This value is essentially equal to the current annual atmospheric increase in CH_4. Therefore, the infrared adsorption characteristics of the atmosphere could be reduced simply by introducing rice cultivars with the lowest CH_4 emission values. From our data, one can predict that the rate of increase of CH_4 in the atmosphere can be decreased by more than 50% by decreasing the CH_4 from rice fields by a factor of 2.5, a ratio similar to that found among existing cultivars.

Other possible candidates suggested by our research for CH_4 mitigation practices are (1) soil consolidation treatments that would cause the soil to become less porous or more clay-like, (2) soil aeration procedures that do not require the use of large amounts of water, and (3) more careful management of organic additions (e.g., prior composting or the use of anaerobic digesters or bioreactors before the organic fertilizers are put on the soil).

Acknowldegments. This work was supported by grants from the United States Department of Agriculture and the United States Department of Energy National Institue for Global Environmental Change and by funding from The Houston Lighting and Power Company.

Role of Methane and Nitrous Oxide in Climate Change

H. Rennenberg, R. Wassmann, H. Papen and W. Seiler

Radiatively active gases are fairly transparent to the incoming shortwave solar radiation, but absorb and reflect long-wave infrared radiation emitted from the surface of the earth. In this way, part of the energy of the long-wave radiation is trapped within the atmosphere. As a consequence, the surface temperature of the earth is kept at the relatively high value. The mean surface temperature of the globe is approximately +15 °C; in the absence of radiatively active gases this temperature would be –18 °C. This natural greenhouse effect of 33 °C is the result of the presence of a combination of about 1% water vapor, 0.04% CO_2, and even smaller quantities of CH_4 and N_2O in the atmosphere (IPCC 1990). An increase in the concentration of these radiatively active gases would enhance global warming. Such an increase has been observed over the last two centuries. This anthropogenic greenhouse effect leads to increased radiative forcing of global climate.

The time course of increase of atmospheric concentrations of the natural greenhouse gases CO_2, CH_4, N_2O and of the anthropogenic greenhouse gas chlorofluorocarbon CFC-11 (CCl_3F) since 1750 are shown in Fig. 1 (IPCC 1990). Data before 1950 are from the analysis of air entrapped in ice cores taken from the Antarctica. Since 1750, there has been a slow linear increase in the concentrations of CO_2, CH_4 and N_2O in the atmosphere. Since the 1950s, there has been an almost exponential increase in the mixing ratios of these radiatively active trace gases in the atmosphere (Fig. 1; IPCC 1990; Duxbury et al. 1993). During the last 240 years, the atmospheric mixing ratio of CO_2 has increased from 280 to 353 ppmv (parts per million in volume) of CH_4 from 0.8 to 1.72 ppmv, and of N_2O from 288 to 310 ppbv (parts per billion in volume; Fig. 1). The residence times, relative absorption capacities, and the relative global warming potential for different time scales are summarized in Table 1 for the natural greenhouse gases CO_2, CH_4, and N_2O. To calculate the global warming potential within different time frames, the effect following addition of 1 kg of each gas was calculated (Shine et al. 1990; Duxbury et al. 1993). Compared with CO_2, CH_4 has a 58 times higher relative absorption capacity per unit mass. Its residence time in the

Fraunhofer Institute for Atmospheric Environmental Research, Kreuzeckbahnstraße 19, 82467 Garmisch-Partenkirchen, Germany. R. Wassmann currently seconded to International Rice Research Institute, P.O. Box 933, 1099 Manila, Philippines

S. Peng et al. (Eds.) Climate Change and Rice

Fig. 1. Atmospheric increases in CO_2 (–·–, 280→353 ppmv), N_2O (···, 288→310 ppbv), CH_4, (–···–, 0.8→1.72 ppmv), and CFC -11 (——, 0→280 ppt) since 1750. (IPCC 1990)

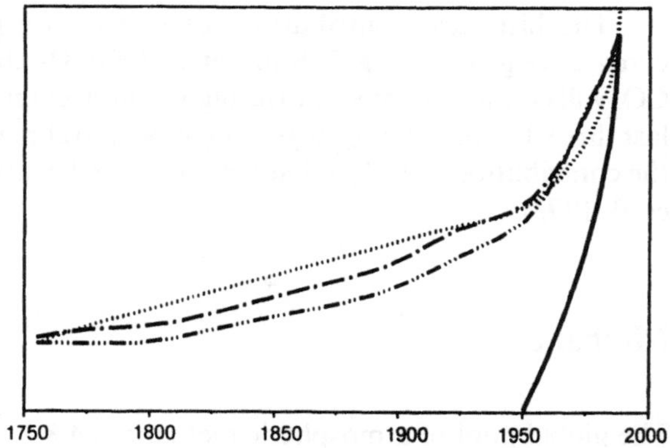

Table 1. Residence time, relative absorption capacity, and global warming potential of CO_2, CH_4, and N_2O. (After Shine et al. 1990; Duxbury et al. 1993)

	Residence time (years)	Relative absorption capacity[a] (unit mass)	Global warming potential[b]	
			20 years	500 years
CO_2	120	1	1	1
CH_4				
Direct[c]	10	58	26	2
Indirect			37[d]	7
N_2O	150	206	270	190

[a] per unit mass change from present concentration.
[b] Following the addition of 1 kg of each gas.
[c] Direct – through direct radiative effect of CH_4; indirect – through the participation of CH_4 in chemical processes that lead to the formation of other radiatively active species.
[d] Estimated contributions of 24, 10, and 3 for generation of tropospheric ozone, stratospheric water, and CO_2, respectively.

atmosphere is only 1/12 that of CO_2. The direct plus indirect relative global warming potential of CH_4 within 20 years was calculated to be 63 times the effect of CO_2; nevertheless, within 500 years the relative global warming potential of CH_4 decreases remarkably to about nine times the value of CO_2, mainly because of the shorter residence time of CH_4 in the atmosphere. Nitrous oxide has a residence time in the atmosphere of 150 years. Its relative absorption capacity is 206 times higher than that of CO_2. Within 20 years the relative global warming potential of N_2O is 270 times higher than that of CO_2 and more than 4 times higher than the global warming potential of CH_4. These values for N_2O decrease only by a factor of 1.42 (compared with CO_2), but increase by a factor of more than 20 (compared with CH_4) if a 500-year scenario is envisaged (Table 1). This indicates the higher effectiveness of N_2O than CH_4 with respect to climate forcing, especially on a long time scale (Shine et al. 1990; Duxbury et al. 1993).

The historical contributions of greenhouse gases to radiative forcing of climate are given in Fig. 2 (Shine et al. 1990; Duxbury et al. 1993). Atmospheric CO_2 will continue to mediate the highest impact on forcing of climate. During the last three decades, the contribution of N_2O to climate forcing increased, whereas, the contribution of CH_4 did not change significantly (Shine et al. 1990; Duxbury et al. 1993).

Methane

The global pool of atmospheric methane is 4.8×10^{15} g (Schlesinger 1992). The highest concentrations of about 1.8 ppm are recorded in the higher latitudes of the Northern Hemisphere; whereas, the lowest concentrations of about 1.6 ppmv are observed in the Southern Hemisphere (Fig. 3). The concentration of CH_4 exhibits a pronounced seasonal variation (Fig. 3), which periodically reduces and enhances the difference between the Northern and Southern Hemisphere to 10 and 4%, respectively. However, the geographical and seasonal patterns can locally be superseded by temporal and spatial variations caused by diurnal cycles of sources, episodic events, and atmospheric mixing patterns.

The abundance of CH_4 in the atmosphere has increased in recent years (1984-1988, Fig. 3). The global CH_4 concentration between 1981 and 1992 increased at a rate 19.8 ± 0.8 ppbv or 1.3% per year (Khalil and Rasmussen 1993). When these values were projected to a longer time scale, the atmospheric CH_4 concentration showed a dramatic potential for increase within the next few decades. However, the recent trend of CH_4 concentration in the atmosphere (1983–1992) reveals that

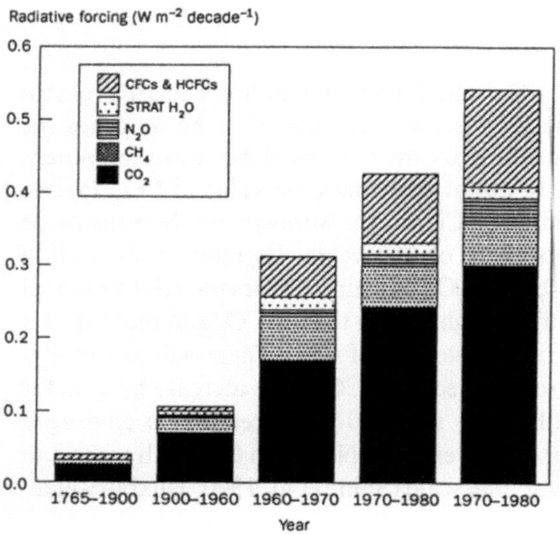

Fig. 2. Historical contributions of greenhouse gases to radiative forcing of climate. (Shine et al. 1990)

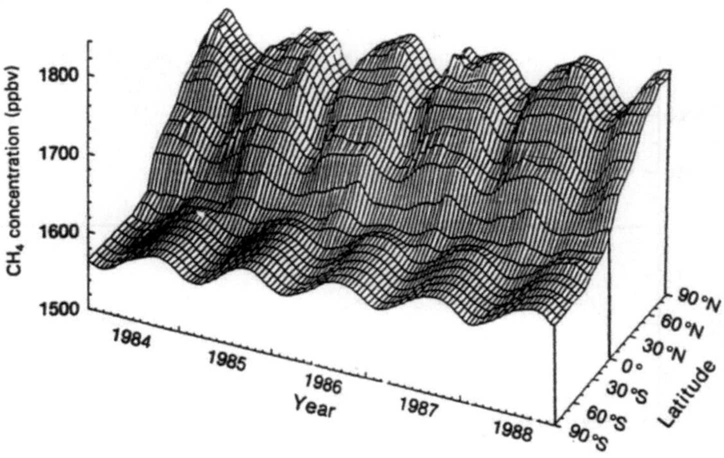

Fig. 3. Global atmospheric distribution of CH_4. (After Watson et al. 1990; Duxbury et al. 1993)

the increase in the abundance of CH_4 in the atmosphere has slowed down (Steele et al. 1992; Khalil and Rasmussen 1993; Dlugokencky et al. 1994; Fig 4). A negative growth rate for atmospheric CH_4 has been described for the Northern Hemisphere during 1992 (Dlugokencky et al. 1994; Fig. 4). The reasons for this tendency are not clear. It seems likely to relate the dramatic economical changes in the former Soviet Union, which has led to a pronounced decrease in exploitation, transport, and consumption of fossil fuels, to this phenomenon (Dlugokencky et al. 1994). Analysis of the isotopic composition of atmospheric CH_4 may be suitable to test this assumption.

The sources and sinks of atmospheric CH_4 as estimated for the 1980s by different authors are compiled in Table 2. Some of the figures for the strength of individual CH_4 sources are connected with a high degree of uncertainty, whereas the predominant sink mechanism as well as the atmospheric accumulation of CH_4 apparently were estimated with reasonable accuracy (Table 2). The chemical oxidation of CH_4 induced by the OH-radical results in an annual scavenging of about 340-500 Tg CH_4. Atmospheric CH_4 is also oxidized in the soil by microbes. This sink is of only minor significance compared with the chemical CH_4 destruction (Table 2). From the increase in the atmospheric concentration during the last decade, an annual difference between sources and sinks of about 40 Tg CH_4 can be calculated.

The figures for the source strengths are derived by extrapolating field data and by using constraints imposed by mass balance and isotopic composition of CH_4. Because accumulation of CH_4 in the atmosphere and the global sinks are relatively well known, the overall source strength can be computed to be about 500 Tg/year. The ^{14}C content of CH_4 indicates a contribution of biogenic sources in the range of 80% of total emissions.

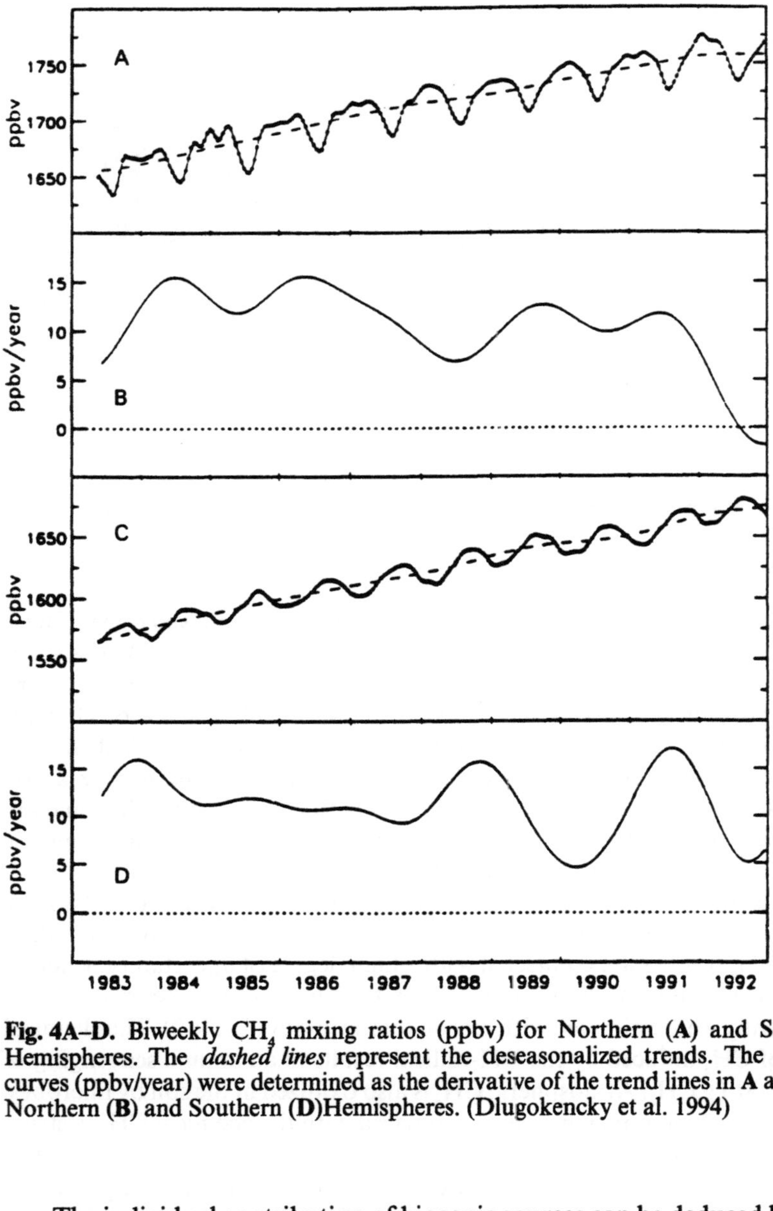

Fig. 4A–D. Biweekly CH_4 mixing ratios (ppbv) for Northern (**A**) and Southern (**C**) Hemispheres. The *dashed lines* represent the deseasonalized trends. The growth rate curves (ppbv/year) were determined as the derivative of the trend lines in **A** and **C** for the Northern (**B**) and Southern (**D**)Hemispheres. (Dlugokencky et al. 1994)

The individual contribution of biogenic sources can be deduced by extrapolating field data. About 70% of the total CH_4 emisssion is attributed to microbial production in waterlogged soils. Natural wetlands are the most important source of CH_4. Large naturally flooded or waterlogged areas are found in tropical and boreal regions. These environments can be distinguished according to their vegetation cover into swamps, marshes, fens, tundras, and various saltwater environments (Bouwman 1990). The pronounced temporal and spatial variabi-

Table 2. Recent estimates of sources and sinks of atmospheric CH_4 (Tg CH_4/year)

	GEIA 1993	IPCC 1990	IPCC 1992
Natural sources			
Wetlands	100–115	100–200	100–200
Termites	20	10–100	10–50
Oceans	10	5–20	5–20
Freshwaters	–	1–25	1–25
CH_4 hydrates	–	0–100	0–5
Anthropogenic sources			
Rice cultivation	40–70	25–170	20–150
Domestic animals	80	65–100	65–150
Sewage treatment	30[a]	–	25(±?)
Landfills	20–25	20–70	20–70
Biomass burning	45	20–80	20–80
Fuel combustion	28		
Coal and gas[b]	60–92	25–50	70–120
Minor	9		
Sinks			
Removal by soils	–	15–45	15–45
Reaction with OH	–	400–600	420–520
Atmospheric increase	–	40–48	28–37

[a] Including animal waste.
[b] Drilling, mining, venting, transmission, and petrol industry.

lity in CH_4 emission rates of natural wetlands is a major bias in trying to develop an accurate estimate of the global source strength of wetlands.

Apart from CH_4 emissions from rice fields, which are discussed in other chapters in this book, another important agricultural source of CH_4 are ruminants. In the rumen, the primary population of fermentative bacteria produce volatile fatty acids, formate, and H_2. The CH_4 production from acetate is of minor importance in the rumen because of its high turnover rate. For animal nutrition, CH_4 emissions represent a significant loss of energy, which accounts for 5–15% of the digestible energy consumed (IAEA 1992). Methane emission rates depend on the diet. Highly digestible diets (e.g., grain) are generally associated with low releases of CH_4 (Johnson et al. 1993).

The most important abiological sources of atmospheric CH_4 are mining, transport, the use of fossil fuels, and biomass burning. However, current estimates of the strengths of these sources are as uncertain as the estimates of biological sources. The total contribution of abiological sources is in the range of 20%. Both biological and abiological sources comprise anthropogenic as well as natural emissions. The contributions of anthropogenic and natural CH_4 emissions to global CH_4 emissions amount to about 70 and 30%, respectively (Hogan 1993).

Nitrous Oxide

From 1977 to 1988, the N_2O concentration in the atmosphere increased at an average annual rate of 0.27 ± 0.01% or 0.8 ± 0.02 ppbv (Khalil and Rasmussen 1992). There are indications that in recent years N_2O has been increasing in the atmosphere at a higher rate than during the mid 1970s (Khalil and Rasmussen 1992). Nitrous oxide accounts for 6–8% of the present radiative forcing of climate (IPCC 1990). In view of its high residence time (about 150 years) in the atmosphere (Khalil and Rasmussen 1992), its high relative absorption capacity, and its high global warming potential (see Table 1), the atmospheric N_2O mixing ratio would further increase even if N_2O emission rates could be stabilized at the present level. It would take several hundred years before a new equilibrium could be established. Therefore, from a long-term perspective, the role of N_2O in climate change may become more important than the role of CH_4.

Individual sources of atmospheric N_2O are even more poorly characterized than those for CH_4. Table 3 gives a compilation of global budgets established by different authors and working groups to give an impression of the dramatic changes that have occurred within the last 7 years with regard to the estimation and identification of N_2O sources that contribute to the global N_2O budget.

In the budget of McElroy and Wofsy (1986), total sources (about 15 Tg N_2O-N/year) and sinks (about 14 Tg N_2O-N) of atmospheric N_2O seemed to be more or less balanced. In this budget, besides natural soils as the dominant source, combustion of fossil fuels was a major source that contributed as much as 4 Tg N_2O-N to the global N_2O source strength (Table 3). These 4 Tg N_2O-N were in good agreement with the annual atmospheric accumulation rate of 3.5 Tg N_2O-N. Other sources that were suggested to contribute significantly to the global N_2O budget (e.g., enhanced biogenic N_2O emissions from soils because of increased N-fertilizer use, and increased deforestation rates in the tropics) seemed to play only marginal roles.

In 1988–1989, it was discovered that an artifact caused by the sampling methods that had been used to quantify N_2O emissions from combustion sources had lead to a heavy overestimation of the contribution of this source to the global N_2O budget (Muzio and Kramlich 1988; Muzio et al. 1989). The strength of this source was now estimated to be less than 0.1 Tg N_2O-N/year. Because the source strength of its photodissociation was reduced to about 7.5 Tg N_2O-N/year, the global N_2O budget was out of balance. The most important sink for N_2O is its photodissociation in the stratosphere (about 11 Tg N_2O-N/year in the budget of McElroy and Wofsy; Table 3). Therefore, the strength of this sink and the accumulation of N_2O in the atmosphere exceeded the source strengths of the identified N_2O sources by about 50%. This imbalance indicated that the emission rates of the identified N_2O sources had been underestimated or that additional sources exist, but had not yet been identified.

Emissions from natural soils make a major contribution to atmospheric N_2O. The GEIA estimate (1993) and the estimate given by McElroy and Wofsy

Table 3. Compilation of different global N_2O budgets (Tg N_2O-N/year) published within the last 7 years

	GEIA 1993[a]	McElroy and Wofsy 1986[b]	IPCC 1992[c]
Sources			
Natural soils	7.6	7.6–8.0[d]	2.75–7.7[e]
Cultivated soils	0.3–2	0.8	0.03–3
Biomass burning	0.1–2	0.7	0.2–1.0
Combustion of fossil fuels	< 0.1	4.0	0.1–0.3
Waste-water treatment*	1–2	–	–
Oceans**	1.4–2.6	2.0	1.4–2.6
Land-use changes	0.4	–	–
Industry	0.4	–	–
Adipic acid production	–	–	0.4–0.6
Nitric acid production	–	–	0.1–0.3
Mobile sources	0.1–1.3	–	0.2–0.6
Aquifers	0.5–1.3	–	–
Sewage	0.2–1.9	–	–
Animal waste	0.2–0.6	–	–
Global warming	0–0.6	–	–
Atmospheric formation	?	–	–
Trash incineration	?	–	–
N-deposition	?	–	–
Coastal and inland waters	?	–	–
Evapotranspiration	?	–	–
Irrigation using groundwater	?	–	–
Total	>12.3–22.8	15.1–15.5	5.2–16.1
Sinks and atmospheric increase			
Stratospheric destruction	–	10.6	7–13
Atmospheric accumulation	–	3.5	3–4.5
Soils	–	–	?
Total	–	14.1	10–17.5

[a] GEIA 1993: Bouwman and van der Hoek referring to Khalil and Rasmussen (1992), Banin (1986)* and Watson et al. (1992)**
[b] Adopted from Davidson (1991)
[c] IPCC 1992, Watson et al. (1992)
[d] Includes temperate grasslands, temperate forests, and tropical forests.
[e] Includes tropical soils (wet forests, dry savannas) and temperate soils (forests and grasslands).

(1986) for the source strength of natural soils agree very well (7.6-8 Tg N_2O-N/year; Table 3), whereas the IPCC estimate (1992) indicates a much higher degree of uncertainty (2.75-7.7 Tg N_2O-N/year). The estimates given for the source strength of cultivated soils show high variation (Table 2). Nitrous oxide is produced in soils by different microbial processes (i.e., denitrification and autotrophic nitrification). Heterotrophic nitrification is also an important source of N_2O (Papen et al. 1993; Anderson et al. 1993) as are other microbial processes in soils, which have not yet been identified (Papen et al. 1993). Microbial production and emission of N_2O to the atmosphere are dramatically enhanced by changes in land use (e.g., deforestation), which may lead to enhanced mineralization of organically bound nitrogen in the soils, and by agricultural activities (e.g., use of mineral and organic-N fertilizers). Severalfold increases in emission rates of N_2O with N-fertilizer application have been widely reported (Eichner 1990). Agricultural practices also have an impact on natural ecosystems (e.g., forests) by indirectly fertilizing these ecosystems with volatile N-compounds such as NH_3. Chronic inputs of nitrogen in the form of ammonia and nitrate may lead to N-saturation, which increases nitrification and denitrification rates and will result in an increase in source strengths of these ecosystems for atmospheric N_2O.

Apart from the major biogenic sources (natural soils and oceans), several anthropogenic sources of N_2O have been identified in recent years. These have not yet been considered as contributors to the global N_2O budget. These anthropogenic sources include industrial sources such as production of adipic acid and nitric acid, mobile sources (IPCC 1992; Table 3), waste-water treatment, aquifers, sewage, global warming, and animal waste (Table 3; GEIA 1993). Each of these sources alone is a small contributor of <0.1-1.9 Tg N_2O-N/year to the present trend of increasing N_2O in the atmosphere. However, it was concluded by Khalil and Rasmussen (1992) that the sum of these sources can account for total anthropogenic emissions of 7 ± 1 Tg N_2O-N/year. Several additional sources with totally unknown source strengths may also contribute to the global N_2O budget: atmospheric formation to N_2O, trash incineration, N-deposition, coastal and inland waters, evapotranspiration, and irrigation with groundwater (GEIA 1993; Table 3).

Methane Emissions from Rice Fields in China

M.X. WANG and X.J. SHANGGUAN

China is the most important rice-producing country in the world. Its harvested area of rice (3.2×10^7 ha) is about 22% of the world total of 14.3×10^7 ha (Neue et al. 1990). According to several authors (Cicerone and Shetter 1981; Seiler et al. 1984; Holzapfel-Pschorn and seiler 1986; M. X. Wang et al. 1987, 1990, 1993a,b; Schütz and seiler 1989; Neue et al. 1990; Schütz et al. 1990; Yagi et al. 1990; Sass et al. 1991c; Bachelet et al. 1993; Shangguan et al. 1993a,b; Wassmann et al. 1993 a,b), Chinese rice fields play a very important role in CH_4 emissions from global rice fields. Wassmann et al. (1993b) estimated that Chinese rice fields may account for 25–35% of total Asian emissions. However, these estimates of CH_4 emission from rice fields in China were based on individual measurements that covered a relatively short time at particular sites. Based on the climate and edaphic parameters, five major regions of rice cultivation can be identified in China: South China Region (SCR), Central China Region (CCR), Southwest China Region (SWCR), Middle and Lower Course of Changjiang River Region (MLCCR), and North China Region (NCR) (Xu 1981). The climate, soil conditions, and cropping systems in these five regions differ greatly. Therefore, CH_4 emission data from all five regions over a long time scale are required. Statistical data on rice cultivation at the province level (China Agriculture Yearbook 1986) were combined with the map given by Xu (1981). This analysis showed that more than 90% of the rice area is in southern China (19% in SCR, 28% in CCR, 30% in MLCCR, and 16% in SWCR). The NCR accounted for only 7%. Since 1987, several measurements have been carried out using the static box technique in some typical rice culture regions in MLCCR, CCR, SWCR, and NCR (Schütz et al. 1990; M.X. Wang et al. 1990, 1993a,b; Dai et al. 1991; Khalil et al. 1991; Z.L. Chen et al. 1993b,c; Shangguan 1993; Shangguan et al. 1993a,b; Wassmann et al. 1993a,c; D.Z. Chen et al. 1994; Toelg et al. 1994). These data have greatly improved our understanding of the magnitudes and characteristics of CH_4 emissions from Chinese rice fields.

Some mechanism studies (e.g., effect of fertilizer and water treatment, CH_4 production and its transport, the role of plants in CH_4 emissions, and temperature effect) have also been conducted. Several practical and economic mitigation measures have been discovered.

Institute of Atmospheric Physics, Chinese Academy of Sciences, 100029 Beijing, China

S. Peng et al. (Eds.) Climate Change and Rice

Temporal Variation in Methane Emissions and Production

Diurnal Variation

Detailed descriptions of diurnal variations in CH_4 emissions are only possible with continuous measurement over a short time interval (Wang et al. 1990; Wassmann et al. 1993a). Four types of diurnal variation have been found in the rice fields in China (Fig. 1). The most common pattern is a single afternoon peak (Fig. 1A), which has been observed in Zhejiang, Sichuan, Hunan, and other sites (M.X. Wang et al. 1990, 1993a,b; Khalil et al. 1991; Shangguan et al. 1993a,b; Wassmann et al. 1993a; Chen et al. 1994; Toelg et al. 1994). This pattern normally occurs during clear weather, with regular diurnal variation of air temperature. Very good correlation has been found between soil (<5 cm), water temperature, and CH_4 emission rate (Shangguan et al. 1993a,b). It was originally thought that the diurnal variation in CH_4 emission might be caused by variations in CH_4 production in the soil (Schütz et al. 1989a). However, contradictory conclusions have been suggested because: (1) CH_4 production in the soil does not show regular diurnal variation, or at least the degree of regularity is much lower than that of CH_4 emission rates (Shangguan et al. 1993a,b), (2) there is a time delay between the variation of air temperature and flooded water temperature and soil temperature (Shangguan et al. 1993 a,b), (3) laboratory experiments have shown that changes in soil temperature affect CH_4 emission; however, there is a 10–11-h phase delay between the change in the two parameters, which indicates that the diurnal change in soil temperature in the rice field cannot be a factor affecting CH_4 emissions (Butterbach-Bahl 1993; Shangguan et al. 1993a,b), and (4) diurnal CH_4 emission rates correlate very well with the temperature of the air and flooded

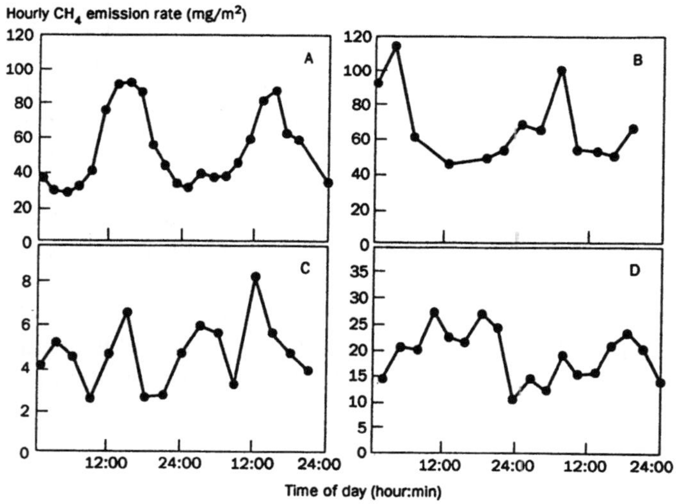

Fig. 1A–D. Diurnal variations in CH_4 emissions from rice fields. **A** Afternoon peak. **B** Night peak. **C** Afternoon-night double peaks. **D** Random peak

water (Fig. 1A), but not with those of deep soil (>5 cm) (Shangguan et al. 1993a,b). Therefore, the most important factor affecting short-term temporal (e.g., diurnal) variation in CH_4 emissions is variation in transport efficiency of CH_4 to the atmosphere. This efficiency depends on changes in air temperature (Shangguan et al. 1993a,b). In continuous measurements of plant emissions and ebulllitions, obvious diurnal variations in the two major transport pathways were observed (Shangguan 1993).

The second pattern of diurnal variation was a night peak, which is rare and was only found occasionally in Zhejiang Province during the summer (Fig. 1B). However, this pattern has never been observed in Hunan Province, where air temperatures are higher than in Zhejiang. The differences are probably due to the different rice varieties planted in the two sites.

The third type of diurnal variation is an afternoon-night double peak (Fig. 1C), which may depend on the mechanisms of both pattern 2 (above) and the efficiency of other transport pathways (e.g., gas bubbles and molecular diffusion) that enhance CH_4 emissions when air temperature is high in the afternoon (M.X. Wang et al. '1993a,b). This pattern of diurnal variation was also rare and was only observed during the summer in Zhejiang Province.

Irregular weather conditions may cause the fourth type of diurnal variation – with CH_4 emissions almost at random during the day (Fig. 1D; Shangguan et al. 1993a,b). In Zhejiang and Hunan Provinces during rainy or overcast periods, or when weather conditions were changing quickly (e.g., clear to rainy), CH_4 emission rates were very irregular. Some weather conditions (e.g., cold or warm fronts) also disrupted regular diurnal variations.

Seasonal Variation

The seasonal variations in CH_4 emissions from different experiment sites differ greatly. These differences may be related to different rice cropping system (e.g., single, early, or late rice), field management (e.g., green manure, wheat, rape, waterlogging), soil properties, weather conditions, water regimes, rice variety, and fertilizer treatment (Fig. 2). In addition to the long-term variation marked by pronounced CH_4 emission peaks, short-term fluctuations (days) may also occur because of local variations in air temperature (M.X. Wang et al. 1993b)

Two pronounced CH_4 emission peaks were normally observed during the tillering and reproductive stages of single-crop rice in Jiangsu and Sichuan Provinces and in Beijing, (Fig. 2A, B; Chen et al. 1993b,c; Shangguan et al. 1993a,b). However, the relative magnitude of the peaks and their timing varied. This phenomenon may be related to field conditions before rice vegetation, the flooded water, and weather conditions. Seasonal variations in CH_4 emisions were different in the wheat-rice fields and single-rice fields in Beijing. In Nanjing, emissions from the winter-waterlogged field were not the same as from the field preplanted with rapeseed. In Sichuan, seasonal variations in CH_4 flux differed slightly between 1987 and 1992. These differences may be related to different

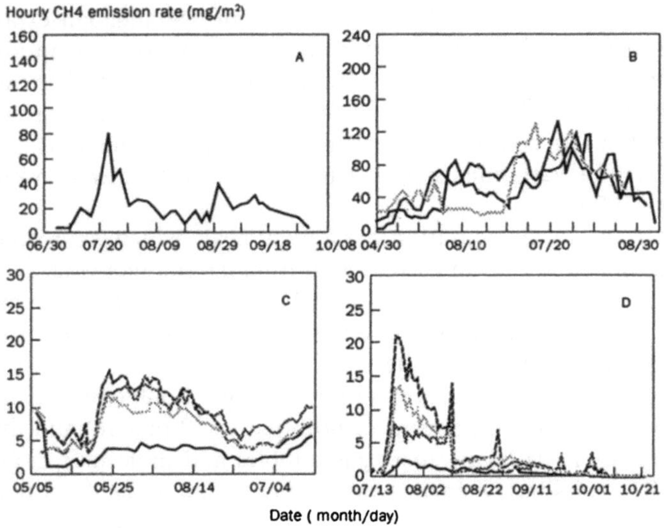

Fig. 2A–D. Seasonal variation in CH_4 emissions from rice fields. **A,B** Single rice crop. **C** Early rice with fertilizer. **D** Late rice with fertilizer. In **B** – – 1988; ·— 1989; ··· 1990; in **C** and **D** ··· normal; –··· organic; +–+ chemical; – – fermented

weather conditions because other factors (e.g., rice variety, fertilization, and field pretreatment) were uniform. Changes in the flooded water may also influence seasonal variations in CH_4 emissions; however, this was not found in the study with single-crop rice.

Weather conditions during the early rice-growing season were similar to the conditions for single-crop rice. The exception was that the single-crop rice grew during the entire period of highest temperature (July and August), whereas the early rice grew only until July. Three peaks in CH_4 emissions were observed in Zhejiang and Hunan Provinces. The magnitudes and time of appearance varied to some extent (Fig. 2C). The first peak appeared at the beginning of rice vegetation, perhaps due to the abundance of organic matter in the soil shortly after the field was flooded. The second and third peaks are thought to be related to root exudates and root litter during the reproductive and ripening stages (M.X. Wang et al. 1990, 1993b). However, the seasonal variations in CH_4 production rates and transport abilities of the rice plant suggest that the transport efficiency of the atmosphere-rice-soil system should also be considered. This may be most important for the second emission peak, when both air temperature and the transport ability of the rice plant favor CH_4 transport to the atmosphere (Shangguan 1993).

Seasonal variations in CH_4 emissions from late rice were very similar in Zhejiang and Hunan Provinces, where CH_4 emissions were the highest a few days after the rice was transplanted and decreased as the rice grew (Fig. 2D). This simple pattern may be related to variations in several parameters involved in CH_4

production and emission. First, air temperatures were highest at the end of July, a few days after the late-rice season begins, and organic matter was abundant in the soil shortly after fertilization and flooding. These factors would explain the high CH_4 production from the soil in the experiment. Second, the rice plants were still too young to play an important role in the transport of CH_4 to the atmosphere. The major transport pathway was gas ebullition, which would also be greatly enhanced by high temperatures. Gas bubbles were observed in the water of the flooded field during this period. According to the calculated ratio of CH_4 emissions to production, the transport efficiency during this period was the highest of the entire late-rice season. Therefore, gas ebullition was the most efficient pathway to avoid reoxidation of the CH_4. Both the highest CH_4 production and the highest transport efficiency occurred at the beginning of the rice vegetation period. Both factors (CH_4 production and transport efficiency) decreased with time and produced a simple pattern of seasonal variation (Fig. 2D). The role of the plant in CH_4 transport is not as important as previously described (Schütz et al. 1989a). At the beginning of the late-rice season, less than 15% of the total emitted CH_4 was through the rice plant.

Annual Variation

Large annual differences (up to severalfold) in average emissions were found at all sites (Chen et al. 1993b,c; M.X. Wang et al. 1993a,b; Wassmann et al. 1933a). For example, in Nanjing, Jiangsu Province, the hourly CH_4 emission rate from a wheat-rice cropping field was 10.8 mg CH_4/m^2 in 1990 and 25.2 mg CH_4/m^2 in 1991. In Zhejiang Province, the average seasonal flux in 1989 was six times higher than in 1988. Wassmann et al. (1993a,b) compared the visible changes in the field site and concluded that changes in rice variety and climate (especially temperature) could not explain this large variation. In Sichuan, a cold summer in 1988 was thought by M.X. Wang et al. (1993a,b) to explain the lowest seasonal emissions of CH_4 between 1987 and 1992. However, our knowledge of the complex effects of factors such as climate variation, different rice cultivars, and soil conditions, precropping systems, and water and fertilizer treatments is insufficient, and these annual variations cannot be clearly explained.

Spatial Variations and Methane Emissions

Methane emission rates from different locations in China are listed in Table 1. The large spatial variation could be caused by soil type and soil properties, climate system, rice varieties, fertilizer treatment, and water management practices.

Chinese rice fields are characterized by a variety of soil types. The five regions for rice culture are classified on the basis of uniform geomorphologic and climatic

Table 1. Methane emissions from various rice fields is China

Location	Period	Field type	OM[a] (%)	CH_4 emission rate E	L	S	Reference
Hangzhou, Zhejiang	1987–89	Double cropping	2.65	0.53	0.77	–	M.X. Wang et al. (1990, 1993a,b); Wassmann et al. (1993b)
Taoyuan, Hunan	1991–92	Double cropping	2.5	0.19	0.37	–	Shangguan et al. (1993a,b); Wassmann et al. (1993c,)
Nanjing, Jiangsu	1990–92	Wheat-rice	2.29	–	–	0.43	Chen et al. (1993b,c)
		Rape-rice	–	–	–	0.34	
		Waterlogged	–	–	–	0.48	
Leshan, Sichuan	1988–92	Single Waterlogged	2.24	–	–	1.41	M.X. Wang et al. (1993a,b), Shangguan et al. (1993a,b)
Beijing	1990–92	Wheat-rice	3.01	–	–	0.42	Chen et al. (1993b,c)
		Single	3.23	–	–	0.22	

[a] OM organic matter, E early, L late, and S single.

conditions and the system of rice cultivation. The soil types in the experimental rice fields in Zhejiang, Hunan, and Sichuan are alluvial meadow soils (clay), red earths (clay), and violet soils (clay), respectively. The soils in Jiangsu Province and Beijing are a clay or sandy loam. However, other physical and chemical parameters must be considered to assess CH_4 production and consumption. Some authors have classified rice soils according to their potential to produce soil CH_4 (Van der Gon pers. comm.). However, measurements conducted in China suggest that organic carbon in the soil is not necessarily related to CH_4 emissions (Table 1). The organic matter (OM) content in Sichuan was among the lowest, but CH_4 emissions were extremely high. In Beijing, the CH_4 flux from a wheat-rice field with 1.33% OM was also much higher than the flux from a single-crop rice field with 3.23% OM. Easily decomposed carbon input from organic fertilizers may be more important for CH_4 production and emission.

Methane emission rates can also be affected by other soil parameters that influences CH_4 consumption and transport. For example, the large effect of percolation rate on CH_4 emissions has been reported by Yagi et al. (1990). The higher the rate of percolation, the lower the level of CH_4 emissions. The percolation rate in the field in Hunan Province was very high (> 10 mm/h) (Shangguan et al. 1993a,b), whereas in Sichuan, the rate was much lower. In Beijing, the percolation rate in the single-crop ricefield was higher than in the wheat-ricefield, which may explain the difference, in CH_4 flux.

The local climate system can also affect CH_4 emissions. At sites with double cropping of rice, the CH_4 flux for late rice is generally higher than for early rice (Table 1). This difference may be due to the trend in weather conditions rather than to absolute differences. For late rice, air temperatures are highest at the beginning of rice vegetation and then decrease. This trend in weather conditions is more coincident with CH_4 emissions from late rice than from early rice. However, similar trends in air temperature at individual locations do not explain large annual and spatial variations in CH_4 emissions. For example, the annual average temperatures in Zhejiang (16.1 °C), Hunan (16.5 °C), and Sichuan (17.1 °C) differ slightly and are unlikely to explain the large spatial variation in CH_4 emissions.

Rice plants can affect the CH_4 flux because they are involved in the processes of CH_4 production, consumption, and transport (Schütz et al. 1989a). However, few studies have investigated this effect. Lower hourly CH_4 emissions were observed in fields in Sichuan Province planted with hybrid rice (59.8 mg CH_4/m^2) compared with the usual variety (75.4 mg CH_4/m^2) from 1988 to 1991. Butterbach-Bahl (1993) found that different rice varieties (Roma and Lido) affected CH_4 emissions in an Italian rice field and in the laboratory; however, Wassmann et al. (1993a) suggested that changes in rice plants were not the reason for the annual variations in CH_4 emissions in Zhejiang. Nevertheless, we believe that rice plants do influence CH_4 emissions and suggest that addititonal laboratory and field sutdies be conducted to quantify the roles of different rice cultivars.

The influences of different fertilizers on CH_4 emission rates have been studied in various parts of China (Table 2). Almost all the studies reported a positive effect of organic manure on CH_4 flux. However, the effect of mineral fertilizers was not clear. In Beijing, CH_4 emissions from wheat-rice fields were much higher than from rice-only fields. The difference may be related to the additional input of organic matter from wheat roots left from the previous harvest. In almost all sites, the CH_4 flux increased with an increase in the amendment of organic manure. In Zhejiang, when only 10% of the normal amount of organic manure was applied to the field, CH_4 emissions were very similar to those from a field that received no additional organic matter. This indicates that the amount of organic manure has an important effect on CH_4 emissions. Fermented organic fertilizer from biogas generators is not as efficient for CH_4 production and emission as fresh manure because the easily decomposed carbon has been removed in the biogas pits.

However, the role of mineral fertilizers is contradictory. In Beijing, the effects of several different chemicals on CH_4 emissions were tested in flooded fields. These studies suggested that $(NH_4)_2SO_4$ and urea inhibited CH_4 flux more efficiently than $(NH_4)_2HCO_3$ and $(NH_4)_2HPO_4$. Seiler et al. (1984) and Schütz et al. (1989b) also suggested a reduction by sulfate and urea, whereas other studies have reported the opposite result (Cicerone and Shetter 1981; Wassmann et al. 1993a). In frequently drained fields in Beijing, the effects of different chemicals did not match these results. In Zhejiang, the effects of minerals were observed by

Table 2. Effect of fertilizer on CH_4 emissions

Location	Treatment	Hourly flux	Comparison (mg CH_4/m^2)	Reference (%)
Zhejiang (1987-1989)	No fertilizer	30.4	100	Wassmann et al. (1993a)
	Mineral[a]	27.5	90	
	Organic[b]	35.1	115	
	Min+Org[c]	28.3	93	
Hunan (1991)	Min+Org[d]	38.3	100	Wassmann et al. (1993c)
	Organic[e]	47.4	124	
	Mineral[f]	10.4	27	
	Ferment[g]	19.2	50[h]	
Jiangsu (1990)	Org+min[i]	10.8	–	Chen et al. (1993c)
	Mineral[j]	2.6	–	
Beijing (1990)	NH_4HCO_3+Org[k]	35.9	100	Chen et al. (1993c)
	NH_4HCO_3+Org[l]	48.9	136	
	NH_4HCO_3[m]	17.5	49	
1992	Organic[n]	23.9	–	Chen et al. (1993c)
	$(NH_4)_2HPO_4$[o]	15.8	–	
	$(NH_4)_2SO_4$[p]	9.6	–	
	$(NH_4)_2SO_4$[q]	9.4	–	
	NH_4HCO_3[r]	15.0	–	
Sichuan (1992)	Norm+min[s]	86.2	100	Chen et al. (1993a)
	Ferm + min[t]	45.3	53	

Note: amounts of fertilizers are per hectare.
[a] Early: 694 kg KC1, late: 694 kg K_2SO_4.
[b] 1024 kg of animal and rapeseed manure.
[c] Combination of [a] plus [b].
[d] Early: 15 t green manure, 38kg KCl, 300 kg mixed min; late: 3 t rice straw, 150 kg urea, 375 kg mixed min.
[e] Early: 15 t green manure, 15 t animal manure; late: 3 t rice straw, 22.5 t animal manure.
[f] Early; 150 kg urea, 60 kg KCl, 450 kg mixed min; late: 90 kg urea, 120 kg KCl, 450 kg mixed min.
[g] Early: 45 t residues from biogas pits; late: 45 t residues from biogas pits.
[h] Fermented fertilizer used in 1991 were stored 1 week in the open.
[i] 15 t animal manure, 188 kg urea.
[j] 600 kg $(NH_4)_2SO_4$.
[k] 3.75 t horse manure, 37.5 kg $(NH_4)_2HCO_3$.
[l] 7.5 t horse manure, 37.5 kg $(NH_4)_2HCO_3$.
[m] 75 kg $(NH_4)_2HCO_3$.
[n] 2.7 t horse manure, some other top dressing.
[o] 144 kg $(NH_4)_2HPO_4$.
[p] 144 kg $(NH_4)_2SO_4$.
[q] 117 kg $(NH_4)_2SO_4$.
[r] 225 kg NH_4HCO_3.
[s] 28.5 t general organic manure.
[t] 30 t fermented manure from biogas pits.

Table 3. Effect of water management on hourly CH_4 emisssions from Chinese rice fields

Location		Treatment	Emission (mg CH_4/m^2)	Comparison[a,b] (%)	Reference
Hunan (1992)					
Early		Const. Moisture	3.1	53	M.X. Wang et al. (1993b)
		Freq. drainage	5.4	92	
		3-cm flooding	5.9	100	
		10-cm flooding	3.0	51	
Late		Const. moisture	5.5	45	
		Freq. drainage	13.6	111	
		3-cm flooding	12.2	100	
		10-cm flooding	15.8	130	
Jiangsu		Flooded	10.8	100	Chen et al. (1993c)
(1990)		Ridge	6.6	61	
(1992)		Norm. flooding	19.8	100	
		Ridge	12.6	64	
Beijing		Flooded	48.9	100	Chen et al. (1993c)
(1990)		Freq. drainage	14.6	30	
(1991)	1.	Flooded	20.4	100	
		Freq. drainage	18.0	88	
	2.	Flooded	10.0	100	
		Freq. drainage	8.7	87	
		Dry land cult.	2.6	26	
(1992)	1.	Flooded	23.9	100	
		Freq. drainage	15.8	66	
	2.	Flooded	8.5	100	
		Freq. drainage	3.6	42	
		Dry land cult.	2.2	26	
Sichuan		Flooded	86.2	100	Chen et al. (1993a)
(1992)		Ridge	44.7	52	

[a] Calculated as [emission (treatment)/emission (normal)] × 100 %.
[b] Comparisons based on same fertilizer treatment.

averaging the 3-year data, but no differences were seen during five individual rice growing seasons (Wassmann et al. 1993a).

It is difficult to quantify the effect of mineral fertilizers because the rice is grown under different conditions with different fertilizers. Because these differences may also influence CH_4 emissions, experiments were conducted in Hunan Province to maintain the NPK levels in the fertilizers at a constant level to quantify the influence of mineral and organic fertilizers (Wassmann et al. 1993c).

Spatial vatiations in CH_4 emissions may be related to the amount of organic manure applied. For example, the input of organic manure in Sichuan was the highest among the fields that were measured, which may be one reason for its high CH_4 emissions. Biogas generators are located throughout China and bring many benefits to farmers. Therefore, the use of fermented fertilizer to reduce CH_4 emissions is very promising.

In addition to the percolation rate of the soil, techniques of water management may affect CH_4 emissions. The rates of CH_4 emission and production in Hunan dropped rapidly during periods of water shortage. The main CH_4-producing layer moved deeper and deeper with decreasing soil moisture (Shangguan 1993). After the fields were reflooded, CH_4 emissions did not resume for a long time (Shangguan et al. 1993a,b; M.X. Wang et al. 1993b). In the CCR, flooded water is normally drained well before (about 20 days) the late rice is harvested to allow green manure seeds to be sowed in the fields. Methane emissions dropped, which reduced total and average flux (M.X. Wang et al. 1993b). Frequent drainage of flooded water has been shown in some places (Beijing, Jiangsu Province) to be a mitigation option; whereas, 3-day frequent drainage is too short to cause significant change (M.X. Wang et al. 1993b; Shangguan 1993; Table 3). Experiments in Sichuan and Jiangsu have shown that the ridge cultivation system used in some parts of China significantly reduced CH_4 emissions (Chen et al. 1993b,c) (Table 3). Frequent drainage and ridge cultivation may become important mitigation options for reducing CH_4 emisssions from rice fields.

Table 4. Estimates of CH_4 emissions from rice fields in different regions of China

Region	Culture	Harvest area	Flooded (days) (10^3 ha)	CH_4 emissions	
				Daily (g/m^2)	Regional total (Tg/year)
South	Early	2871	70–90	0.19	
	Late	2971	80–100	0.34	1.3–1.6
	Wheat-rice	235	80–110	0.43	
Central	Early	3984	70–90	0.19	
	Late	3943	80–100	0.34	2.0–2.5
	Wheat-rice	1061	80–110	0.43	
East	Early	2614	70–90	0.53	
	Late	2707	80–100	0.77	4.2–5.4
	Wheat-rice	4446	80–110	0.43	
Southwest	Early	107	70–90	0.53	
	Late	86	80–100	0.77	3.4–4.1
	Waterlogged	1603	100–120	1.41	
	Wheat-rice	3182	100–120	0.34	
North	Single	2261	90–110	0.24	0.5–0.6
Total		33 495			11.4–14.2

Total Methane Emissions from Chinese Rice fields

The soil map of Xu (1981) and statistics on provincial rice production (China Agriculture Yearbook 1986) have been used to calculate the area of rice grown in the five regions of rice cultivation in China (Table 4). Methane emission rates from representative provinces were used to extrapolate the total emssions from rice fields in China. The individual emission data that were used covered the entire season. Based on these calculations, it is estimated that a total of 12.3 (11.4–14.2) Tg CH_4 is emitted for Chinese rice fields each year.

Soil Characteristics Affecting Methane Production and Emission in Flooded Rice

Z. Wang, H. Kludze, C.R. Crozier, and W.H. Patrick, Jr

Flooded soil is a unique physiobiochemical environment. A thin aerobic surface layer and an underlying reduced soil layer develop after submergence. The thickness of the oxidized surface is determined by the net soil O_2 consumption rate and the O_2 transport rate through the flooded water. The anaerobic layer develops after the consumption of the original O_2 dissolved in the soil and is followed by the sequential reduction of soil oxidants (e.g., NO_3^-, Mn^{4+}, Fe^{3+}, SO_4^{2-}). Methane production occurs later. Methane produced in this anaerobic layer must pass through the aerobic surface layer, where it may be oxidized by methanotrophs. When aquatic plants are present, CH_4 may be emitted through the aerenchyma of these plants (Nouchi et al. 1990).

Methane production and oxidation in flooded rice soils are both microbiological processes controlled by many biological, chemical, and physical factors in the environment. In the presence of rice plants, the rhizosphere will affect both production and oxidation of CH_4. During the growth cycle of rice, soil environmental conditions fluctuate because of changes in water level, temperature, plant-root extension, and fertilizer applications. In such a dynamic system, it is important to understand the controlling factors to characterize and eventually reduce CH_4 emission to the atmosphere.

Methane Production

Carbon Sources

Studies using C-isotope techniques demonstrate the two distinct metabolic pathways of biological CH_4 formation: CO_2 reduction that uses hydrogen gas, fatty acids, or alcohols as a hydrogen donor; and transmethylation of acetic acid or methyl alcohol, which does not involve CO_2 as an intermediate (Vogels et al. 1988):

Wetland Biogeochemistry Institute, Louisiana State University, Baton Rouge, LA 70803-7511, USA

S. Peng et al. (Eds.) Climate Change and Rice

$$CO_2 + 4H_2 \rightarrow CH_4 + 2H_2O \quad (1)$$

$$CH_3COO^- + H^+ \rightarrow CH_4 + CO_2. \quad (2)$$

A limited number of simple compounds provide energy and carbon for the growth of methanogenic bacteria. These compounds are mainly various organic acids formed during organic matter fermentation. Recognized substrates include H_2 plus CO_2, formate, acetate, methanol, and methylated amines (Oremland 1988). Three main sources supply carbon for methanogenic bacteria in flooded rice soil: the original soil organic matter, the exogenous supply of organic material to the soil, and the root litter and exudates from growing rice plants.

The exogenous supply of organic material to the soil, whether it be for the disposal of crop residues or as a source of fertilizer, appears to be the single most important contributor to CH_4 production. Yagi and Minami (1991) found that application of rice straw at rates of 6–9 t/ha increased CH_4 emission rates 1.8–3.5-fold. However, application of compost prepared from rice straw did not appreciably enhance CH_4 fluxes. This indicates the importance of labile carbon in CH_4 formation. Methane production rates have been shown to be linearly correlated with water-soluble carbon in the soil or with readily mineralizable carbon in both laboratory study and field experiments (Yagi and Minami 1990a; Van Cleemput et al. 1991; Crozier et al. 1994). In addition, CH_4 production also depends on the soil oxidant content (Z. Wang et al. 1993a). The method of application (i.e., depth of placement also affects methanogenesis (Sass et al. 1991a). More CH_4 is released and the emission is more sporadic when organic matter is placed at a greater depth. Wang et al. (1992) showed that addition of rice straw increased CH_4 production in a Crowley rice soil. A linear relationship between CH_4 production and rates of application of rice straw was observed in this experiment.

Although organic matter addition generally increases CH_4 production, CH_4 formation was reduced in a laboratory experiment in which 1% of organic matter was added to an acidic soil (Morgan 1968). This reduction lasted about 3 weeks before CH_4 production increased. This initial decrease in CH_4 production was caused by large amounts of acetic acid and smaller amounts of propionic and n-butyric acids, which caused a soil pH decrease. However, the same amount of organic matter added to an alkaline soil stimulated CH_4 production. These results indicate the importance of soil pH to CH_4 production.

Root exudates and plant litter may provide substrates for both methanogenic and methanotrophic bacteria. Root exudates consist mainly of carbohydrates, organic acids, amino acids, and phenolic compounds (Martin 1977). Root exudation enhances CH_4 emissions from 3-week-old rice plants in laboratory experiments (Raimbault et al. 1977). Kludze et al. (1993) reported a small decrease in CH_4 production in unvegetated pots compared with vegetated pots. This disparity was attributed to the additional contribution made by root exudates and root autolysis products to the organic substances in the soil medium. A similar explanation was offered by Holzapfel-Pschorn and Seiler (1986).

In addition to root exudation, deteriorating plants contribute to the organic carbon pool from above-ground plant and root litter (Schütz et al. 1991). These additional sources of substrates are the cause of temporal and spatial variation in CH_4 production and emission (Schütz et al. 1989a).

Soil pH

Methanogens are pH-sensitive populations. Most of them grow over a relatively narrow pH range (about 6–8) and the optimal pH is about 7 (Alexander 1977; Oremland 1988). Nevertheless, CH_4 production does occur in acidic environments such as peat bogs (Crawford 1984). There are also a few strains of alkaliphilic (pH optima 8. 1–9.7) bacteria that produce CH_4 (Oremland et al. 1982).

Introduction of acidic materials frequently results in a decrease in CH_4 production (Z. Wang et al. 1993b). Figure 1 shows the effect on CH_4 production of introducing different amounts of HCl or NaOH to soil suspensions. The maximum rate of CH_4 production was shown in the pH range 6.9–7.1, which is the natural soil pH or slightly higher. At pH values lower than that of the natural soil, CH_4 production was reduced sharply. Below pH 5.8 and above 8.8, CH_4 production in the soil suspension was almost completely inhibited.

The relationship between CH_4 emission and air-dried soil pH is not strong (Z. Wang et al. 1993a). Some acidic and alkaline soils produce more CH_4 than soils with near natural pH after the convergence of soil pH to 7 during anaerobic incubation. This might be related to differences in other controlling factors in the different soils, because pH is only one of the factors that affect CH_4 production.

The application of rice straw and chemical fertilizers might have some influence on soil pH, and results in a temporary change in the rate of CH_4 production. Fertilizer application, especially urea, changes soil pH significantly.

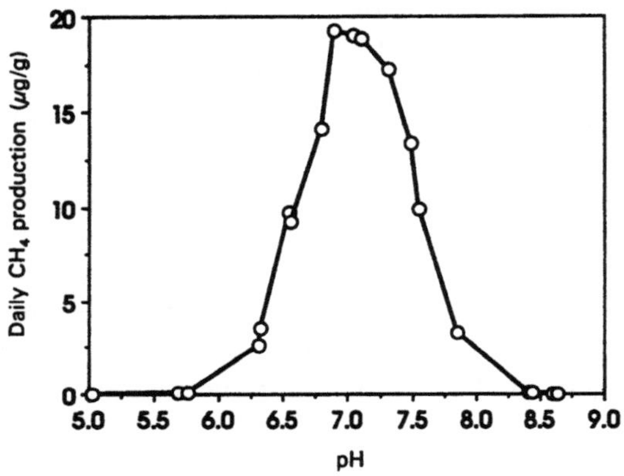

Fig. 1. The optimum pH for CH_4 production in a Crowley silt loam rice soil. (Z. Wang et al. 1993b)

Conflicting observations on the effect of urea fertilization on CH_4 emissions have been reported. Schütz et al. (1989b) reported a reduction in CH_4 emissions after the incorporation of $(NH_4)_2SO_4$ or urea into the soil. Another report showed a stimulatory effect of urea fertilization on CH_4 emissions (Lindau et al. 1991). These conflicting measurements may be related to the effect of urea fertilizer on soil pH. Z. Wang et al. (1993a) studied 16 flooded rice soils with a pH range of between 4.8 and 8.1. In most acidic soils, the addition of urea stimulated CH_4 emission. In all the nonacidic and alkaline soils, soil pH also increased, but CH_4 emission was inhibited. Apparently, a critical effect of fertilizer applications is to shift the soil pH toward or away from the range for CH_4 production.

Similar pH effects were also obtained when organic matter was added to anaerobic soils (Morgan 1968). There was a decrease in soil pH and a reduction in CH_4 production for the first 3 weeks after addition of 1% organic matter to acidic soil. Addition of organic matter to an anaerobically alkaline soil also decreased soil pH, but CH_4 formation markedly increased.

Soil Temperature

Various reports show that soil temperature and CH_4 production or emission generally have a positive relationship. In field experiments, a significant positive correlation between soil temperature and CH_4 emission rate was not always observed because of other limiting factors. Seiler et al. (1984) found a positive correlation between CH_4 emissions and soil temperature at a depth of 0.5 cm. A positive correlation between CH_4 flux and soil temperature was also observed in a field study conducted in China. The Q_{10} value was found to be 3 (Khalil et al. 1991).

A study conducted in peat slurry concluded that the microflora involved in CH_4 metabolism are not well adapted to low temperature (Dunfield et al. 1993). They incubated soil samples at different temperature (0–35 °C) and found that optimum temperature for both CH_4 production and consumption was 25 °C. However, CH_4 production was much more temperature-sensitive (activation energy 123–271 kJ/mol, Q_{10}=5.3–16) than was CH_4 consumption (20–80 kJ/mol, Q_{10}=1.4–2.1). In the 0–10 °C range, CH_4 production was negligible, but CH_4 consumption was 13–38% of maximum.

Soil Redox Potential and Soil Oxidants

Flooded rice fields are characterized by lack of sufficient oxygen in the soil atmosphere to act as the sole electron acceptor for microbial, plant, and animal respiration (Reddy and Patrick 1984). Redox potential measures the ability of a soil environment to supply electrons to an oxidizing agent, or to take up electrons from a reducing agent. In a complex environment such as a soil ecosystem, redox potential is determined by a number of reactions, and is a comprehensive measurement of soil chemical and biochemical oxidation-reduction process. Because methanogenesis occurs only under strict anaerobic conditions, CH_4 production is closely related to soil oxidation-reduction status.

Reductions Before the Formation of Methane

As long as oxygen is present in the soil, other oxidized components of the soil (e.g., NO_3^-, Mn^{4+}, Fe^{3+}, SO_4^{2-} and CO_2) will generally not be used as electron acceptors in biological reductions. However, after submergences, O_2 dissolved in the flooded water and in the soil will be consumed quickly. The need for electron acceptors by facultative anaerobic and true anaerobic organisms results in the reduction of several oxidized components. Reductions of NO_3^- to NO_2^- and N_2O to N_2, Mn^{4+} to Mn^{2+}, Fe^{3+} to Fe^{2+}, SO_4^{2-} to S^{2-} and CO_2 to CH_4 will occur sequentially in the soil (because of thermodynamic principles) as long as available C sources exist (Patrick and Delaune 1977). A corresponding decrease in soil Eh indicates the depletion of the subsequent oxidants (Fig. 2). For example, nitrate is reduced to N_2O and N_2 in the Eh range of +250 to +350 mV. Manganic forms are reduced in a slightly lower range. Ferric iron reduction occurs in the range of +120 to +180 mV. Sulfate reduction was reported to occur when the soil Eh was as low as –150 mV (Connell and Patrick 1969; Jakobsen et al. 1981).

Critical Initial Eh of Methanogenesis

Methane production does not reach appreciable rates until most of the sulfate is removed from the soil and water systems by sulfate-reducing bacteria (Martens and Berner 1974; Jakobsen et al. 1981). Although the initiation of methanogenesis has been reported at an Eh as low as –300 mV (Cicerone and Oremland 1988), methanogenesis has been shown by Connell and Patrick (1969) to initiate at a higher Eh (about –150 mV). However, the rate of CH_4 production was very low until all sulfate in the soil was reduced and soil Eh reached a certain low level.

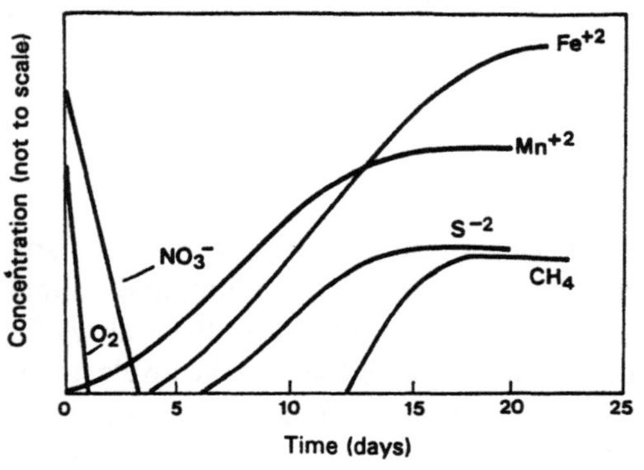

Fig. 2. Sequence of reduction of inorganic redox components in a soil after flooding. (Patrick and Delaune 1977)

The critical Eh for CH_4 production has recently been observed in the range of –140 to –160 mV (Masscheleyn et al. 1993; Z. Wang et al. 1993b). An exponential relationship between CH_4 production and soil Eh was observed when soil Eh was lower than –150 mV (Fig. 3).

Fertilization

Fertilization, especially application of N, is essential for intensive rice cultivation, Because fertilizer application may alter soil pH, microbial populations, and plant litter and root exudate inputs, changes in CH_4 production are also likely. Variable results following the application of chemical fertilizer have been obtained by different researchers. Yagi and Minami (1991) did not observe a significant difference in CH_4 emissions between fertilized and control treatments in a field experiment. Schütz et al. (1989b) found a significant reduction in CH_4 emission when $(NH_4)_2SO_4$ or urea was incorporated into the soil.

The most common N fertilizer used in flooded rice fields is urea, which constituted 75% of the total (Maene et al. 1987). Much attention has been paid to the effect of urea application on CH_4 production in flooded rice. In a field experiment, Kimura et al. (1992a) compared the effects of broadcasting three kinds of nitrogen fertilizers [$(NH_4)_2SO_4$, NH_4Cl, and urea] on rates of CH_4 emission. Urea-treated rice fields produced the most CH_4, followed by the NH_4Cl and $(NH_4)_2SO_4$ treatments. The effect of urea fertilization on CH_4 production may be related to its effects on changing microbial activity, increasing plant litter and roots exudates, and altering soil chemical characteristics (e.g., pH) in the short term.

The application of $(NH_4)_2SO_4$ was shown to produce a gradual suppression of CH_4 production in a rice soil suspension (Z. Wang et al. 1992). In a field study, application of $(NH_4)_2SO_4$ to flooded rice fields was found to reduce CH_4 emission (Schütz et al. 1989b). The degree to which CH_4 emission rates declined with application of $(NH_4)_2SO_4$ was related to both depth of fertilizer incorporation,

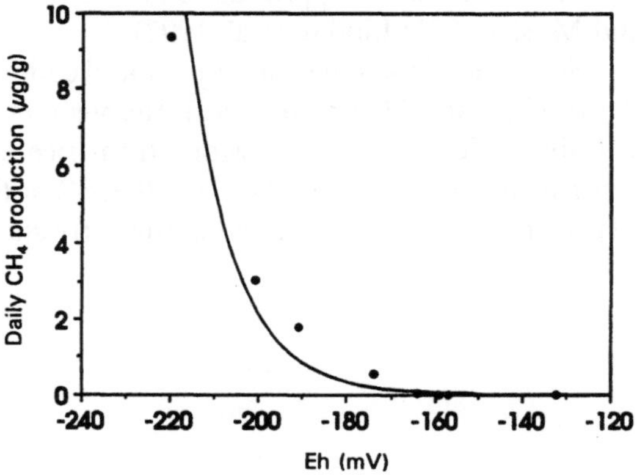

Fig. 3. The exponential relationship between soil redox potential and CH_4 production. (Wang et al. 1993b)

and whether or not rice straw was also applied. When $(NH_4)_2SO_4$ was applied as the sole amendment, incorporation to a depth of 20 cm reduced CH_4 emissions to only 42% of control levels, which was less than the reduction that occurred when the fertilizer was raked into the soil (82% of control) or surface applied (91% of control). However, when rice straw was applied in addition to $(NH_4)_2SO_4$, the reduction effect was not significant.

The importance of (NO_3^- as an electron acceptor may affect CH_4 production. A much lower CH_4 emission rate was measured by Lindau et al. (1990) in NH_4NO_3 treated plots compared with plots treated with urea and $(NH_4)_2SO_4$. Nitrate tends to buffer soil Eh and slow CH_4 production (Patrick and Delaune 1977). An increase in soil Eh from −200 to +200 mV was observed in a laboratory study after nitrate was applied to a highly reduced soil. This increased soil Eh to much higher than the critical soil Eh for CH_4 production (Z. Wang et al. 1992).

Inhibitors of Methanogenesis

Oxygen, alternate electron acceptors, and some chemicals inhibit methanogenesis by different mechanisms. A central electron carrier in CH_4 production biochemistry is coenzyme F_{420}. Oxygen causes an irreversible disassociation of this coenzyme (Vogels et al. 1988). Exposure to low levels of O_2 (e.g., a few ppm) lowers the adenylate charge of methanogens and causes death (Roberton and Wolfe 1970).

Alternate electron acceptors other than O_2 (e.g., NO_3^-, Fe^{3+}, Mn^{4+}, and SO_4^{2-}) inhibit methanogenesis in mixed microbial ecosystems by channelling electron flow to microorganisms that are thermodynamically more efficient than methanogens (e.g., denitrifiers or sulfate reducers).

The chemical 2-bromoethanesulfonic acid is a structural analogue of coenzyme M that is found in methanogens. Application of this chemical as an inhibitor of methanogenic bacteria has recently gained popularity. Chlorinated CH_4 (e.g., chloroform and methylene chloride) is a competitive inhibitor of CH_4 formation (Cicerone and Oremland 1988). Nitrification inhibitors were also reported to inhibit CH_4 production and emission in some field studies (Bronson and Mosier 1991; Lindau et al. 1993).

Some metal ions, such as copper and cadmium, may inhibit methanogenesis (Drauschke and Neumann 1992). Inubushi et al. (1990b) studied the effects of cadmium (Cd) on CH_4 emissions from rice soils in both laboratory and pot experiments. By addition of Cd at 1.9 (Cd^+) or 6.9 ppm (Cd^{2+}) on a dry-soil basis, CH_4 formation was suppressed in the early stage of anaerobic incubation.

Methane Oxidation

Emissions of CH_4 from a particular source to the atmosphere are the result of both production and consumption of CH_4 in the ecosystem. Some ecosystems that produce CH_4 can also function as sinks of atmospheric CH_4 when O_2 is present (Seiler 1984; Schütz et al. 1990; Frenzel et al. 1992). Bacteria that consume CH_4 for growth are known as methanotrophs and are part of a larger grouping of organisms termed methylotrophs. All methanotrophs isolated and studied to date are obligate aerobes because the enzyme responsible for the initial step in CH_4 oxidation is a monooxygenase enzyme that requires molecular O_2.

Surface Layer

Water saturation for an extended time usually results in changes in chemical properties of soil as well as in microbial populations. In flooded rice soils, the dissolved O_2 may be depleted within the top 3–5 mm of surface layer (Frenzel et al. 1992). It is assumed that CH_4 generated by methanogens in anaerobic zones of the soil profile can be used by methanotrophic bacteria as it passes through the upper oxidized layer (Wagatsuma et al. 1992). Methane oxidation in the soil surface of a flooded rice field consumes about 80% of the potential diffusive CH_4 flux (Conrad and Rothfuss 1991).

Rhizosphere

Methane oxidation in the rhizosphere is considered the most important internal sink for CH_4 produced in the soil profile (Holzapfel-Pschorn and Seiler 1986). Methane fluxes increased after the gas phase of an incubation chamber was changed from air to N_2 (Frenzel et al. 1992). This indicates that CH_4 oxidation decreases because the O_2 supply transported through plants from the atmosphere into the rhizosphere was cut off.

Aquatic macrophytes can either enhance or reduce CH_4 emission. In rice and other hydrophytes, atmospheric O_2 that is required for respiration is transported to the submerged organs from the leaves through the plant aerenchyma by diffusion or by mass flow. Because these internal air spaces of plants are particularly well developed in culm and roots, the ventilation system plays an important role in gas exchange between the soil, the plant, and the atmosphere (Nouchi et al. 1990). In addition, rice plants enhance CH_4 emissions by providing substrates for methanogenesis in the form of root and leaf litter and root exudates. The vascular system decreases CH_4 emissions because it allows O_2 transport into the rhizosphere. The presence of O_2 in the rhizosphere reduces CH_4 production and increases CH_4 oxidation. Decrease in CH_4 concentration in the rhizosphere of some aquatic plants has been reported by Wagatsuma et al. (1992). Because the supply of oxygen is increased in the rhizosphere, the population and activity of CH_4 oxidizers are enhanced, which lead to consumption of CH_4 before it is emitted (Hanson 1980). An increase in soil Eh from –250 to +100 mV as a single root moved along a microelectrode was reported by Flessa and Fischer (1992). This Eh level is much higher than the initial Eh for methanogenesis. Methane-producing bacteria are inhibited in such conditions. Methane emissions will be lower in zones where rice roots are present. Schütz et al. (1989a) reported that as rice plants grew, the contribution of plant-mediated CH_4 emissions increased as did the percentage of produced CH_4 that was oxidized. Similarly, Kludze et al. (1993) found strong correlations between the development of aerenchyma in rice roots, rhizosphere oxidation, and CH_4 emissions.

Algae Effect

The presence of algae on the surface of a flooded soil may also affect CH_4 emissions because O_2 is released during algal photosynthesis. Harrison (1914) observed that the gases that occur in rice soils consist mainly of CH_4 and N_2 together with small amounts of CH_2 and H_2. Aiyer (1920) showed that, with the exception of N_2, the other gases found in rice soil were almost undetectable as long as the surface of the soil was not disturbed. An *organized film* that covered the soil surface was considered to possess the power of arresting and assimilating these gases and of leading to an increased output of O_2 at the surface of the soils.

The composition of this organized film may have been the thin layer of algae that is commonly observed in most rice fields. The relationship between CH_4 emissions and the photosynthetic activity of algae in rice soil has not been demonstrated. It is possible that the O_2 released during algal photosynthesis decreases CH_4 emissions by promoting CH_4 oxidation.

Methane emissions decreased substantially in the presence of a thin layer of algae in microcosms without rice plants (Z. Wang et al. 1994). Decreased CH_4 emissions may have been the result of either a physical barrier to diffusion or to O_2 release and subsequent stimulation of methanotrophs or inhibition of methanogens. In the presence of rice, an algal layer did not reduce CH_4 emission significantly. This indicated that the CH_4 was mainly emitted through plant aerenchyma in the presence of rice.

Factors Affecting Methane Transport to the Atmosphere

Net CH_4 emission is the difference between production and consumption. In addition to CH_4 production and oxidation, the processes that regulate the flux of CH_4 from rice fields into the atmosphere include ebullition, molecular diffusion, and vascular transport by plants.

Ebullition

Ebullition of CH_4 gas occurs when the partial pressure of entrapped CH_4 within the soil exceeds the hydrostatic pressure. The result is an upward surge of the gas into the atmosphere. Bartlett et al. (1988) attributed 49–64% of the total CH_4 flux to ebullition; whereas, Crill et al. (1988) reported values of up to 70%. The presence of vegetation moderates ebullition. For example, Takai and Wada (1990) observed that CH_4 ebullition was important during the early stage of flooding when rice plants were small; whereas, vascular transport became more important as the rice plants grew. Mattson and Likens (1990) listed wind speed, water temperature, atmospheric pressure, solar radiation, water level, and local water table as the factors that may influence the ebullition process.

Molecular Diffusion

Dissolved CH_4 gas may diffuse from rice soils through the soil-water and air-water interface (Bartlett et al. 1985). The diffusion of gases in water is about 10^4 times slower than in air; therefore, the diffusive exchange of gases almost stops when soils are waterlogged and most of the nonplant supported emission is by ebullition. However, some CH_4 emission through diffusion does occur. Conrad and Rothfuss (1991), and Whalen and Reeburgh (1990) observed that subsurface microbial CH_4 oxidation is important in controlling CH_4 emission by diffusion.

Vascular Transport

Rice plants develop aerenchyma, an intercellular gas-space system, that provides the roots with O_2. This gas-space system enables the transport of other gases, including CH_4 and CO_2, from the soil and sediment to the atmosphere. Much higher CH_4 emissions were found in vegetated rice fields than in those that were not vegetated. A seasonal pattern of highest CH_4 fluxes in the last 2–3 weeks before harvest was observed by Cicerone et al. (1983). This suggests either greater gas permeability at the ripening stage or a greater input of C because of root death or leaf-litter input.

The aerenchyma tissue of rice plants facilitates the emission of CH_4 produced in the anaerobic zones of flooded soil. About 95% of the CH_4 emitted by rice soils may be transported through the aerenchyma system (Seiler et al. 1984). Inubushi et al. (1990c) reported that soil-pore CH_4 and plant-stem CH_4 concentrations are correlated with the distribution of both CH_4 production and root biomass in the soil. In this context, Kludze and DeLaune (1995) found that stressful conditions that curtailed root growth resulted in significant reductions in CH_4 emission, although CH_4 production rates and individual root air-space formation were high. This is believed to be caused by a reduction in the total pathway for the escape of the gas. The growth stage of the plant (Inubushi et al. 1990c) and diurnal fluctuations in photosynthesis and respiration rates (Bouwman 1990) have also been reported to affect plant-mediated CH_4 emissions. However, Seiler et al. (1984) noted that CH_4 emissions from plants do not seem to be under stomatal control.

Methane Emission and Rice Cultivars

Variations exist among rice cultivars in root aerenchyma formation (Kludze et al. 1994) and root density (Kludze and DeLaune 1994). Both factors collectively determine the total pathway for the transport and oxidation capabilities of CH_4 gas. In addition, cultivar differences in the formation of root exudates and litter would influence CH_4 formation and its net emission. Therefore, it is imperative that morphological differences in the gas-conduction path and variations in root exudation and litter formation among rice cultivars be considered in any breeding program aimed at developing rice cultivars with enhanced CH_4 oxidation and reduced CH_4 production.

Acknowledgments. The research and preparation of this chapter was funded by the United States Environmental Agency Agreement CA817426 to the International Rice Research Institute. The manuscript has not been subject to review by the Agency, and therefore does not necessarily reflect the views of Agency. No official endorsement should be inferred.

Comparison of Micrometeorological and Chamber Measurements of Methane Emissions from Rice Fields

E.T. Kanemasu,[1] I.D. Flitcroft,[1] T.D.H. Shah,[2] D. Nie,[3] G.W. Thurtell,[4] G. Kidd,[4] I. Simpson,[4] M. Lin,[4] H.-U. Neue,[3] and K. Bronson[5]

Estimates of global CH_4 emissions from rice fields have been derived entirely from the results of experiments with chamber (Watson et al. 1990), which are known to have several shortcomings (Hutchinson and Livingston 1993). Therefore, there is a need to validate the accuracy of the chamber measurements using an independent measuring technique (Denmead and Raupach 1993). Micrometeorological measurements of trace gas fluxes do not disturb the plant microclimate and are integrations of fluxes over much larger areas than can be obtained with chambers. Although a number of factors favor the use of chambers, micrometeorological measurements could be used to test the accuracy of the results obtained from chambers.

The use of micrometeorological techniques to measure CH_4 fluxes has been limited by the shortage of CH_4 detectors suitable for such applications (Mosier 1990), but a tunable diode laser (TDL) system has been developed at the University of Guelph specifically for micrometeorological applications. The TDL system permits precise measurements of CH_4 concentrations and was used to measure CH_4 gradients above a rice field. These measurements were then combined with adjacent measurements of the eddy diffusivity to provide estimates of CH_4 flux.

Materials and Method

The study was carried out in flooded rice fields at the IRRI, Philippines (14.2°N, 121.1°E), during March 1992. Four fields (25 × 100 m) were aligned in rows that were oriented southeast-northwest. The soil in each field was classified according to the soil Taxonomy guide (Soil Survey Staff 1992). Soil characteristics are given in Table 1.

[1] Department of Crop and Soil Sciences, University of Georgia, Georgia Experiment Station, Griffin, GA 30223-1797, USA
[2] Department of Plant and Soil Science, Alabama A & M University, Normal, AL 35762, USA
[3] Man-Tech Environ, Tech Inc., Research Triangle Park, NC 27709, USA
[4] Department of Land Resources Science, University of Guelph, Guelph, ON Canada
[5] International Rice Research Institute, P.O. Box 933, 1099 Manila, Philippines

S. Peng et al. (Eds.) Climate Change and Rice

Table 1. Characteristics of rice soil used for CH_4-emission studies (IRRI farm)

pH	6.4
OC%	1.57
Total N%	0.174
Avail. P (Olsen) mg/kg	10
CEC cmol/kg	37.3
Texture	Clay
Soil order	Aquandic Epiaqualf
Soil series	Guadalupe

Field preparation (wet plowing, pudding, and harrowing) of both fields began on 9 December 1991 and rice (IR-72) was transplanted on 8 January 1992. Fertilizer treatment for the fields was 87 kg N/ha broadcast and incorporated at final harrowing and 30 kg N/ha top-dressed both at panicle initiation and at flowering.

By 10 March, when this comparison started, about 50% of the plants were heading, the estimated leaf area index was 4.5, and the canopy height was 0.65 m. Micrometeorological measurements continued until 25 March, but data collected after 15 March were used to examine the effect of cultural practices on CH_4 emission and are not reported.

The plots for the micrometeorological measurements lay fallow for one year prior to the planting of rice in December 1991. The plots of the chamber measurements were double-cropped with rice during this period. Consequently, the organic carbon content of the previously fallow field was 2.1% compared with 1.6% in the field where the chambers were situated. During the comparison period both fields were flooded to a depth of about 5 cm, although the depth varied considerably (± 3 cm) between fields and from day to day.

Chamber Measurements

Four chambers were located in an adjacent field. Each chamber consisted of a transparent Plexiglas box with aluminum profiles and a hinged cover lid. The box rested on posts in the floodwater. Each box enclosed an area of 1 m^2, which covered 25 rice hills. The CH_4 flux inside a box was measured every 2 h by closing the lid for 24 min and monitoring the rate of increase of CH_4 concentration over 6-min intervals, using a Flame ionization Detector gas chromatograph housed next to the rice field (Schütz and Seiler 1989). Fans were used to mix the air in the chamber to reduce the disturbance of the microclimate of the plants, and the chamber lids remained open when measurements were not being taken. Measurements were taken continuously during the comparison using the four chambers.

Micrometeorological Techniques

The micrometeorological sensors were located at the southwest edge of the third plot, which was 75 m from a paved road. During the data collection period, the predominant wind direction was northeast, and in this direction the fetch distance was about 90 m.

In this study, the CH_4 flux was determined by simultaneous measurements of the CH_4 concentration gradient above the rice crop and the eddy diffusion coefficient for scalars. The CH_4 flux was obtained from:

$$F = -K_c \frac{\partial c}{\partial z}, \tag{1}$$

where F is the vertical CH_4 flux (g m^{-2} s^{-1}), c is the CH_4 density (g/m^3), and K_c is the eddy diffusion coefficient for CH_4 (m^2/s). We use the convention that fluxes away from the surface are positive.

The eddy diffusion coefficient for CH_4 was determined separately by three micrometeorological methods: Bowen ratio energy balance, eddy correlation, and wind profile. These three techniques depend on flux-gradient theory, but each derives the transfer coefficient from an independent measurement. The three sets of eddy diffusion coefficients were not entirely independent because the wind profile and eddy correlation methods use latent and sensible heat fluxes from the other apparatus to correct for stability. In practice, the corrections for stability were small (less than 10% of the eddy diffusivity).

Tunable Diode Laser

The TDL system has been developed to provide highly sensitive, real-time measurements in the field or the concentration of trace gases. In this application, the laser is used to determine the CH_4 concentration of sampled air by measuring the profile of a single, selected, CH_4 absorption line. The measured profiles is integrated to yield the total absorbed laser intensity, which, when compared with the absorbed laser intensity of a reference gas, gives the CH_4 concentration of the sampled air.

The laser is mounted in a Dewar and cooled to liquid nitrogen temperatures. A heater mounted inside the Dewar enables precise modulation of the laser temperature, which in turn determines the frequency of the laser emissions. The profile of the absorption line is determined by scanning the laser center frequency across the width of the line and measuring the absorption at over 200 points across the absorption line.

The laser beam is collimated into a 20-mm diameter, 1.4-m-long sample cell, into which the sample air is drawn. The emerging beam is then split (8/92) and focused onto the 1 × 1 mm elements of the sample and reference detectors, both of which are Peltier cooled Hg-Cd-Te infrared detectors. The reference cell contains a known mixture of CH_4. Comparisons of the absorption by the sample gas and by the reference gas enable the concentration of CH_4 in the sampled air to be determined.

Mean gradients of CH_4 concentration were obtained by alternately sampling air drawn from intakes placed 0.125 and 1.0 m above the crop. A solenoid was used to switch between intakes every 4 s. These measurements were averaged to give 30-min concentration gradients.

Bowen ratio energy balance

The transfer coefficient for water vapor, K_v, was determined by:

$$K_v = \frac{LE}{\lambda \varepsilon \frac{\Delta e}{\Delta z}}, \tag{2}$$

where LE is the latent heat flux, $\Delta e/\Delta z$ is the vertical vapor pressure gradient, λ is the latent heat of vaporization of water, and ε is the ratio of the molecular weights of water vapor and dry air. This transfer coefficient is assumed to be equal to that of CH_4 and is substituted into Eq. (1) to calculate the half-hourly CH_4 flux. LE is calculated from the energy balance equation and the Bowen ratio:

$$LE = -\frac{Q-G}{1+\beta}, \tag{3}$$

where Q is the net radiation, G is the soil heat flux, and β is the Bowen ratio (the ratio of sensible heat flux to latent heat flux). The Bowen ratio was determined from:

$$\beta = \gamma \frac{\Delta T}{\Delta e} \tag{4}$$

where γ is the psychrometric constant and ΔT and Δe are the vertical gradients for temperature and vapor pressure.

Net radiation was measured with a net radiometer (Model Q6, Radiation Energy Balance Systems, Seattle, WA) mounted 1 m above the crop and raised as the crop grew. Ground heat flux was estimated with a combination of two heat flux plates buried 0.02 m below the soil surface and two sets of thermocouples installed above the plate, one in the soil and one in the water.

Unfortunately, the sensors could not be placed far enough from the edge of the field to eliminate unrepresentative effects, which caused the measured ground heat flux to vary rapidly between 30-min periods. An attempt to remedy this was made by fitting the ground heat flux values to a sine function to smooth short-term variations.

The vertical gradients of temperature and vapor pressure were measured with a Bowen ratio apparatus (Gay and Greenberg 1985) built at the University of Georgia. When the Bowen ratio approaches –1, Eq. (4) becomes ill-conditioned and values of LE (and therefore K_v) are not reliable. We discarded the data when the value of the Bowen ratio was between –0.6 and –1.4. This eliminated most of the nighttime measurements.

Eddy Correlation

A three-dimensional sonic anemometer (Model SWS-211/3V, Applied Technologies Inc, Boulder, CO) was mounted 1 m above the rice crop, and was rotated as required to point into the wind. The sonic anemometer measured wind speeds in three orthogonal directions, at a rate of 100 Hz. Temperatures were also measured at the same rate.

Values were block averaged into one-tenth of a second intervals, which were then output to a portable computer. A moving average filter was applied to the output in real time, and deviations from the average were obtained (u′,v′,w′, and T′). All time-averaged variances and covariances ($\overline{u'u'}$, $\overline{u'w'}$, $\overline{w'T'}$) were subjected to a coordinate rotation transformation and the friction velocity (u_*) was calculated using:

$$u_* = \sqrt{\overline{u'w'}}\,. \tag{5}$$

The average sensible heat flux was calculated from the mean of the covariance of w and T values over 30-min periods.

Wind Profile

The wind profile above the canopy was measured using four cup anemometers (Model 10B, Met One, Grants Pass, OR). Calibrations were provided by the manufacturer, but the anemometers were tested by installing them at the same height and comparing their outputs. The results indicated that they agreed to within 0.1 m/s. The anemometers were installed at heights of 0.125, 0.25, 0.5 and 1.0 m above the canopy and were raised during the experiment to maintain this distance as the rice grew.

Selected wind profiles were used to find the best estimates of u_* and the roughness length (z_0) for each 30-min period by regressing u(z) against $\log(z-d) - \psi_m$ using:

$$u(z) = \frac{u_*}{k}\left[\ln\frac{(z-d)}{z_0} - \psi_m\left(\frac{z}{L}\right)\right], \tag{6}$$

where u(z) is the mean wind at height z, d is the zero-plane displacement height for the crop, k is von Karman's constant (0.4), and ψ_m is the Businger form of the dimensionless stability correction factor (Arya 1988).

The Monin Obukhov length (L) was calculated using values of u_* and the sensible heat flux (H) from the sonic anemometer system. Throughout this experiment, daytime stability corrections altered the estimates of K by less than 10%.

The zero-plane displacement was estimated by varying d to find the value giving the minimum least square error of regression between u(z) and the stability corrected heights. Periods when the mean wind speed measured by the sonic anemometer was less than 1 m/s and periods when the best estimate of z_0 lay outside the range 0.02–0.1 m were excluded. The resulting values of d showed a small linear increase through the measurement period with a large amount of scatter, and was modeled as a linearly increasing function of time. Having determined the zero-plane displacement, regressions were again performed to determine u_* for each 30-min period. The values of friction velocity from the wind profile data are compared with those from the sonic anemometer in Fig. 1, which shows that the two systems agree closely and that there is little bias.

Calculation of Methane Flux

Having determined the friction velocity using both the sonic anemometer and the wind profile measurements, the CH_4 fluxes were calculated using the integrated form of the flux-profile relation for scalars:

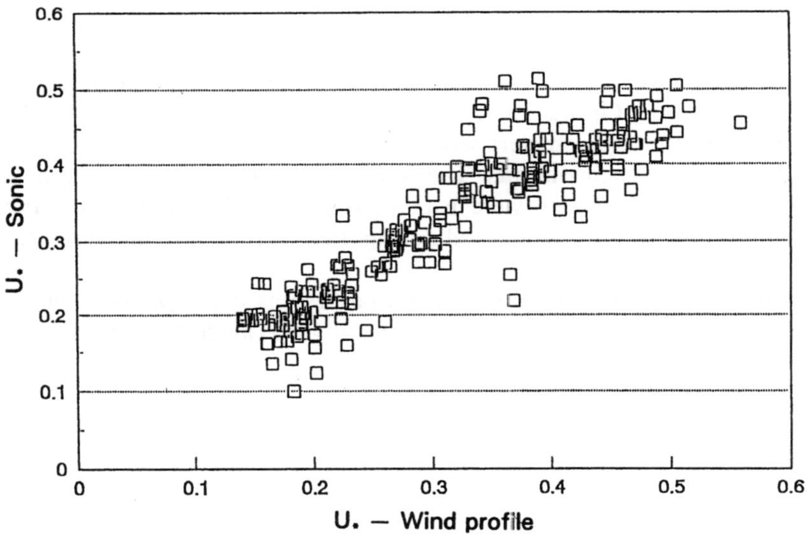

Fig. 1. A comparison of the mean 30-min friction velocities calculated from the wind profile and from the sonic anemometer. (IRRI, Philippines, 11–23 March 1992)

$$F = \frac{u_* k (c_2 - c_1)}{0.74 \left(\log \frac{(z_2 - d)}{(z_1 - d)} + \psi_h(z_1) - \psi_h(z_2) \right)} \tag{7}$$

$$\psi_h(z_i) = 2 \log \left(\frac{1 + x_i^2}{2} \right) \tag{8}$$

$$x_i = \left(1 - \frac{9 z_i}{L} \right)^{\frac{1}{4}} \quad \frac{z}{L} < 0 \tag{9}$$

$$\psi_h(z_i) = \frac{-4.7 z_i}{0.74 L} \quad \frac{z}{L} > 0. \tag{10}$$

The heights z_1 and z_2 are those at which the CH_4 concentration difference (c_1 and c_2) is calculated (z_2 is the upper height).

Error Analysis

Methane Gradients

The tunable diode laser system has been designed to measure very low concentrations and when sampling was conducted over 30-min periods, a resolution of 0.1 ppbv was achieved. Typical daytime concentration difference were 60 ppbv, and nighttime values were an order of magnitude larger. Measurement errors from this system are, therefore, at least an order of magnitude smaller than those associated with the determination K. Therefore, we assumed that errors in the estimates of CH_4 flux arose solely from the errors in calculating K.

Bowen Ratio

Measurement errors in this system are caused mainly by uncertainty in the measurements of net radiation, which we estimated at 5%. Systematic errors in the measurements of the wet and dry bulb temperature gradients were minimized by exchanging the positions of the psychrometers every 5 min, and we assumed that radiation errors were also minimal. The other source of error is the measurement of ground heatflux, which has an estimated uncertainty in the order of 50%.

Taking typical daytime values of net radiation and ground heat flux, we calculated percentage errors to be:

$$R_n - G \equiv 700(\pm 35) - 70(\pm 35) = 630 \pm 70 \quad (11\%) \ \mathrm{W/m^2} \tag{11}$$

Adding estimated errors from the psychrometry of 5%, we derived a total error in this system of 16%.

Eddy Correlation

Errors in the calculation of u_* arise from the measurement accuracy of the vertical and horizontal wind speeds, and errors associated with the statistical processing of the data, which include coordinate rotations, averaging periods, and aliasing errors. Errors in measurements of wind speed are particularly acute at low wind speeds, when distortion of the flow around the sonic head becomes significant (Wyngaard 1991). We conservatively estimate that errors from these sources are less than 10% when mean horizontal wind speeds exceed 1 m/s. Periods when the mean wind speed was less than 1 m/s were discarded.

Other quality controls were performed on the 30-min statistics to remove some poor quality measurements. Uncertainty in the zero-plane displacement height was also included in the error analysis, and this was found to dominate the total error in the CH_4 flux.

Wind Profile

The cup anemometers are inaccurate at low wind speeds because the anemometers have a threshold of 0.22 m/s. Therefore, measurements were discarded when wind speeds of less than 1 m/s were recorded by the sonic anemometer. Unfortunately, winds were commonly light, intermittent, and variable in direction at night.

The principal source of error was the error in the zero-plane displacement height. By comparison, the errors of regression were negligible. Errors from this source were found to be approximately 30% of the flux.

Fetch

Another potential uncertainty in our measurements arises from the limited fetch. During the comparison period, daytime wind directions were almost exclusively from the direction of the longest fetch. To test the possibility that upwind obstructions were altering the turbulent eddy structures, the variance of vertical and horizontal velocity (uw) was plotted against a dimensionless stability parameter (z/L) (not shown). The covariance was first normalized by dividing by the vertical velocity variance (ww). The data were in good agreement with arguments (Arya 1988) that for conditions approaching neutral, the ratio uw/ww should approach a constant of about 1.3, and in more unstable conditions the ratio should increase. Most of the outlying points were obtained at low wind speeds.

The adequacy of the fetch was also confirmed by the very close fit of the wind profile to the stability-corrected log profile. There were no signs of distortion of the profile because of inadequate fetch. In addition, the mean wind speed was compared with the mean wind speed measured by the top cup anemometer (1 m above the canopy). The agreement was good, except for wind speeds of less than 1 m/s.

The data set was also categorized by wind direction to investigate possible differences in fetch wind direction. For conditions approaching neutrality, there was no discernable effect from wind direction, and we conclude that any inadequacies in the fetch did not affect the measurements.

Chambers

The variation in average daily fluxes between the two chambers was significant. Chamber 2 gave consistently lower fluxes than chamber 5. Because only two chambers were available on 2 out of 4 days, from which to calculate an average, we have not calculated standard errors. To show the variability of the measurements, we plotted the raw values of the fluxes in Fig. 2.

Results

Figure 3 shows the daytime average fluxes of CH_4 (with error bars) as measured by all the techniques used. In general, the sonic anemometer (sonic) and the wind profile (WP) methods gave almost identical average values between 5 and 6 $\mu g\ m^{-2}\ s^{-1}$. The energy balance (EB) method gave rather lower average fluxes compared with the other two micrometeorological methods, except on day 73, when there were considerable amounts of missing data from the sonic and wind

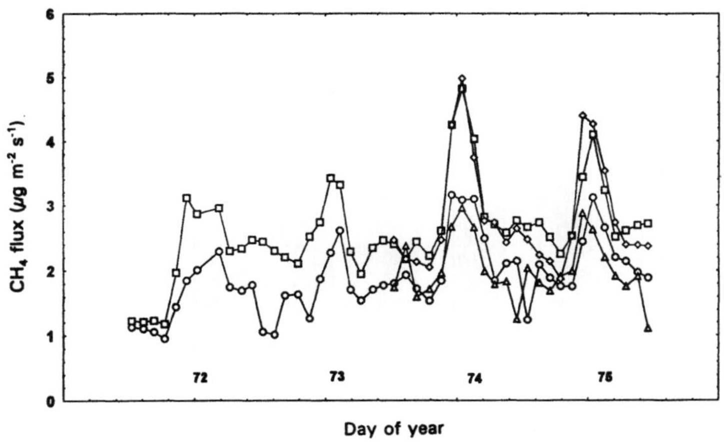

Fig. 2. Diurnal variation of the CH_4 flux from 12 to 15 March 1992 at IRRI, Philippines. TDL-based measurements are 30-min averages and chamber measurements are samples every 2 h. (O chamber 2; □ chamber 5; Δ chamber 8; ◊ chamber 13)

profile systems. These differences are all within the measurement errors of any one of the three systems, and we conclude that the estimates of K from the three methods were not significantly different. The average fluxes obtained from the chambers were lower than those measured by the micrometeorological techniques by a factor of between 2 and 3, although there was often a factor of 2 in variations between chambers.

The measured diurnal variations of the fluxes measured in chambers from 12 to 15 December are shown in Fig. 3. The three micrometeorological methods gave very similar flux patterns during the 5 days (the hourly fluxes derived from micrometeorological measurements are not shown). All measurements show a diurnal cycle in the emissions, with maximum flux at about solar noon. This is to be expected because the production of CH_4 in soil results from biological activity and is, therefore, strongly temperature-dependent.

The maximum flux measured for a single half-hour was approximately 9 μg m^{-2} s^{-1} (12 March, wind-profile method). During the 5 days, most of the measured values were between 5 and 8 μg m^{-2} s^{-1}.

Summary and Conclusion

There were substantial difference between the fluxes measured by the micrometeorological and chamber methods. The fluxes obtained by the micrometeorological techniques were consistently higher than those obtained by the chamber method. Both measurement sites had the same soil properties and received

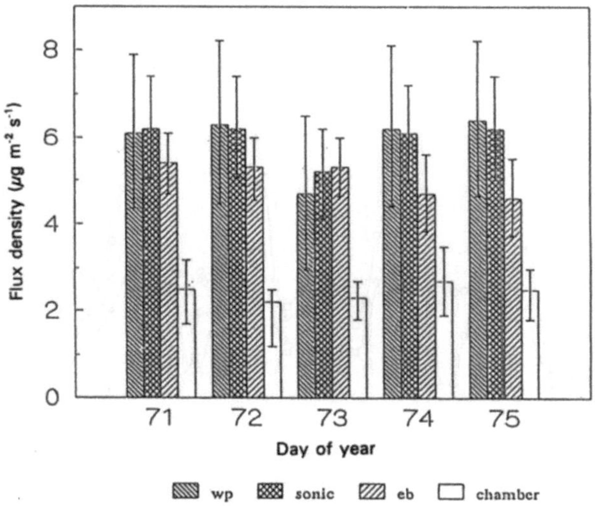

Fig. 3. Average daytime CH_4 fluxes calculated by the four different methods. Average values and error bars are shown for each method. (IRRI, Philippines)

similar management during that season. However, the preceding weed fallow in the micrometeorological fields resulted in higher level of organic inputs and of soil organic carbon in this field than in the chamber field (2.1% compared with 1.6%). Production and emission of CH_4 is strongly affected by organic inputs (Van der Gon et al. 1992). Addition of rice straw to rice fields increases CH_4 production more than threefold (Yagi and Minami 1990a).

The water depth varied in each field independently during the course of the experiment, and this may also have contributed to the differences because of resulting differences in soil temperature between the two fields. Probably the discrepancies in easily decomposable carbon between the fields was mainly responsible for the observed differences in CH_4 emissions, although we cannot exclude the possibility that the differences were partly due to the measurement techniques. Although enclosures offer many advantages over micrometeorological methods, including relatively low cost, portability, and the minimal land area requirements, they do suffer from two main drawbacks. First, chambers measure the flux from a small area (in this case, 1 m^2); whereas, trace-gas emissions have been known to vary by a factor of 2 over distances of a few meters (Matthias et al. 1980). The chambers used in this experiment may have given biased results because of insufficient spatial sampling. Second, permanently installed chambers alter the environment inside the enclosure. Straw yield was 6% less inside the chamber and grain yield was reduced by 20%. Yields outside the chambers were similar to those from the micrometeorological field.

The micrometeorological measurements reported here were the first attempt to measure CH_4 fluxes from rice fields using new gas-detection systems based on a tunable diode laser. The eddy diffusivity for CH_4 was measured using three separate apparatus and the results agreed within the measurement errors of each apparatus. Further intercalibration between different techniques of CH_4 flux measurements are required in which all measurements are done in the same fetch of emissions.

World Inventory of Soil Emission Potentials: Development of a Global Soil Data Base of Process-Controlling Factors

N.H. Batjes[1], E.M. Bridges[1], and F.O. Nachtergaele[2]

Climatologists, oceanographers, and atmospheric scientists have long appreciated the many links between their fields of study (Bolin et al. 1986; Trabalka and Reichle 1986; Prinn 1994). The consequent increase in the atmospheric concentration of trace gases (e.g., CO_2, CH_4, N_2O, and the chlorofluorocarbons) since the industrial revolution is predicted to change climate and induce sea level rise (Houghton et al. 1992). This may have critical effects on natural and managed biotic systems, on which the world food, fuel, fiber, and timber supplies depend, and profound environmental and socioeconomic repercussions (Idso 1989; Brouwer et al. 1991; Walker 1994).

The importance of soils and their biota in the greenhouse effect is becoming better understood (Matson and Ojima 1990; Bouwman 1990). Nonetheless, there are still significant areas of uncertainty in current scientific knowledge. This applies, for instance, to the chemical, physical, and biological processes that control production, absorption, and emission of natural greenhouse gases in different terrestrial ecosystems. Practical difficulties are associated both with the measurement of trace-gas fluxes in the field (Mosier 1990) and subsequent extrapolation or scaling-up of these values to regional and global scales through modeling (Rosswall et al. 1988; Bliss 1990; Matson and Ojima 1990; Burrough 1993).

Georeferenced data bases linked to a geographical information system (GIS) can provide a spatial framework for scaling results of flux measurements and fundamental process studies to ecosystems at the global level (Bliss 1990; Oldeman and Sombroek 1990; Matthews 1993). It is in this context that ISRIC implemented the project entitled World Inventory of Soil Emission Potentials (WISE), which was a sequel to the conference on Soils and the Greenhouse Effect (Bouwman 1990). The aim of the WISE project, sponsored by the Dutch National Research Programme on Global Air Pollution and Climate Change (NOP), is to arrive at a geographic quantification at the global level of soil factors and processes that regulate fluxes of CO_2, CH_4, and N_2O between soils and the atmosphere. The two main objectives in this project were to compile a global 1:5

[1] International Soil Reference and Information Centre (ISRIC), P.O. Box 353, 6700 AJ Wageningen, The Netherlands
[2] Land and Water Development Division, Food and Agriculture Organization (FAO), Viale delle Terma di Caracalla, 00100 Rome, Italy

S. Peng et al. (Eds.) Climate Change and Rice

M, 0.5° latitude by 0.5° longitude soil data base, and to make a useful contribution to the understanding and quantification of gaseous emission potentials of soils, with special reference to CH_4.

Methane and Climate Change

Globally, CH_4 is the second most important greenhouse gas (Houghton et al. 1992). Methane is important to the control of global warming because of its strong ability to absorb infrared radiation and its relatively short atmospheric lifetime of about 10 years. Methane also plays an important role in photochemical reactions in the troposphere and stratosphere, and changes in its concentration distinctly affect the chemistry of the atmosphere (Thompson and Cicerone 1986; Khalil and Rasmussen 1987). In recent years, there has been an appreciable deceleration in the annual rate of increase in atmospheric CH_4 concentrations, from 1.2% per year in the late 1970s to about 0.3% per year in 1992. A number of explanations have been proposed (Dlugokencky et al. 1994; Prinn 1994; Rudolph 1994).

The major sources and sinks of atmospheric CH_4 are known (Crutzen 1991; Houghton et al. 1992). Natural wetlands and irrigated rice fields account for about one-third of the total estimated source strength of 515 Tg CH_4/year (Houghton et al. 1992). The ranges listed in a review of 19 calculated global emissions of CH_4 from rice cultivation vary widely, from 59 to 200 Tg CH_4/year (Minami 1993). This reflects the problems encountered in extrapolation of local emission rates to the global scale, and figures remain uncertain. The current best estimate is 60 Tg CH_4/year, with a range of 20–150 Tg CH_4/year (Houghton et al. 1992). The reported differences in global emissions are the consequence of: the large hourly, daily, and seasonal variation in measured CH_4 fluxes, uncertainty about the relative importance of the process-controlling factors (e.g., water management, organic matter applications, fertilizer treatment, and soil type), and the extrapolation of a limited number of flux measurements, often obtained only for short periods, to areas and time spans for which they may not necessarily be representative.

Rice requirements of the world population are predicted to increase at the compound rate of 1.7% per year between 1990 and 2025 (IRRI 1993). This increase will mainly be achieved through an increase in the harvested area and by higher rice yields. Methane emissions from irrigated rice fields are likely to increase as a result (Anastasi et al. 1992), unless sound mitigation options to reduce CH_4 emissions are introduced.

A comprehensive discussion of the CH_4 literature is beyond the scope of this paper because many reviews have been published (Cicerone and Oremland 1988; Bouwman 1990; Neue and Roger 1993; Batjes 1992a). The balance between the microbial processes of methanogenesis and CH_4 consumption controls to a great extent whether a certain terrestrial ecosystem will function as a sink or source of atmospheric CH_4. Methanogenesis occurs in all anaerobic environments in which organic matter undergoes decomposition (e.g., natural wetlands, ricefields, and in the digestive tracts of ruminants and termites). In general, processes of CH_4 oxidation in soils and their controls remain less understood than those of methanogenesis (King and Adamsen 1992; Oremland and Culbertson 1992; Koschorreck and Conrad 1993; Panikov et al. 1993; Topp 1993).

The main transfer process of CH_4 to the atmosphere is ebullition, diffusion, and transport through the aerenchyma of vascular plants (Seiler et al. 1984; Nouchi et al. 1990; Miura et al. 1992a; Delwiche and Cirecone 1993). Uncertainties remain in assessing the relative importance of soil factors, plant community or rice cultivars, and other environmental and agricultural controls of CH_4 production, consumption, and emission (Sass et al. 1990; Bouwman 1991; Parashar et al. 1990; Batjes 1992a). Laboratory and field experiments are needed to refine our understanding of the regulatory processes and their controls to provide a better basis for developing mechanistic CH_4 models.

Various studies have shown that soil type influences CH_4 emission from ricefields (Inubushi et al. 1990c; Neue et al. 1990; Yagi and Minami 1990b; Denier van der Gon et al. 1992; Kimura 1992; Neue and Roger 1993). However, the effects of soil type may be overruled by other factors (e.g., water regime, temperature, fertilizers, type of organic matter applications, and rice cultivars because of their different capability to transport CH_4 and different root exudation rates).

The minimum set of soil properties required for reliable modeling of CH_4 production in soils is still open to debate. Nevertheless, a basic data set will certainly include: hydrologic regime, measures of the Eh/pH-buffer capacity, organic carbon content, C/N ratio, percentage clay, mineralogy, free-Fe content, sulfate content, and salinity (Neue et al. 1990; Batjes 1992a; Denier van der Gon et al. 1992; Kimura 1992). Whereas a large number of these soil factors can be taken directly from profile descriptions, others (e.g., Eh/pH-buffer relationships) must be derived from pedotransfer functions. These considerations point at the usefulness of developing a soil data base that can: show where potential areas of CH_4-producing soils occur in the world, and provide the basic quantitative data necessary for deriving pedotransfer functions.

Development of the Data Base

Rationale

The basic data layers for a Global Environmental Geographic Information System to calculate gaseous emission include climate, seasonality of water regime and inundation, vegetation or land use, agricultural practices, and soil type (Bouwman 1990; Sombroek 1990; Matthews 1993). During a workshop held at ISRIC (Batjes and Bridges 1992), it was concluded that the Soil Map of the World (FAO-Unesco 1971–1981) is the only viable geographic base map of soil resources. Although known to be partly out of date, this 1:5 M scale map will remain the sole uniform map available for compiling the area data on world soils pending completion of the 1:1 M World Soils and Terrain (SOTER) Data Base (Oldeman and Van Engelen 1993). Unlike Zobler (1986), who used the original printed Soil Map of the World to compile a 1° × 1° resolution data base for global modeling purposes, the WISE project makes use of an edited, digital version of the Soil Map of the World on which errors have been corrected, some boundary changes included, and complete information on the composition of the soil mapping units is used (FAO 1991). The area data of the WISE data base can be updated gradually as sections of the Soil Map of the World are revised by FAO and collaborating institutes in the context of regional SOTER activities (Fig. 1). The ultimate WISE data base will consist of: a GIS file with information on the component FAO soil units of each 0.5° × 0.5° grid cell, that corresponds with the area-data, a suite of soil profile data for the respective FAO soil units (with associated subfiles listing the analytical methods used and source of primary data), and a file of derived characteristics for the topsoil and subsoil of the respective FAO soil units to provide a uniform basis for subsequent modeling activities.

Soil Area Data

Small-scale maps encompass a marked degree of data integration, the aim being to simplify the geographical distribution of soils to a regionally representative pattern of spatially dominant soils. In the context of the WISE project, the predictive use of the digitized, 1:5 M scale Soil Map of the World becomes of

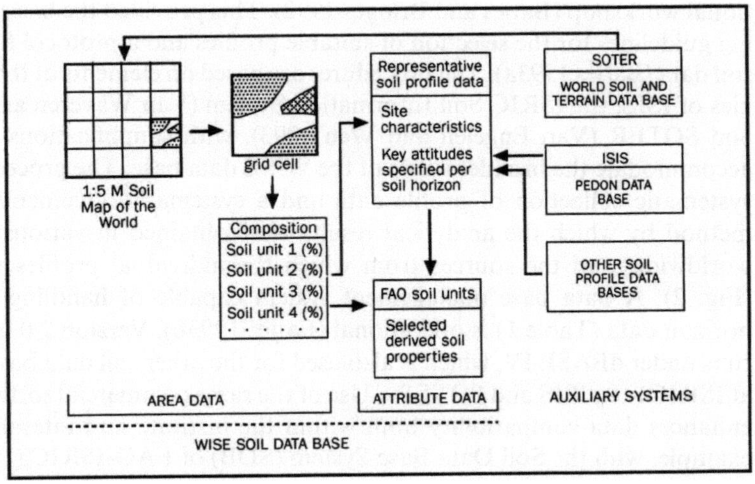

Fig. 1. Schematic representation of the World Inventory of Soil Emission Potentials (WISE) data base, with its area data and attribute data. (Batjes 1992b)

significant importance. The areas of soil units depicted on this map are used as a cartographic basis for the preparation of a 0.5° × 0.5° resolution data base, corresponding with about 55 × 55 km at the equator. This grid size is commonly used in global change research and has been requested by the sponsors of the WISE project.

First, the type and relative area of the component soil units occurring at the center of each 5′ × 5′grid-cell of the digitized Soil Map of the World have been identified and characterized, using composition rules developed by FAO (1991). Next, the information for the 36 component 5′ × 5′ cells of a particular 0.5° × 0.5° grid have been computed to provide the definitive area data (Nachtergaele, unpubl. data). The algorithms for this part of the work have been prepared by staff of the FAO Land and Water Development Division (AGLS), with whom WISE personnel are closely cooperating.

Prototypes of area data for Africa and the Near East have been tested for possible inconsistencies in the griding algorithms. The area data for the remaining sections of the world, prepared by FAO-AGLS, will ultimately be transferred to one global file at ISRIC. To each grid on this map will be attached, through the FAO-Unesco legend descriptors, a suite of representative soil profile; this forms the soil-point data (Fig. 1). This linkage with actual profile data in WISE forms a significant improvement on the information held by the world soil file (Zobler 1986) and on the still limited profile data set of Webb and Rosenzweig (1993).

Soil Attributes

The attributes that would be needed to develop the WISE data base were identified in a literature survey (Batjes 1992b) and discussed during an international workshop (Batjes and Bridges 1992). This provided the basis for developing guidelines for the selection of suitable profiles and a protocol for coding the soil data (Batjes 1993a). The procedures are based on elements of the methodologies of ISIS, the ISRIC Soil Information System (Van Waveren and Bos 1988), and SOTER (Van Engelen and Wen 1993), with simplifications necessary to accommodate the broader scope of the WISE data base. The procedures allow a systematic collection of profile data and a systematic documentation of: the method by which the analytical results were obtained in various laboratories worldwide, and the sources from which the individual profiles were derived (Fig. 2). A data base management system capable of handling the site and horizon data (Table 1) is operational (Batjes 1993b). Version 2.0 of this system runs under dBASE IV, which is also used for the other soil data bases developed at ISRIC (i.e., ISIS and SOTER). Use of the same commercial software package enhances data compatibility both within the institute and internationally, for example, with the Soil Data Base System (SDB) of FAO-ISRIC (1989).

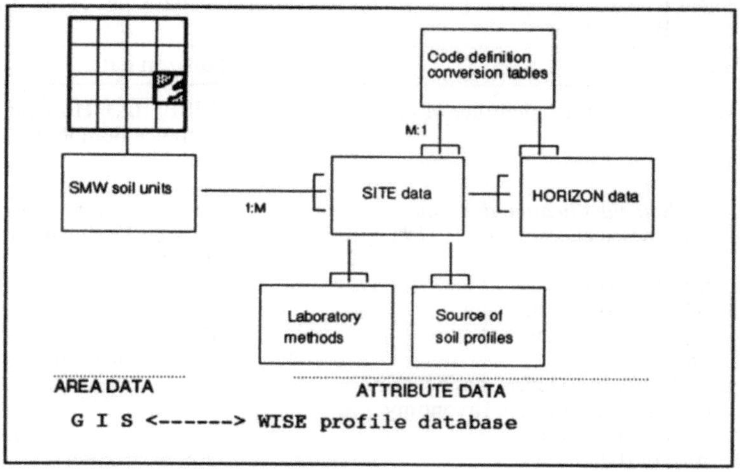

Fig. 2. Schematic representation of the SITE and HORIZON data files and the related key-attribute files. (Batjes 1993b). (M:1 Many to one relations; 1:M one to many relations)

Collection of Profile Data

Collection and processing of profiles is done according to the guidelines developed for the WISE data base. In the data collection activity, much attention is being paid to the necessity for global distribution of soil profiles in the data base. Responses to an international data collection activity requesting descriptions of regionally representative profiles for the various countries of the world, on tape or in manuscript form, have been limited. Problems of data compatibility and comparability between disparate data bases have had to be addressed, for example where different criteria have been used to describe similar features. In this context, Vogel (unpubl. data) reviewed the possibility of correlating soil attributes obtained with different analytical procedures in the various laboratories of the world. For any soil, results obtained for a particular attribute (e.g., cation exchange capacity or bulk density) often vary widely with the sampling procedure, sample pretreatment, and analytical methods used (Driessen 1986; Pleijsier 1989; Burrough 1993).

Staff at ISRIC have developed software to mechanically transfer profile data held by the major international data base holders from one system to another (Tempel, unpubl. data). Protocols for transferring data held in ISIS and SDB to WISE are being tested, and similar protocols for the transfer of profile data obtained from the Soil Conservation Service (SCS) at Lincoln are being developed. They will be used to transfer about 390 ISIS profiles, about 200 SDB profiles, and about 1300 SCS profiles to the profile component of the WISE data base.

Table 1. Attribute data for the SITE and HORIZON data files of the WISE profile data base. (Batjes 1993a)

Site data	Horizon data
WISE_ID (unique identifier of profile)	WISE_ID + Horizon_NO (unique reference number for horizon within a profile)
Soil classification and source	*General Attributes*
FAO-Unesco classifcation (1974 Legend)	Horizon Designation
phase	Depth, top
topsoil texture class	Depth, bottom
FAO-Unesco classification (1990 Revised Legend)	matrix color (dry and moist)
	Mottling
phase	Presence of roots
USDA subgroup level classification	
edition (year) of Soil Taxonomy	
Local classification	*Chemical Attributes*[a]
Source of data	Organic carbon
Name of Laboratory where analyses were made	Total N
Soil profile description status	Available P
Data of Description	pH-H_2O
	pH-KCI
Location	pH-$CaCI_2$
Country	Electrical conductivity (EC)
Location of soil profile, descriptive	Free $CaCO_3$
Latitude (deg/min/s)	$CaSO_4$
Longitude (deg/min/s)	Exchangeable Ca^{+2}
Altitude	Exchangeable Mg^{+2}
	Exchangeable Na^+
General site data	Exchangeable K^+
Major landform	Exchangeable Al^{+3} + H+ (exchangeable acidity)
Landscape position	
Aspect	Exchangeable Al^{+3} (exchangeable aluminium)
Slope	
Drainage class	Cation exchange capacity (CEC)
Groundwater depth	Effective CEC (at field pH)
Effective soil depth	Base saturation (as percentage of CEC)
Parent material	
Köppen climate classification	
Land use	*Physical attributes*[a]
Natural vegetation	Structure type
	Particle size distribution:
	Weight percent sand
	Weight percent silt
	Weight percent clay
	Stone and gravel content
	Bulk density
	Volume percent water held at specified sections
	Hydraulic conductivity at specified suctions

[a] Analytical methods are specified in a separate key-attribute file (see Fig. 2).

When digital files were unavailable, for example, when unresolved copyright issues prevented their release, profile data had to be entered manually into the WISE data base. In such situations, greater reliance than had been expected fell on the collection of data present in the ISRIC library. So far, about 1650 profiles have been manually transcribed to data entry forms and entered in the computer. All profiles in WISE are given an indication of their inferred quality, ranging from good to poor, based on the nature of the original materials (Batjes 1993a). This type of subjective quality rating is likely to be of considerable value to the end-user, but it is difficult to establish in a consistent fashion (Henderson-Sellers 1992). Data quality is further ensured by a computerized check on possible inconsistencies in all coded data and in selected numerical data (e.g., pH, C/N ratio, and clay content). This computerized procedure supplements the compulsory, visual checks of all data sets stored in WISE.

Pedotransfer Functions

Certain physical soil properties are often needed yet seldom collected on a routine basis, for example, water retention versus pressure head relationships, and pressure head-hydraulic conductivity relationships. Surrogate values can be calculated from soil data held in the profile data base through use of pedotransfer functions (Bouma and Van Lanen 1987; Gregson et al. 1987; Vereecken et al. 1992; Bohne et al. 1993; Tietje and Tapkenhinrichs 1993). Examples in information that can be derived from soil parameters held in the WISE data base are listed in Table 2. The accuracy of the prediction of individual pedotransfer functions varies with the functions used (Vereecken et al. 1992; Tietje and Tapkenhinrichs 1993).

Table 2. Examples of information that can be derived from soil parameters held in the WISE data base. (After Ingram 1993)[a]

Soil parameters	Derived data
Soil surface color	Albedo
Organic C, total N, and bulk density per horizon	Organic C and N pools (e.g., as kg/m^2 to 1-m depth)
θ (h) Bulk density and rooting depth	Water-holding capacity
Heat capacity of soil constituents	Heat conductance
CEC, OC content, clay content, and θ (h)	Mineralogy
CEC and exchangeable bases	Soil nutrient status
pH	Soil acidity
Particle size distribution, bulk density, and θ (h)	K_{sat}, K (θ)

[a] θ Volumetric water content, h soil water pressure head, ρ bulk density, K hydraulic conductivity, CEC cation exchange capacity, and OC organic carbon content.

Applications

Methane Production Potentials

The WISE data base can be used for a range of applications, each of which corresponds to a well-defined scenario (Table 3). Because each attribute has a different effect on a particular process it must be weighted in a different manner depending on the topics under investigation (e.g., studies of crop production, soil gaseous emission potentials, or soil pollution). In the context of the WISE project, the focus is on CH_4. First, the data base will be used to make an inventory of global wetland soils, and to present refined estimates of CH_4 production potentials of these soils using an expert system (Fig. 3). This will require refinement of a qualtitative scheme to describe the different aspects of CH_4 production.

During the WISE workshop, held in August 1992, it was recommended that ISRIC and the International Rice Research Institute (IRRI) make best use of past and current field measurements of CH_4 production, oxidation, and emission for model development and verification. In this context, a joint ISRIC-IRRI questionnaire was prepared and sent to research groups involved in field measurements of CH_4 emissions. One aim of the questionnaire was to obtain data to derive an expert system to permit spatial extrapolation with a geographical information system linked to WISE and auxiliary data bases. Unfortunately, responses to the questionnaire provided less information useful for developing a modeling approach than had been anticipated by the WISE workshop.

Table 3. Selected examples of soil attributes needed for different types of environmental studies[a]

Topics	Soil processes	Relevant soil factors
Crop production	Water release, weathering, cation exchange	Fertility status, salinity, soil moisture characteristics, and rootable depth
Pollution by heavy metals	Adsorption, solubility	Organic matter content, pH, $CaCO_3$ content, water balance, and salinity
Acidification	Weathering, base exchange	$CaCO_3$, exchangeable bases, cation exchange capacity, and mineralogy
CH_4 emission	Methanogenesis and CH_4 oxidation	Organic matter content and quality, pH, sulfate content, texture, N-content, nutrient status, drainage class (redox-potential), and soil porosity
N_2O emission	Denitrification and nitrification	organic matter content, texture, pH, soil moisture characteristics, soil drainage, nutrient status, and soil porosity

[a] Also needed from auxiliary sources are climate, land use and vegetation, and management practices (e.g., fertilizer application, organic matter amendments, irrigation and drainage practices, and loading with pollutants).

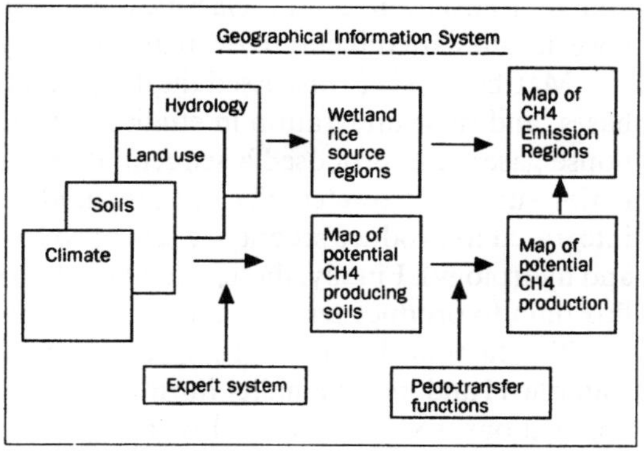

Fig. 3. Schematic representation of projected use of WISE and auxiliary data bases to identify potential CH_4-producing soils and, ultimately, to make refined calculations of CH_4 emissions from wetland rice soils

Consequently, a collaborative activity with colleagues from the Soil Science and Geology Department, Wageningen Agricultural University, was initiated. This work was done in the framework of another NOP-sponsored project on the influence of soil parameters on the production and emission of CH_4 in and by wetland rice soils. In this effort, incubation experiments on potential CH_4 production of rice soils were analyzed to develop pedotransfer functions that can relate soil characteristics, held in the WISE data base, with potential CH_4 production from soil units present on the Soil Map of the World. After suitable relationships have been found, there will be a need for supporting field experiments. Contacts with research institutes specializing in rice and CH_4, established in the framework of a joint ISRIC-IRRI questionnaire activity, should prove useful to identify research groups that could participate in possible follow up experiments; this is an activity that would extend beyond the current WISE project. IRRI will further cooperate in the interregional research program on CH_4 emissions from ricefields in China, Indonesia, Thailand, and the Philippines under a contract funded by the United Nations Development Programme (UNDP). Future collaborative activities may include the use of the WISE data base and other auxiliary data bases to propose refined global estimates of CH_4 emissions from wetland rice soils.

Global Digital Data Bases

Geographic data bases of parameters (e.g., vegetation cover, land use, and soil properties) determine extrapolation and modeling capabilities at the global scale. These data are needed at scales that reflect their effect on regional and global processes and that are relevant to modeling goals (Wessman 1992). Bouwman (1990), Sombroek (1990), and Matthews (1993) reviewed common data sets for global emission inventories. Uncertainty in the data bases can arise from different sources, including errors in the original materials used in data base compilation and in translation of the classification systems used to record data. These uncertainties are difficult to evaluate and vary among the data sets.

Matthews (1993) discussed the design and development of global digital data bases, and their integration in studies of terrestrial sources and sinks of greenhouse gases, and proposed a sequential approach. First, primary data bases on major surface characteristics are prepared. These data bases are subsequently integrated to produce secondary sets of source categories (e.g., soils, vegetation, and hydrology). Finally, these secondary data bases are combined with measured flux data to produce tertiary data sets on emissions.

The factors that have the greatest effect on CH_4 emissions are probably: water level and its seasonality during the growing season, temperature, fertilizer applications, soil type, rice cultivar or natural vegetation type, and agricultural practices, including direct seeding or transplanting (Khalil 1993; Matthews 1993). At present, there are insufficient data to incorporate all these factors in a study of the global budget. For example, there is still no global information on the application of organic matter to rice fields, whereas the importance of organic substrate quality and quantity to CH_4 emissions is well known (Minami 1993). Similarly, the nature of regional preplanting activities (e.g., puddling, direct seeding, and transplanting), drainage, and water percolation, which through their effect on soil redox potentials and crop growth have a significant effect on CH_4 emission, remains difficult to quantify in a global model. Nonetheless, global estimates of CH_4 emission can substantially be improved by incorporating the current knowledge on water levels and temperature (Khalil 1993; Minami 1993). Bachelet and Neue (1993) proposed an expert classification of soils to group rice-growing regions according to their CH_4-emission potentials, for which additional experimental backing is now being sought. An important contribution of the WISE project to this work will be in providing data on the extent and characteristics of the world soils using currently available knowledge.

Modeling Approaches

Most global modeling approaches for CH_4 emissions from natural wetlands (Matthews and Fung 1987; Aselmann and Crützen 1990) and irrigated rice (Matthews et al. 1991; Bachelet and Neue 1993) use a bottom-up approach, are empirical, and are based on simple assumptions. They rely on published information on the type, location, and extent of wetlands, and on still small numbers of measurements of CH_4 flux. An inventory approach, which includes the spatial distribution and characteristics of the wetlands, is preferred in these studies (Roulet and Matthews 1993). Bachelet and Neue (1993), for example, showed the importance of considering soil characteristics in global studies of CH_4 emissions from irrigated ricefields. An alternative, top-down approach to constrain the global CH_4 budget is through inverse modeling (Fung et al. 1991).

There is a fairly general agreement that the present understanding of the controlling factors and processes is not yet adequate to implement process models of CH_4 emissions from rice cultivation (Japan Environmental Agency-US Environmental Protection Agency 1990; Matson and Ojima 1990; Van Breemen and Feijtel 1990). Deciphering mechanisms of CH_4 formation, oxidation, and transport from the rhizosphere to the atmosphere calls for fundamental research that is beyond the mandate of ISRIC. Work related to model development must take place in cooperation with scientists from field-research institutes. The task of the WISE project is to identify soil-sensitivity factors that play a crucial role in determining the magnitude of CH_4 emissions. Links have been established with researchers at IRRI, Nagoya University, and Wageningen Agricultural University for these aspects of the work.

Development of mechanistic CH_4 models is still not a realistic option on a global scale. Research priorities for developing such models, at field level, include studies on: the biogeochemistry of CH_4, both in wetland rice fields and natural wetlands, including methanogenesis, methanotrophy, and regulating factors; the effects of climate, soil, water management, cultivars, fertilizer application, and cultural practices on CH_4 fluxes; temporal variation in CH_4 fluxes between sites; field-level measurements to assess spatial variability, and mechanistic simulation models to synthesize the process-level and field-level data; and effects of mitigation techniques for CH_4 emissions, including their impact on N_2O emissions (Japan Environmental Agency-US Environmental Protection Agency 1990). Braatz and Hogan (1991) propose to develop three major, interactive modules describing: soil redox processes, the development of the rice plant, and the dynamics of the physical and microbial processes. Only a limited number of the soil-data requirements identified by Braatz and Hogan (1991) are collected during routine soil surveys (Msanya 1987; WISE data collection activity). A coordinated effort is needed to collect the basic data sets at regionally representative sites, using uniform and standardized procedures, to test and validate process-based CH_4 models. These deterministic models, by definition, will produce results pertaining to single events. The environmental and land-use factors, however, will always be subject to spatial and temporal variability, and there will be a degree of stochastic uncertainty in the results of deterministic process models. Future research should consider the possibility of combining a mechanistic and stochastic approach in a single CH_4 model.

Discussion and Conclusion

Although attemps have been made in the past to use the printed FAO-Unesco Soil Map of the World as a basis for determining the nature of the soil cover for modeling purposes (Zobler 1986), the resolution used was coarse and the potential of the map as a source of information was only partially exploited. Algorithms developed by FAO-AGLS as part of the WISE project have greatly expanded the amount of information on the world soil pattern that can be used in global models. Nonetheless, the spatial and temporal variablility, and quality of soil profile data, remain controversial issues (Driessen 1986; Pleijsier 1989; Mausbach and Wilding 1991; Burrough 1993). A full-scale update of information on world soil resources using the SOTER procedures is a separate and much longer-term project (Oldeman and Van Engelen 1993). As sections of the Soil Map of the World are updated in the framework of regional SOTER activities, new spatial information can readily be incorporated in the WISE data base. Simultaneously, soil profiles from updated areas can be incorporated into the profile component of the WISE data base. The gathering in a uniform manner of soil-profile data in WISE represents a major task in soil science. As the number of soil profiles for a particular region increases, it will become possible to improve the knowledge about the range in characteristics of each FAO-Unesco soil unit (e.g., stratified by climate and parent material).

There seems to be a good agreement that the list of attributes considered in the WISE data base (Table 1) will help meet the needs of terrestrial and atmospheric modelers, including those of the IGBP Digital Information System (DIS) working group with which ISRIC is cooperating (Ingram 1993). Likely applications of the WISE data base in the context of trace-gas studies include (Batjes and Bridges 1994); simple extrapolation of process studies for similar soils to estimate regional or zonal fluxes of a trace gas; process model studies for regional and global flux estimates; and interpretation of inverse-modeling estimates of terrestrial source and sink terms. At the time of writing, the WISE project is nearing completion and requests for the data base are already being received. Some immediate applications in the context of global climate change are being investigated, but many of the opportunities provided by the existence of such a data base remain to be exploited both by ISRIC and the international modeling community. A range of auxiliary data bases, for example, on climate and land use, of a similar resolution are also needed for these studies. Several of the necessary global data bases still need to be developed or to be released in the public domain for wider use.

It still remains difficult to establish to what degree the respective process-controlling factors quantitatively affect the production, oxidation, and emission of CH_4 in a certain terrestrial ecosystem, at a particular time. This is why proposals for developing mechanistic CH_4 models have only recently been formulated, initially with respect to ricefields. The preliminary data requirements for these models in terms of soil conditions, as formulated by Braatz and Hogan (1991), already exceed the level of information held in most high-resolution soil survey reports. New data-collection activities should be initiated at the level of the experimental site.

Refined estimates of CH_4 emission from wetlands can already be obtained by extrapolating seasonally averaged, field-measured CH_4 fluxes (considered representative for well-described natural wetlands or irrigated rice ecosystems) to the corresponding global ecosystem. Such an extrapolation will not only require additional data on CH_4 fluxes, but also more detailed and up-to-date information about the geographic and seasonal distribution of the various types of natural and anthropogenic wetlands and their soils. The WISE data base will be useful in the context of these activities. Other anticipated uses are to refine estimates of N_2O emissions from soils (Bouwman et al. 1993), improve the soil information in climate-vegetation succession models (Walker 1994), and assess the vulnerability of soils to chemical pollution.

Acknowledgments. Niels Batjes acknowledges sponsorship by the Netherlands National Research Programme on Global Air Pollution and Climate Change (Project 851039) to participate in the International Symposium on Climate Change and Rice.

Nitrous Oxide Emissions from Flooded Rice

K.F. BRONSON and U. SINGH

Flooded soils, such as those used for rice production, are assumed to be sites of denitrification reactions that lead to the emission of nitrous oxide (N_2O) and dinitrogen (N_2). Nitrification is believed to occur in the thin oxidized surface layer of soil and in the oxidized rhizosphere (Patrick 1982). Nitrate that diffuses from the oxidized to the reduced zones of soils is immediately denitrified. Soil and fertilizer nitrogen (N) lost to the atmosphere represents an economic loss for the rice farmer. Ammonia losses from flooded rice systems have been reported to be as high as 56% of urea-N applied (Freney et al. 1990). Direct measurements of economically significant denitrification of fertilizer N have not been observed; the maximum is 4% of applied N (Mosier et al. 1990). However, few measurements have been made of denitrification of the appreciable amounts of soil NO_3 that accumulate during fallow periods in rice soils. These losses should be highest immediately after the permanent flood that is imposed for puddling and establishment of transplanted flooded rice.

In addition to the economic loss that denitrification represents, the denitrification product (N_2O) is a very potent greenhouse gas (300 times more radiatively active than CO_2 on a mass basis; Rodhe 1990) with a very long residence period in the atmosphere (150 years Rodhe 1990). The N_2O oxidation products NO and NO_2 also participate in the destruction of stratospheric ozone (Crützen 1976). The concentration of N_2O in the atmosphere has been increasing at the rate of 0.3% per year (IPCC 1990)

Nitrification-Denitrification in Flooded Rice Soils

Several excellent reviews of N transformations in flooded soils have been published (Patrick 1982; Keeney and Sahrawat 1986; Reddy and Patrick 1986; Reddy et al. 1986). The focus of this chapter is the nitrification-denitrification reactions that produce N_2O.

Soils and Water Sciences Division, International Rice Research Institute, P.O. Box 933, 1099 Manila, Philippines, and International Fertilizer Development Center

S. Peng et al. (Eds.) Climate Change and Rice

Flooded soils are characterized by the absence of O_2 in the profile (Reddy and Patrick 1986). Diffusion of molecular O_2 is 10 000 times slower in the water-filled pores of flooded soils than when air is present. After O_2 is displaced, the requirement of facultative and strict anaerobic organisms for electron acceptors is met by the reduction of oxidized compounds (e.g., NO_3, MnO_2, Fe_2O_3, and SO_4) (Patrick 1982). These sequential reductions result in a redox status in flooded soils as low as −240 mV.

Dissolved O_2 in the overlaying floodwater diffuses down into the top few millimeters of soil, where it is rapidly consumed. This thin surface layer in flooded soils is brown in color, whereas the reduced zone below is gray. The redox potential at the interface of the aerobic soil layer and the reduced zone below is about 300 mV (Reddy and Patrick 1986). The thickness of this aerobic layer is controlled by organic matter content (which reduces the thickness), the presence of NH_4 (which can increase the thickness), and the percolation rate of the floodwater (higher rates can increase thickness).

Nitrification of NH_4 that has diffused upward can occur in the aerobic layer. When the NO_3 that is formed diffuses down into the reduced zone, it is rapidly denitrified to N_2O and N_2. The amount of N_2O evolved during nitrification (Bremner and Blackmer 1978) in the aerobic layers of flooded soils is unknown. Ammonium diffusion and nitrification are the rate-limiting steps in these reactions, not NO_3 diffusion and denitrification, which occur rapidly (Reddy and Patrick 1986).

Soil properties that affect nitrification in the aerobic soil layer include pH and pO_2 (Katyal et al. 1988). The presence of light can inhibit nitrification in flooded soils, possibly because of direct effects on nitrifiers or because of increased immobilization of N by algae (Katyal et al. 1988).

Magnitude of Nitrous Oxide Losses in Flooded Rice Soils

In one of the first studies of N_2O emission in a flooded rice field, Denmead et al. (1979) reported that 1.4% of NO_3 lost was lost as N_2O in Australia. Freney et al. (1981) reported that just 0.1% ammonium sulfate N was lost as N_2O in flooded rice in the Philippines. Smith et al. (1982) observed that 0.01–0.05% of applied urea-N was evolved as N_2O in Louisiana. Lindau et al. (1990), also in a Louisiana rice field, found that daily N_2O emissions never exceeded 3 g N/ha whether urea or KNO_3 was applied. In the Philippines, Bronson et al. (1993) found that N_2O emissions accounted for 0.06 and 0.15% of urea and $(NH_4)_2SO_4$-N, repectively. The main reason for low N_2O fluxes from flooded rice soil is slow rates of ammonium diffusion and nitrification, which tend to keep rates of denitrification low. Additionally, factors controlling the $N_2O/(N_2O + N_2)$ ratio must be considered.

The proportion of N_2O to $(N_2O + N_2)$ that is produced and evolved in soils is influenced by O_2 and NO_3 concentration (Firestone 1982). Weier et al. (1993)

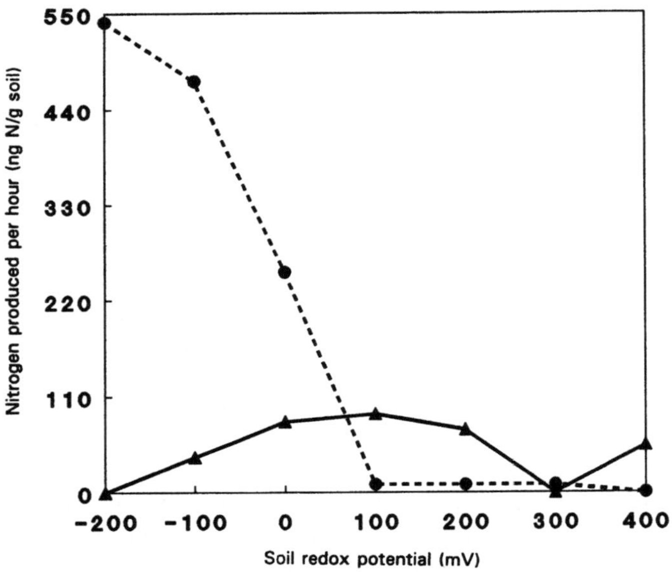

Fig. 1. Nitrous oxide (▲) and dinitrogen (●) emission rates as a function of soil redox status (average hourly rates over a 4-day period). (After Masscheleyn et al. 1993)

conducted laboratory studies and reported that the $N_2O/(N_2O + N_2)$ ratio decreased as the percentage of water-filled pore space increased, and that the ratio increased as NO_3 concentration increased. Masscheleyn et al. (1993) measured N_2O and N_2 production in laboratory studies as a function of redox status (Fig. 1). Dinitrogen began to appear at 0 mV and increased as the Eh fell to –200 mV. No N_2O was produced below –100 mV.

In flooded rice soils where the pore space is essentially 100% filled with water, and only low amounts of NO_3 are produced, the $N_2O/(N_2O + N_2)$ ratio should always be small. Lindau et al. (1990) found that N_2O evolved in field studies was < 1% of the N_2 evolved. Bronson and Mosier (1991) reported that in greenhouse studies of straw-amended rice soil, the proportion of $N_2O/(N_2O + N_2O)$ was less than 0.01%. The redox potential in this study was less than –200 mV.

Nitrous Oxide Emissions during Rice Fallow and from Permanent Flood

During fallow periods between rice crops, aerobic conditions allow nitrification of NH_4 mineralized from soil organic matter. When the soil is permanently flooded for rice production at the end of these fallow periods, the NO_3 may all be denitrified. Buresh et al. (1989) reported that during dry season fallow, 52–77 kg

NO_3-N accumulated. All of the NO_3-N disappeared after flooding for rainy season rice, presumably from denitrification. George et al. (1993) observed that NO_3, which accumulated during dry season fallow or during the dry to wet season transition, declined near the end of these seasons and completely disappeared when flood for rice was imposed. George et al. (1993) also cited evidence for NO_3 leaching at this time. Few direct measurements of denitrification have been made during these periods when the soil is flooded and rice has not yet been transplanted. In the Philippines, Bronson et al. (1993) observed N_2O fluxes between 25 and 70 g N_2O-N/day upon flooding after 6–8 weeks of rainfed fallow (Fig. 2B,C). They measured small N_2O fluxes after permanent flooding if the preceding fallow period was excessively wet.

Measurements of N_2O emissions during the generally dry fallow periods have been made by Bronson et al. (1993). They observed high N_2O fluxes (30–100 mg N_2O-N/day) immediately after rainfalls of > 2 cm during fallow periods, apparently due entirely to denitrification of accumulated NO_3 (Fig. 2A,C). Between large rain events in the fallow periods, N_2O emissions were about 1–2 mg N_2O-N/day, or about ten times higher than during the flooded rice season. During these times, N_2O was presumably produced both from nitrification of NH_4 in the dry surface layer of soil and denitrification in the wet subsoil.

Mitigating Emissions of Nitrous Oxide

Split applications of N fertilizer (i.e., breaking application up into several doses) in flooded rice culture is probably the simplest way to further reduce the already low rates of N_2O emission. Nitrification inhibitors are another option. Bronson and Mosier (1991) showed in greenhouse studies that the nitrification inhibitor wax-coated calcium carbide reduced N_2O (as well as N_2 and CH_4) emissions in flooded rice, although the magnitude of the N_2O loss was already quite low. In dry-seeded rice culture in Australia, intermittent irrigations prior to the establishment of permanent flood, result in significant losses of N_2O to denitrification (Keerthisinghe et al. 1993). The N_2O loses in this field study were effectively controlled with the application of nitrification inhibitors with N fertilizer. However, general use of nitrification inhibitors in the transplanted rice-growing regions of Asia, where fertilizer use is often suboptimal, is unlikely.

Soil management during fallow periods is more important to conserve NO_3 that accumulates for the following rice crop and to prevent large N_2O fluxes after flooding. The traditional weedy fallow is effective in this way (Buresh et al. 1989; George et al. 1993), but the planting of a crop is even more efficient in cycling inorganic N. A negative aspect of cycling native N from the fallow periods with weeds or crops is that this added C will likely stimulate CH_4 production and emission during the flooded rice season.

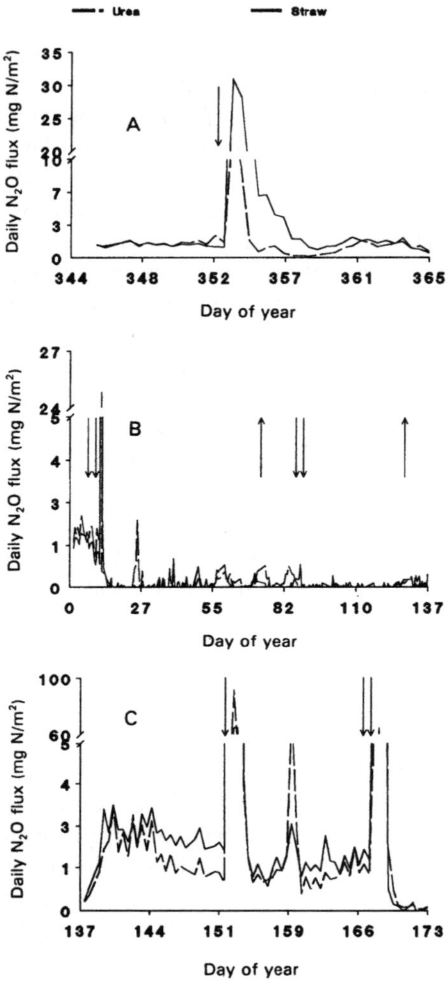

Fig. 2A–C. Nitrous oxide fluxes in Aquandic Epiaqualf as affected by N source. **A** Wet-dry season fallow (1992). **B** Wet-dry season fallow (days 0–27) and dry season (days 27–137) (1993). **C** Dry-wet season fallow (1993). *Single downward-facing arrows* represent rainfalls > 2 cm, *double downward-facing arrows* represent flood irrigation, and *single upward-facing arrows* represent drainage. (Bronson et al. 1993)

Nitrous Oxide-Methane Interaction

It is clear that the redox potentials in soil where N_2O and CH_4 are produced are quite different (Masscheleyn et al. 1993). An interaction between N_2O and CH_4 was reported by Bronson et al. (1993). Additions of organic amendments stimulated CH_4 emissions but resulted in lower N_2O fluxes. The added organic matter probably led to a thinner aerobic surface layer, which limited nitrification and reduced N_2O emissions. Management strategies for CH_4 mitigation that involve changing redox status (e.g., midseason drainage) have the potential to increase N_2O emissions. However, Bronson et al. (1993) observed in one dry season experiment, in which four N applications of 50 kg N/ha were applied, that midseason drainage suppressed CH_4 emissions, without affecting the N_2O flux (Fig. 2B).

Future Research Needs on Nitrous Oxide Emissions from Rice

Several large gaps in knowledge still exist in N_2O production and emission from flooded rice.

The amount of $N_2O + N_2$ entrapped in rice soil is largely unknown, and it is not recovered in standard techniques of flux measurement. A few attempts have been made to quantify entrapped $N_2O + N_2$ (Lindau et al. 1990; Samson et al. 1990). Nitrous oxide entrapped in a heavy-textured rice soil for long periods would probably get reduced to N_2 or come out in drainage water. During the growing season, $N_2O + N_2$ flux measurements are higher when rice plants are included in the chamber. This indicates the importance of plant-mediated transport (Mosier et al. 1990).

Little or no denitrification measurements have been made in rainfed ecosystems. In these systems, the water regime is highly variable, and nitrification-denitrification reactions probably occur during alternate wetting and drying cycles. It is entirely possible that loss of fertilizer N to denitrification in rainfed systems is as important as NH_3 volatilization losses.

Finally, given the amounts of N_2O that Bronson et al. (1993) observed during dry rainfed fallow periods in rice soils, it is likely that NO losses at those times are also agronomically important (Davidson 1991).

Global Climate Change: Role of Rice in Methane Emissions and Prospects for Mitigation

R. RANGANATHAN, H.-U. NEUE, and P.L. PINGALI

If atmospheric CH_4 continues to increase at the current rate of 1% per year, it is expected to contribute more to future climate change than any other gas, with the exception of CO_2. Methane has so far received little attention because CO_2 emissions are considered a greater threat. In developing economic responses to global warming, Schelling (1991) argues that although CH_4 is a serious greenhouse gas, its low residence time gives researchers two to three decades to decide on strategies of reduction.

The Intergovernmental Panel on Climate Change has recommended immediate reductions in anthropogenic emissions of 15–20% to stabilize CH_4 concentrations at current levels (IPCC 1990). But at the core of the issue on CH_4 mitigation from rice are the questions: Is global warming occurring? and: Can emissions from rice cultivation be reduced without affecting rice production? The vast amounts of literature on the ongoing changes in atmospheric composition suggest that global warming will occur, but so far there is no evidence of imminent rapid change. There is little scientific consensus on the magnitude and effects on human society because of changes in precipitation, soil moisture, rises in sea level, and increased severity of storms (IPCC 1990). However, if the higher projections from climate models prove to be correct, the stress on the environment, agriculture, and other human enterprises could be grave (Parry 1990; Pittock 1991; Adams et al. 1992; Kaiser et al. 1992; Rosenzweig et al. 1993).

Options that limit rates of emission of greenhouse gases are being evaluated (National Research Council 1992; Reilly and Anderson 1992; Dornbusch and Poterba 1991); but questions remain. What economic tools are available to control anthropogenic emissions of greenhouse gases? Should reductions come from the industrialized sector or agricultural? In the case of CH_4, the issue is inextricably linked with agricultural practices and revolves around equity. Can agriculture-dependent economies adopt climate-control policies and management techniques that may reduce agricultural productivity? Can developing nations implement costly policies and techniques when even progressive economies find much of their economic activity falling short of demonstrated best practices?

International Rice Research Institute, P.O. Box 933, 1099 Manila, Philippines

S. Peng et al. (Eds.) Climate Change and Rice

Trade-Offs Between Rice Production and Methane Mitigation

A review of the literature on factors that control CH_4 emissions from rice fields suggests that economic factors that contribute to increasing rice production also enhance emissions. Therefore, CH_4 mitigation may have a negative impact on rice production. The potential trade-off in CH_4 mitigation and rice production is a particularly sensitive issue because of the already declining productivity of rice in the irrigated rice ecosystems of Asia (Pingali et al. 1990; Cassman and Pingali 1994), which contribute 76% of the current rice production in the region (IRRI 1993). Despite impressive gains in rice production in many Asian countries, there is little cause for complacency about food security. In many countries, growing populations and rising incomes are contributing significantly to increasing demands for rice (IRRI 1989). Rice production must increase at a compound rate of 1.7% per year (IRRI 1989, 1993) and future output growth is expected to come from increased rice yield per unit of land because opportunities for area expansion have been exhausted (Pingali and Rosegrant 1993).

Methane mitigation options have been discussed by many researchers (Sass et al. 1992a; Neue 1993; Wassmann et al. 1993e). Adams et al. (1992) calculated the cost associated with achieving diminished CH_4 through a 50% reduction in fertilizer use from a spatial-equilibrium model of US agriculture. They estimated a cost of USD590/10^3 kg of CH_4 because of reduced rice yields as a consequence of reduced nitrogen applications. Reductions in nitrogen also require an increase in total area in rice production. The analysis, although only an initial estimate of the cost-effectiveness of the option considered, suggests that the trade-off is high. Variations in CH_4 production and emission are caused by a complex interaction of soil physical and chemical properties and water management, modified by organic and inorganic fertilizer amendments and practices of crop management (Neue and Roger 1993). A calculation of the trade-off between rice production and CH_4 mitigation is difficult to make because of the many interactive processes. Not all these processes are well understood. Even if it were possible to estimate the loss in rice production from a particular strategy for CH_4 abatement, the benefit purely in terms of a reduction in CH_4 is hard to judge because in situ measurements of CH_4 emissions from the diverse subecosystems can be conducted at only a limited number of sites.

Attempts to estimate global CH_4 emissions from rice fields have been made. Bachelet and Neue (1993) replicated three published techniques to estimate CH_4 emissions from rice fields and included soil-related factors to arrive at a revised estimate. Their method was an improvement over the earlier ones because it used a classification of soils by their potential for CH_4 emissions. The classification itself was based on still-to-be verified features of the soil that affect CH_4 production, but it showed that the inclusion of an additional variable in the estimation procedure reduced earlier estimates of global emission by close to 25%.

Some of the major economic factors (e.g., water management, intensive rice monoculture systems, input use, crop establishment methods, and rice cultivars) that contribute to CH_4 production and emission are discussed along with options, if any, for CH_4 mitigation. The possible trade-offs between rice production and CH_4 mitigation in the different rice ecosystems are discussed.

Water Regime and Management

Because of their distinct flood-water regimes, rice ecologies are major discriminators for potential CH_4 production (Neue and Roger 1993). Many lowlands are naturally flooded in the rainy season. Excess water in river and coastal floodplains is controlled by terracing, leveling, and leveeing of rice fields. In areas with irrigation infrastructure, the availability of water as a supplement in the wet season and assured irrigation facilities in the dry season have allowed for intensive (two or three consecutive rice crops on the same piece of land) rice cultivation. The irrigated rice ecosystem has the highest potential to produce CH_4 not only because flooding is assured and controlled but also because of the intensity of rice cultivation. Readily mineralized organic matter, the substrate for CH_4 formation in wetland soils, is obtained from rice straw, weed, and aquatic biomass incorporated into the soil during land preparation and before rice seedlings are transplanted.

Reductions in CH_4 can be effected either by decreasing the period of inundation and draining fields during those times in the season when plants are least susceptible to water stress or by moving away from rice cultivation. Multiple aeration, feasible only with complete water control and when drainage is possible, was found to reduce CH_4 emissions by 88% without any loss in yield but needed 2.7 times more water than that required by the normal flood-water treatment (Sass et al. 1992a). Midseason drainage, similar to that practiced in Japan (Yagi and Minami 1990a), resulted in a seasonal emission that was 52% of the CH_4 flux from a normal uninterrupted flood situation. No significant loss in grain yield was recorded.

The intensive use of irrigation water has led to a buildup of soil salinity in the semiarid and arid zones, whereas in the humid zones, waterlogged fields have lower productivity levels from low rates of organic-matter decomposition, lower nitrogen availability, and accumulation of soil toxins. Increasing the efficiency of water use in the irrigated lowlands is expected to reduce degradation in the rice fields and conserve water. Tabbal et al. (1992) showed that intermittent flooding to keep the soil saturated reduced water use by 40% without a significant effect on yield. The possibility of intermittent flooding without significantly affecting yield depends on the soil characteristics. Although there are differences in the reported quantities of water used and saved using intermittent flooding and drying, controlled multiple or midseason aeration is technically feasible and could contribute to both increased water use efficiency and CH_4 mitigation in the irrigated rice ecosystem.

Unlike in the irrigated ricelands, CH_4 emissions have not been measured in rainfed areas that account for midseason droughts and year to year rainfall variations. Unpredictable rainfall makes the rainfed ecosystem less favorable for continuous CH_4 production and emission. However, depending on rainfall amount and distribution and soil-related factors, the potential for CH_4 emission in some rainfed environments could be close to that of the irrigated ecosystem. A major factor limiting the yield potential of the rainfed lowlands is a reliable and timely supply of water. Methane abatement using water management is, perhaps, not a vaiable strategy for such environments.

Improved efficiency of water use in the irrigated ecosystem is a possible strategy to both limit CH_4 emissions and alleviate the stress on the soils of this environment with no significant losses in rice yield.

Intensive Systems of Rice Monoculture

Rice farming, to a large extent, is controlled by the availability of water and the physical conditions of the soil. Many lowlands are naturally flooded in the rainy season, and the rice plant is unique in its ability to thrive under flooded conditions. The potential for CH_4 emission under such conditions has been established. The availability of water in the dry season makes the irrigated lowlands the most intensively cultivated ecosystem. Two, sometimes three, rice crops are grown each year. With the second and third rice crops, CH_4 emissions from the irrigated lowlands is greatly increased.

The CH_4 emission potential of rice-rice systems is raised further by the high yields obtained in dry season irrigated rice as a result of assured water supply combined with higher solar radiation and high input levels. Neue et al. (1994) recorded higher emissions in the dry season compared with the wet. High flux rates associated with high yields are attributed to increased rice tillers and roots which provide more emission pathways. Improved root growth also enriches the carbon source for methanogenesis. Higher soil temperature is yet another factor contributing to higher emissions in the dry season. Holzapfel-Pschorn and Seiler (1986) found that CH_4 fluxes doubled when soil temperature increased from 20 to 25 °C.

Long-term trials at experimental stations show that even with the best available cultivars and management, yields from continuous cropping tend to decline over time when input levels are held constant (Cassman et al. 1994; Cassman and Pingali 1994). The decline in productivity is attributed to a degradation of the resource base. Reducing the intensity of rice cultivation has been suggested as a possible means of arresting this long-term degradation of the resource base. It has obvious benefits from the point of view of mitigating CH_4 emissions. Neue and Roger (1993) have suggested a shift from wetland to upland rice cultivation. Wassmann et al. (1993e) briefly mention it too but shy away from suggesting an attempt to implement such a proposal in the light of an increasing demand for rice.

Moving away from a rice-rice farming system is feasible in areas where drainage will not be a problem for the second crop. In a sequential rice-nonrice cropping system, crops would need to be cultivated in distinct anaerobic and aerobic phases. Establishment of upland crops following wetland rice could be problematic because of soil compaction. In areas where drainage is required in the dry season, upland crops stand the risk of being waterlogged. The trade-off between rice production and crop diversification in an intensely cultivated irrigated ecosystem is high.

Crop Establishment

Transplanting and direct seeding (wet and dry) are the two methods of establishing a rice crop. Although transplanting is practiced widely in the irrigated environments, in the rainfed environment uncertain rainfall is a major factor affecting the choice of the method of crop establishment. The decision to transplant or directly sow seeds generally depends on farmer assessment of drought risk and other socioeconomic conditions. Soil texture and topography also play a role in the decision. Under optimum conditions, the yield of rice is unaffected by the method used.

Transplanted and wet-seeded rice require puddled soils and have similar requirements with regard to land preparation. Depending on the rainfall distribution, the time lag between land preparation and transplanting or wet seeding might be 6–8 weeks. During this time, anaerobic soil conditions might prevail. Dry-seeded rice reduces this flooding period and eliminates the need to puddle the soil and weed during land preparation. Although the direct effects of methods of crop establishment are likely to be insignificant, water regimes and cultural practices associated with these procedures do have an impact on CH_4 emission. In the irrigated ecosystem dry-seeded rice may mitigate CH_4 emissions because of a shortened flood period and minimal soil disturbances. Land preparation, puddling, and weeding cause the flooded soil to be disturbed, a physical disturbance of flooded soils releases large amounts of soil-entrapped CH_4. Dry seeding reduces this disturbance. Dry-seeded rice however, requires high levels of herbicide use because manual weed control is difficult. The effects of herbicides on CH_4 production and emission are not known. Crop losses from dry-seeded rice have been reported, perhaps because of competition from weeds.

In an early attempt at modeling (Ranganathan et al., in prep.) a simple algorithm was developed to estimate CH_4 emissions from rainfed sites in a given season and year, based solely on the soil aeration status of the rice[1]. From a sample of farms in the Philippines where the water status of the field was measured regularly over the rice-growing season (Saleh et al. 1993), aerobic soil conditions prevailed 60-80 days after transplanting. This implies that the potential for CH_4 emission in the latter half of the rice-growing season is negligible. Estimates of cumulative CH_4 emitted over the season, derived from the use of the algorithm, ranged from 25 to 75% of that emitted from irrigated rice fields

(Wassmann, in prep.). This illustrates the variability that exists within a season and within a small geographic spread.

In rainfed areas where soil moisture is favourable for a second crop or the availability of labor is a constraint, dry seeding of rice is practiced. Dry-seeded rice is sown and harvested earlier than a transplanted crop. The probability of standing water early in the rice-growing season is low; whereas, late-season anaerobic conditions are a real possibility. This was borne out by a sample of dry-seeded farms where the water level (depth of standing water) was regularly measured by Saleh et al. (1993). Anoxic soil conditions started 60–70 days after seeding and often continued until well after harvest. In the study by Sass et al. (1992a), CH_4 emission from the late-season flood treatment was 1.6 times higher than that observed in the treatment where the crop was continuously flooded. Easily decomposable organic matter accumulated during the nonflooded period and fermentation took place when the emission potential of the rice plants was highest. Rice yields were lower by 25%. A significant factor mediating late-season CH_4 production and emission is soil type. In acid soils, it may take weeks after anoxic conditions prevail for CH_4 production; whereas, in calcareous and alkaline soils, CH_4 production can occur within hours after flooding (Neue and Roger 1993).

Determination of the trade-off between CH_4 emission resulting from different methods of crop establishment and rice production in rainfed areas requires consideration of factors such as crop loss from weeds, availability of labor, wages for labor, the opportunity cost of family labor, the cost of chemical weed control and other socioeconomic costs that may accrue from a change in method. The eventual trade-off will vary from site to site; nevertheless, it is considered to be low for both irrigated and rainfed environments.

Use of Organic and Inorganic Fertilizers

Incorporation of crop residues is a common practice in the tropics. Rice straw, green mulches, and weed biomass are the only nutrient sources for many resource-poor farmers. The quality and quantity of organic matter added to the soil influences CH_4 production. Methane emissions increase in direct proportion to the application rate of rice straw (Schütz et al. 1989b). Denier van der Gon et al. (1992) found that the addition of 20 t/ha (fresh weight) of *Sesbania rostrata*

[1] The algorithm cannot be used unless information is available on field-saturation conditions and on the time sequence of the aerated spells during the rice-growing season. Daily rainfall alone is not a good predictor of the water level status of a field. Soil texture, slope, position of the field in relation to the watershed, depth of the water table, height of the bunds, and management of the levees are among some of the variables that affect the level of standing water on a field. To generate information on field-saturation conditions, a model that accounts for these variables is required. Even as a first step in approximating the emission potential of an area, simple models require vast amounts of data and make it imperative to establish the minimum data set required for a reliable estimate of CH_4 emissions.

strongly enhanced CH_4 emissions. Methane mitigation requires that the quantities of organic fertilizer used be minimized. The trade-off in resource-poor areas is clear and is higher for rainfed than for irrigated environments.

Irrigated rice is characterized by high use of nitrogen fertilizers. Although the plant-mediated effect of increased nitrogen fertilizers on CH_4 production and emission is known, the chemical processes that directly effect CH_4 formation and emission are still not clear. The addition of nitrogen fertilizers result in higher CH_4 emission because the increased number of rice tillers and roots provide more emission pathways (Neue et al. 1994; Denier van der Gon et al. 1992). An increase in rice-crop biomass has also been assumed to stimulate CH_4 production by an increase in root exudates and sloughing off of roots and vegetative matter, which are all easily decomposable organic material. CH_4 emissions are higher in the later growth stages because of well-developed aerenchymal passages and an accumulation of organic matter (Sass et al. 1990).

The direct impact of chemical fertilizers on CH_4 formation and emission is still not clear. In some experiments, urea applications reduced CH_4 emissions as compared with unfertilized plots (Schütz et al. 1989b), in others (Wang et al. 1992), urea increased CH_4 emissions, perhaps as a consequence of increased soil pH after urea hydrolysis. Similar inconsistent results have been reported for ammonium sulfate applications (Schütz et al. 1989b; Wassmann, pers. comm.). The inhibition of CH_4 emissions with the addition of a sulfate fertilizer is attributed to sulfate-reducing bacteria competing for hydrogen.

Nitrogen fertilizer is generally applied in two or three split applications. To minimize losses due to volatilization, deep placement of fertilizer granules is recommended. The source and mode of application of fertilizers is anticipated to have some direct impact on the processes that control CH_4 emissions (Schütz et al. 1989b; Braatz and Hogan 1991). However, results from trials on deep placement of fertilizers are inconsistent (Denier van der Gon et al. 1992; Wassmann et al. 1994). If deep placement of fertilizer granules was found to reduce CH_4 formation or emission, the objective of CH_4 abatement would be served and the efficiency of fertilizer use would be improved. However, the chemical processes in the soil that underlie the effects of mineral fertilizers on CH_4 emissions must still be understood.

Rainfed ecosystems are characterized by low input use. The introduction of improved varieties is expected to increase the use of mineral fertilizers and raise the productivity of these regions. Significant progress is expected in the coming years to increase the use of both farm-grown and commercial fertilizers. Along with improved rainwater management, the potential for CH_4 emissions will increase. Until more is known about the chemical processes related to fertilizer application, recommendations on an optimum use of organic and inorganic fertilizers that minimize loss in production will be difficult.[2]

Rice Cultivars

Wide variations in the amount of CH_4 emitted by different cultivars have been observed (Butterbach-Bahl 1993). The structure of the aerenchyma, particularly the intersection between root and shoot aerenchyma, controls the diffusion of CH_4. Other studies have shown two or three maxima in CH_4 emissions over the growing season, which are believed to coincide with periods when plants provide organic root exudates and litter to CH_4-producing bacteria in the soil. Ladha et al. (1986) demonstrated different patterns of root exudation from rice varieties. The development of new rice cultivars with a low potential for CH_4 emissions is an attractive mitigation option, but the inheritance of these traits and their relationship with yield potential must still be explored. Soil and environmental factors (e.g., soil type, redox potential, and soil moisture) may have an impact on the ability of the rice plant to emit CH_4, which adds another facet to the selection of varieties with a low potential to emit CH_4.

Methane Mitigation Strategies and Policy Options

Several kinds of uncertainties characterize climate change – the future mix of greenhouse gases, their impact on climate, and regional distribution of climatic change. On the CH_4 issue, there is a debate on whether CH_4 mitigation is necessary at all because of its relatively low atmospheric lifetime, the scientific uncertainty about underlying processes, and the efficacy of mitigation options. Table 1 summarizes information on the determinants of CH_4 emissions, the level of scientific uncertainty associated with mitigation strategies, and the trade-off in rice production by ecosystem. A conventional cost-benefit analysis of the CH_4 mitigation strategies is considered premature because of the uncertainties in calculating the cost of measures to slow CH_4 emissions and determining the value of the benefits of abatement. Much of the economic analyses in the global-change debate are still initial ideas and are not definitive solutions to the problem. Methodological issues such as uncertainty, discounting, and measuring costs and benefits of a reduction in emission are still being settled. Sustainability and environmental concerns have raised new questions about the efficiency criteria used in conventional economic analysis. Norgaard and Howarth (1992) talk of intergenerational rights, the rights of future generations to natural resources and environmental services, and the inadequacy of lower discount rates and non-

[2] In a greenhouse experiment, encapsulated calcium carbide was used as a nitrification inhibitor and found to severely limit CH_4 production while increasing rice production (Bronson and Mosier 1991). However, little is known about the environmental consequences of prolonged use of such inhibitors, the safety and health aspects related to their use, the commercial availability of such chemicals and the costs asociated with farmer extension services and application methods. The interaction, if any, with different soils and chemical fertilizers must also be determined.

Table 1. Determinants of CH_4 emissions from rice cultivation, their influence on emissions from the irrigated and rainfed ecosystems, and potential trade-offs with rice production

Determinants of CH_4 emissions	Degree of influence[a] on emissions from ecosystems		State of knowledge about application[b] for mitigation		Potential trade-off between mitigation and rice production	
	Irrigated	Rainfed	Irrigated	Rainfed	Irrigated	Rainfed
Economic factors						
Water regime and management	High	Moderate-low	Moderate	?	Low	?
Rice farming systems						
Rice-rice	High	High-Moderate	High	High	High	High
Rice-nonrice	High	Moderate-low	High	High	High	High
Rice-fallow	High	Moderate-low	High	High	High	High
Crop management						
Transplanting	Moderate	Moderate	Low	?	Low	Low
Wet seeding	Moderate	Moderate	Low	?	Low	Low
Dry seeding	Moderate	Moderate	Low	?	Low	Low
Weeding	Moderate	Moderate	Low	?	Low	Low
Land preparation	Moderate	Moderate	Low	?	Low	Low
Input use						
Organic matter	High	High	High	High	High	High
Inorganic fertilizers[c]						
Type	Moderate	Moderate	Moderate	Moderate	Low	Low
Quantity	Moderate	Moderate	Low	Low	Low-moderate	Low-moderate
Placement	Low	Low	Low	Low	Low	Low
Timing	Moderate	Moderate	Moderate	Moderate	Low	Low
Nitrification inhibitors	High	High	Low	?	Low	Low
Varieties and cultivars						
Traditional	?	?	Low	Low	?	?
High-yielding varieties	?	?	Low	Low	?	?
Hybrid	?	?	Low	Low	?	?
Root depth,biomass, and distribution	Moderate	Moderate	Low	Low	?	?

Physical and Environmental factors						
Major soil characteristics						
Organic C content	High	High	High	High	Moderate	High
Texture	Moderate	Moderate	High	High	None	None
pH	High	High	Moderate	Moderate	Low	Low
Redox potential	High	High	Moderate	Moderate	Low	Low
Climate						
Solar radiation	Low	Low	High	High	None	None
Temperature	High	High	High	High	None	None

[a] The interactive effect of the emission-determining factors is not included in this analysis.
[b] The physical and environmental factors have little or no application for mitigation.
[c] Only direct effects of inorganic fertilizers, and not plant-mediated effects, are considered.

market valuation of resources in the classical analysis to develop sustainable economic and policy guidelines. The importance of biological and physical data to economic assessments cannot be overemphasized.

Lave and Dowlatabadi (1993) believe scientific uncertainty is not the most important obstacle to decision-making (on emission control programs). Rather, they suggest that the degree of optimism of the decision-maker and the decision criterion itself are most important. However, it is desirable to develop strategies that optimize economic development by investing options for CH_4 mitigation that, subject to all the current uncertainties, appear to contribute more to development in the future. At the current level of scientific uncertainty of the benefits and methods of mitigating CH_4 and considering the importance of maintaining food security, two options exist that not only meet the needs of mitigating CH_4 but also help to arrest the long-term productivity decline of rice. Because they combine the issues of sustainability and environmental degradation of the rice ecosystems with CH_4 mitigation, rice cultivation with CH_4 abatement strategies could be viewed as economic growth with environmental impact taken into account. These best-bet options of CH_4 containment are crop diversification and increased efficiency of water use.

Crop Diversification

Crop diversification has merits both for CH_4 abatement and conservation of the resource base. It has been cited as a possible strategy by Pingali (1992) and Pingali and Rosegrant (1993) to arrest the long-term degradation of the most productive rice environment in Asia. Diversification is most feasible and profitable in areas with good market access and in locations where drainage is not required in the dry season. Thailand provides a good example of a country in which the development of markets for nonrice products led to diversification away from rice. Thailand is one of the largest exporters of rice, most of which comes from rainfed areas. The prospects for, and constraints to, crop diversification are discussed by Pingali (1992).

If the current trend of declining productivity persists, Pingali and Rosegrant (1993) expect a decline in employment opportunities and income levels. Studies on food consumption have shown that per capita rice intake largely depends on income level: at low income levels rice is a luxury commodity, as income levels rise rice is replaced by vegetables, bread, fish and meat. Per capita consumption of rice has declined substantially since the mid-1970s in the high and middle income countries of Japan, South Korea, Malaysia, and Thailand. With the anticipated economic growth in Indonesia, India, China and Vietnam, it is expected that the demand for rice in these countries will, in time, decrease. As markets operate more freely in these countries, the opportunities for crop and income diversification will increase. The growth in demand for rice because of expanded populations is likely to be higher than the decline caused by changes in food preference. However, if other policies are in place to address this matter, demand for rice may be met.

Efficiency of Water Use

If the time when anaerobic soil conditions prevail is reduced, the production of CH_4 will be arrested and the quantity emitted will be limited. Degradation of the rice soils through the intensive use of water and increased competition for water for nonagricultural uses also make it imperative to improve the efficeincy of water use in irrigated rice. In public irrigation systems, controlled aeration or water control can be acheived through modifications and improvements in physical facilities and water supply, and through new rules of operation for better efficiency of water distribution. Rosegrant and Svendson (1993) report a substantial cutback in public investment on irrigation during the 1980s because of low rates of return and income payoffs. However, to maintain productivity levels (the impact of declining investments in irrigation on productivity is a delayed one) through improved water use, selective rehabilitation of irrigation systems cannot be avoided. The environmental benefits of this rehabilitation must be better accounted for when payoffs are evaluated to justify renewed public investment.

With regard to incentives to farmers to increase the efficiency of water use, Pingali and Rosegrant (1993) recommend volumetric water fees rather than irrigation fees based on the area irrigated. The true value of irrigation water is often not reflected in irrigation fees for most irrigation projects. Rosegrant and Binswanger (1994) recommend the implementation of a pricing policy for water in which the social opportunity cost of water is charged to all users. Their recommended efficiency pricing of water includes cost of water based on the volume of water used, a price for the duration of delivery, and a tax on the productivity of the land for users of surface water. In the case of private groundwater users, the price would include the cost of water pumped, the time the pump is in operation, and a tax on the electricity or fuel used to pump the water.

Discussion

A noteworthy feature of these strategies is their particular relevance to irrigated environments. The irrigated lowlands are the most productive of all rice ecosystems, have the highest potential for CH_4 emissions, and, in the face of growing demand for rice, are of greatest importance. The importance of rice in most countries of Asia has led governments to institute policies detrimental to efficient growth – striving for self-sufficiency, protecting internal markets against price fluctuations of the international market, controlling domestic markets, setting floor prices and ceilings, importing large quantities of rice in election years, and providing rice rations to government servants (Barker et al. 1985). Given the trade-off in rice production with CH_4 mitigation, the key to successful implementation of a mitigation program rests on taking into consideration the twin issues: profitability of rice through improved methods of cultivation, and income generation and distribution.

The benefits of CH_4 mitigation are global, but the costs will be borne locally. Long-term control of greenhouse gases requires that the technological capabilities of the developing countries be improved and that the infrastructure for research and absorption of the new technologies be created. The poorer rice-growing nations may participate in the reduction of CH_4 if some sort of international agreement is made to provide them with additional financial and technical resources to make the necessary changes. There is little accord on such agreements. As long as the North-South divide on which countries bear the largest burden of responsibility to curb global warming lasts, it is politic to recommend mitigation options that do not involve technologies that impair the profitability of rice cultivation and that currently do not propose institutional and political changes that are hard to make.

The profitability of rice cultivation is already on the decline because of degradation of the resource base, long-term declines in real rice prices, and a stagnant yield frontier (Pingali 1993). Against this backdrop and that of uncertain benefits from CH_4 mitigation, abatement strategies that lower rice production are hard to justify. Mitigation achieved as a spin-off from policies that address sustainability issues have a greater chance of acceptance. Options for CH_4 mitigation have already been shown to be categorically indistinct from strategies that seek to halt the productivity decline of ricelands.

Future technology will have a role to play in the debate on climate change. A major focus in rice research is to increase the yield frontier while maintaining or improving the resource base. Research trials at the International Rice Research Institute and national research centers have demonstrated the potential of tropical hybrid rices to increase yield in the irrigated ecosystem (Virmani et al. 1993). With policies and incentives that improve the efficiency of water use and allow for less intensive rice cultivation in the irrigated lowlands, the trade-offs in rice production though the implementation of strategies for CH_4 mitigation can be minimized. The research agenda on cultivar selection and breeding for increased yields must include a search for varieties with low potential for CH_4 emissions. The rainfed ecosystem poses a different set of problems. New technologies and new varieties will use the resources more efficiently but may also increase the use of inputs. If ongoing research were to establish a positive relationship between inorganic fertilizers and CH_4 emissions, the loss in production from a reduced use of fertilizer would be unacceptable. The transfer of mitigation technologies developed for irrigated rice may help limit future increases in CH_4 emissions from rainfed rice. Nevertheless, until better methods to estimate emissions from rainfed areas, and the trade-offs of CH_4 mitigation with rice production are known, few recommendations can be made. Research efforts to understand the complex interactions between the controlling factors of CH_4 emissions must be continued with the objective of developing feasible mitigation options for unacceptably high CH_4 emissions from rice production. Research is relatively inexpensive compared with the cost of implementing policies that are seemingly in conflict with the aim of increasing production.

In conclusion, scientific literature suggests that despite the great uncertainties, greenhouse warming is a potential threat sufficient to justify action now. Substitution in production and consumption of rice and technological adaptations could reduce the economic and social impact of physical changes in the environment. Of course, these plans can only be justified by the current credibility of the threat. However, they set the stage for further action, if needed. Better husbandry of resources in the rice environment may be a start toward meeting the threat of a changed climate. A strong scientific program is needed to understand the phenomena of CH_4 production and emission from rice fields and to develop technologies that reduce the trade-off between CH_4 emissions and rice production. Joint socioeconomic and biological research is essential because economic indicators offer information on the technologies that are most likely to lead to efficient results.

Mitigation Options for Methane Emissions from Rice Fields

H.-U. NEUE,[1] R. WASSMANN,[1,2] and R.S. LANTIN[1]

Recent estimates of global CH_4 emissions provide a fairly good overall balance between sources, sinks and atmospheric accumulation of CH_4; but global emission rates from many individual sources, including rice fields, are still quite uncertain. The global growth rate of the atmospheric concentration of CH_4 has slowed during recent years from 20 ppbv/year (about 1.3% per year) in the late 1970s to a rate of about 13 ppbv (about 0.75% per year) in 1989 (IPCC 1992) and 4.7 ppbv in 1992. The Northern Hemisphere trend for 1992 is 1.8 ppbv and the Southern Hemisphere trend is 7.7 ppbv (Dlugokencky et al. 1994). It is not known whether the declining growth rate is due to decreasing emissions or increasing oxidation of CH_4. More information and mechanistic understanding of fluxes of CH_4 from individual sources and the fate of CH_4 in the atmosphere are needed to reduce uncertainties of current and future emission estimates.

Projected global population levels indicate that the demand for rice will increase by 65% over the next 30 years, from 460 million t/year today to 760 million t/year in 2020 (IRRI 1989). The growing demand must be met by intensified rice production on the existing 144 million ha of harvested riceland, especially in the irrigated and rainfed rice ecologies of Asia. This will most likely increase CH_4 fluxes from wetland rice fields if current technologies continue. Regardless of uncertainties in global emission rates from rice fields, the achieved understanding of factors and processes controlling CH_4 fluxes provides various technical opportunities to mitigate emissions without reducing the production potentials of rice.

Variation of Methane Emission from Rice Fields

When rice fields are flooded during the entire growing season, CH_4 emissions increase and generally show three distinct seasonal maxima (Schütz et al. 1989b; Yagi and Minami 1990a; Wang et al. 1990). The first maxima develops shortly after flooding, the second during the vegetative stage of the rice plant, and the

[1] International Rice Research Institute, P.O. Box 933, 1099 Manila, Philippines
[2] Fraunhofer Institut für atmospherische Umweltforschung, Garmisch-Partenkirchen, FRG

S. Peng et al. (Eds.) Climate Change and Rice

third during the grain filling and maturity stage (Neue and Sass 1994). If the rice field is flooded only after crop establishment the first two maxima may coincide (Sass et al. 1990; Lindau et al. 1991). The first maxima of CH_4 emission is apparently caused by fermentation of easily degradable soil organic matter in the soil.

If the amount of easily degradable carbon is low at the beginning of the season, no initial peak of CH_4 emission develops. The difference in CH_4 emission rates between fields receiving mineral fertilizer only and those receiving mineral fertilizer in combination with organic amendments decreases over time and may become marginal at the end of the growing season (Neue and Sass 1994). This clearly indicates the increasing impact of the rice plant over the season, not only in mediating CH_4 emission but also in providing carbon for CH_4 production. Plants also affect CH_4 ebullition. After an initial peak, CH_4 ebullition declines because of increasing plant-mediated emissions. Most CH_4 generated from root exudates and decaying roots seems to be emitted through the rice plant during this period of vigorous vegetative growth. After flowering, ebullition increases again until maturity. The general pattern of CH_4 fluxes is similar between the wet and dry seasons. Because of better plant growth, CH_4 emission rates are higher in the dry season except at the maturity stage.

Diel variation of CH_4 fluxes is highly correlated with temperature fluctuation. The temperature dependence of the seasonal emission pattern is not as obvious. Sass et al. (1991b) observed maximum CH_4 production in rice soils at about 37 °C irrespective of differences in CH_4 production potentials. The temperature optimum for the production of methanogenic substrates by fermenting bacteria may not concur with methanogenesis. Similar activation energies for CH_4 production and emission suggest that both processes may be limited by the same production step. This would suggest that emission is a faster process than production. Therefore, a large buildup of CH_4 in a flooded soil grown to rice should not be expected (Sass et al. 1991b).

Effect of Soil Properties on Methane Fluxes

Because CH_4 is produced only by strictly anaerobic bacteria (methanogens), a sufficiently low redox potential (Eh) is required. Z. Wang et al. (1993a) showed that the critical soil Eh for initiation of CH_4 production is approximately –150 to –160 mV. The magnitude of soil reduction is determined by the amount of easily degradable organic substrates, their rate of decomposition, and the amounts and kinds of reducible nitrates, iron and manganese oxides, sulfates, and organic compounds. The NO_3^-, Mn^{4+}, Fe^{3+} and SO_4^{2-} must be reduced in the soil before CH_4 production occurs (Patrick and Reddy 1978). As long as a flooded soil contains significant amounts of active oxidants, CH_4 formation is restricted and carbon is mineralized to CO_2. In most soils, rapid initial decrease of Eh after flooding is caused by rapid decomposition rates of organic substrates and the low buffer capacity of nitrates and Mn oxides. Subsequently, the most important redox buffer systems are Fe(III) oxyhydroxide/Fe(II) and organic compounds.

The formation of CH_4 after soil submergence is preceded by the production of volatile acids. Short-term H_2 evolution immediately follows the disappearance of O_2. This is followed by CO_2 production increases and CH_4 formation increases (Takai et al. 1956; Neue and Scharpenseel 1984). According to Takai (1970), the bulk of CH_4 in rice fields is formed through decarboxylation of acetic acid, which results in a 1:1 ratio of CO_2 and CH_4 formation. Z. Wang et al. (1993a) found highest CH_4 production rates at a pH of 6.9–7.1 in an acid rice soil. Small changes in pH sharply lowered CH_4 production. Below pH 5.75 and above 8.75, CH_4 production was inhibited completely. Acharya (1935) reported that, during decomposition of organic substrates, the preliminary stage of acid formation is more tolerant to pH reactions, but CH_4 formation is greatly impeded outside the range of pH 7.5–8.0. Parashar et al. (1990) found highest CH_4 emission rates at a pH of 8.2 in some Indian rice soils. The actual optimum pH for CH_4 production may differ slightly among soils because of differences in substrate supply and the complex linkage between pH and Eh. In calcareous and alkaline soils, CH_4 production may occur within hours after flooding an air-dried soil. In acid soils, it may take weeks before CH_4 is formed. In very acid soils, CH_4 may not be formed at any time (Neue and Roger 1993).

Soils can be classified into four groups according to CH_4 produced during anaerobic incubation for some weeks (Z. Wang et al. 1993a; Neue et al. 1994). Each group has a distinct pattern of CH_4 formation and differs significantly in the total amount of CH_4 produced. Differences are related to substrate (hydrogen, acetate) limitation and to soil conditions. No simple correlation has been found between stable soil properties and CH_4 production. The rate of reduction and the Eh and pH of reduced soils are significantly correlated to CH_4 production (Neue and Roger 1993; Z. Wang et al. 1993b; Neue et al. 1994). Total organic carbon and water-soluble carbon are correlated with CH_4 production if other limiting factors are mitigated (Z. Wang et al. 1993b). Reduced sandy soils high in organic carbon produce more CH_4 than clay soils with similar carbon contents (Neue and Roger 1993; Neue et al. 1994). The negative impact of clayey texture on CH_4 production may be caused by the formation of organomineral complexes. Sandy soils show lower entrapped CH_4 (Z. Wang et al. 1993b) because the pore-size distribution enhances ebullition and diffusion (Neue and Roger 1993). Methane fluxes in clayey soils may also be lower because entrapped CH_4 may be oxidized before it can escape to the atmosphere. Methane production is limited in sandy soils if water percolation and the resultant redox potential is high. Disturbances of anaerobic soils by cultural practices (e.g., puddling, transplanting, fertilization, and weeding) release soil-entrapped CH_4 to the atmosphere. Denier van der Gon et al. (1992) estimated that these soil disturbances contribute about 10% to the total CH_4 emission. Rice soils that are prone to CH_4 production mainly belong to the orders of Entisols, Inceptisols, Alfisols, Vertisols, and Mollisols. Oxisols, most of the Ultisols, and some of the Aridisols, Entisols, and Inceptisols are less favorable to CH_4 production when flooded.

Mitigation Options

Variations in emission rates because of climate, soil properties, duration and pattern of flooding, rice cultivars and crop growth, organic amendments, fertilization, and cultural practices cause the high uncertainties in current global estimates of CH_4 emission from rice fields. A geographic information system of these factors and mechanistic modeling of CH_4 fluxes are required to improve the estimates and predictions. Irrigated rice has obviously the highest potential to produce and emit CH_4 because of assured and controlled flooding, high fertilization, and good rice growth. Rice production and CH_4 emission in rainfed rice varies widely in time and space. Rainfall within the watershed primarily controls floodwater regimes with periods of droughts and floods common during the growing season. Tidal wetlands and deep-water rice comprise less than 10% of ricelands and their CH_4 emission potential may be low because of salinity and deep floodwater, respectively. Upland rice is not a source of CH_4 because it is never flooded for a significant period.

Water Management

Methane emission rates vary markedly with water regimes. A single midseason drainage may reduce seasonal emission rates by about 50% (Sass et al. 1992a; Kimura 1992). Multiple aeration for 2–3 days at 3, 6, and 9 weeks after initial flooding reduced CH_4 emission by 88% and did not reduce rice yields compared with the normal irrigation in southern United States (where dry-seeded rice is flooded 4–5 weeks after seedling emergence until 10–14 days before harvest) (Sass et al. 1992a). Short aeration periods at the end of the tillering stage and just before heading improve wetland rice yields (Wang 1986). However, multiple aeration requires 2.7 times more water than the normal floodwater treatment (Sass et al. 1992a). When soil macropores start to drain during drying, large amounts of soil-entrapped CH_4 may be released (Neue et al. 1994). A net reduction of CH_4 emissions will only be achieved if soils become fully aerated. Increasing percolating rates of water may supply enough oxygen to the soil to raise the Eh, decrease CH_4 production, and increase CH_4 oxidation. Percolating water also transports organic solutes and dissolved gases into the subsoil or groundwater where leached CH_4 may be oxidized or released to the atmosphere elsewhere (Kimura 1992). In soils with high CH_4 production, Inubushi et al. (1992) found that a percolation rate of about 4 mm/day reduced CH_4 emissions up to 58%.

Where possible, fields are commonly drained 14 days before maturity to allow easier harvesting of the crop; but there are constraints to implementing delayed flooding and midseason drying. The timing must be correct and water must be available for reflooding. During the monsoonal rainy season ricelands are naturally flooded in tropical Asia and fully controlled drainage is often impossible. Holdback of precipitation in rice fields is also needed to conserve water on site and prevent flash floods elsewhere. In the dry season, irrigation water may become scarce and expensive. Water deficit during the reproductive stage can cause a high percentage of sterility of the rice plants (Yoshida 1981). Water deficits during the vegetative stage may reduce plant height, tiller number, and leaf area, and this may also reduce yields if plants do not recover before flowering.

In the United States, Australia, and Europe where large power units can be employed, dryland tillage and dryland seeding of rice is common. Fields are flooded only after crop establishment. The shortest flooding period of dry-seeded rice consistent with high yields of grain is 30 days from 75 to 105 days after seeding (Sass et al. 1992a). This requires high control of soil moisture because a stress slightly above field capacity may reduce grain yield by 20–25% compared with continually flooded treatments (De Datta 1981). In most Asian countries, 40% of rice soils have at least 45% clay (Kawaguchi and Kyuma 1977). Farmers are unable to keep soil moisture at field capacity at high evapotranspiration. Clayey soils also have a poor structure and are hard when dry. Wet tillage is the preferred land-preparation method in tropical Asia where hand and animal power is still common and the principal form of mechanization is a 10–15-hp hand tractor. The advantages of wet tillage are lower draft requirements, reduced water percolation, ease of transplanting, and improved weed control. Wet field preparation requires at least 2 weeks: then land soaking until the soil is saturated, plowed, puddled, leveled, and harrowed. One third of the total 1240 mm of water required for an irrigated, transplanted rice crop is needed for the wet field preparation (Yoshida 1981). Transplanting rice is still the major practice of rice culture in most of tropical Asia. Less water is required when pregerminated seeds are directly seeded on puddled fields without standing water. This practice is becoming popular in areas with good water control. Wet-seeded rice fields are reflooded 2–3 weeks after the crop is established.

Fertilization

Fertilization improves plant growth and generally increases CH_4 emissions. Effects of different N fertilizer on CH_4 production from soil incubation studies have been reported by Wang et al. (1992). Addition of urea did not increase total production of CH_4 but stimulated CH_4 production in the acid soil, possibly because of a short-term increase in pH after urea hydrolysis and a resulting decrease in Eh. Addition of ammonium to the floodwater can inhibit CH_4 oxidation in the soil-floodwater interface and reversibly increased the CH_4 flux (Conrad and Rothfuss 1991). Application of nitrate-containing fertilizer increases the soil Eh and results in a decrease in both the rate and the total amount of CH_4 produced. Sulfate-containing fertilizer decreases CH_4 production, especially when applied in large amounts. Addition of sulfate-containing N fertilizer has little effect on soil Eh and pH. Competition of sulfate-reducing bacteria for hydrogen and hydrogen-sulfide toxicity are the most likely mechanisms. Schütz et al. (1989b) reported a decrease of 6% in CH_4 emissions when ammonium sulfate was applied to the surface and up to 62% when incorporated into the soil. Lindau et al. (1993) found that sodium sulfate was more effective than ammonium sulfate in reducing CH_4 emissions. Brackish water inhibits methanogenesis (Garcia et al. 1774; De Laune et al. 1983). Suppression of methanogenesis by sulfate reducers in freshwater sediments at in situ concentrations of as low as 60 μM has been found (Lovely and Klug 1983). Application of 100 kg/ha ammonium sulfate will provide sulfate concentrations in the soil solution that may reduce CH_4 production. Sulfate reduction will rapidly deplete the sulfate pool. The magnitude of decrease of CH_4 emissions by sulfate-containing N fertilizer depends on the reoxidation of sulfides. Inhibitors of nitrification such as acetylene (Raimbault 1975), nitrapyrin (Salvas and Taylor 1980), and dicyanidiamide (Lindau et al. 1993) also inhibit methanogenesis and CH_4 oxidation. Slow release of acetylene from calcium carbide, encapsulated in fertilizer granules greatly reduced CH_4 emissions (Bronson and Mosier 1991).

The most deficient nutrient for high wetland rice yields is nitrogen, followed by P, K, and Zn. The most common source of N fertilizer in wetland rice is urea. Most farmers apply nitrogen fertilizer in two or three split applications. The first split is applied during final land preparation or shortly after transplanting. The remainder is top dressed at later growth stages, especially at the early panicle stage. Studies on fertilizer use and rice-crop management to minimize nitrogen losses and to increase the efficiency of fertilizer have been reviewed (De Datta 1981, 1987; De Datta and Patrick 1986). Up to 60% of basal-applied nitrogen may be lost by volatilization of NH_3 and by nitrification and denitrification. Soil incorporation or deep placement of N fertilizer without standing water at final harrowing reduces volatilization losses. Broadcasting basal N fertilizer into floodwater at the early growth stage of rice (low sink strength) results in extensive N losses (as ammonia) to the atmosphere because high pH values result from algal assimilation or alkaline irrigation water (Fillery and Vlek 1986). Potassium chloride is the principal fertilizer source of K and superphosphate is the primary source of P fertilizer. Both K and P are basically applied during final land preparation. Zinc may be added as a seed treatment by dipping seedling roots in a ZnO solution or by broadcasting Zn salts at the time symptoms of Zn deficiency occur. The major constraints to the use of sulfate-containing fertilizer or CH_4 inhibitors are availability and price.

Organic Amendments

Methanogens constitute the last step in the electron-transfer chain generated by the anaerobic degradation of organic matter. Readily mineralizable soil organic matter is the main source for the fermentation products that finally drive CH_4 formation in wetland rice soils. Generally, there is a strong correlation between water-soluble carbon and CH_4 production (Vermoesen et al. 1991). Organic amendments to flooded soils increase CH_4 production and emission (Schütz et al.

1989b; Yagi and Minami 1990a; Sass et al. 1991a; Cicerone et al. 1992; Neue et al. 1994) by enhancing the reduction of soils and providing carbon sources. Based on the content of readily mineralizable carbon, rice straw or green manures produce more CH_4 and humified substrates such as compost produce less CH_4 per unit carbon. Wang et al. (1992) showed that CH_4 production increased in proportion to the application rate of rice straw.

Incorporation of 5 t/ha of straw increased CH_4 emissions 2.5 times compared with mineral fertilization in experiments at the IRRI research farm where shallow flooding was maintained during the whole growing season. Methane production and emission, especially ebullition, increased during early crop growth. Emissions were similar at the late growth stage irrespective of organic or mineral fertilizer inputs. Organic amendments at Taoyuan, Hunan Province, China, raised emission rates compared with mineral fertilization by a factor of three (Wassmann et al. 1994). Incorporation of crop residues yielded emission rates in Texas, that were two times as high as emission rates after removal of plant stubble (Sass et al. 1991a).

Not only the quantity, but the quality of organic material influences CH_4 production and emission. Application of composted material that has a higher degree of humification only slightly increases CH_4 formation (Yagi and Minami 1990a; Minami and Neue 1994). Application of 12 t/ha compost to a Gleysol and Andosol resulted in 62 and 40% lower CH_4 emission rates, respectively, when compared with incorporation of 6 t/ha rice straw. In Taoyuan, Hunan Province, China, the incorporation of sludge from biogas generators into ricefields reduced CH_4 emission by 60% compared with application of unfermented manure (Wassmann et al. 1994). There are approximately 7 million small-scale biogas generators in rural areas of China (Ma 1981) and the fermented sludges are frequently used as fertilizer in Chinese ricefields. The recycling of crop residues in rice fields through fermentation in biogas generators limits increases in CH_4 emissions as well as aerobic composting and provides an additional energy source for households.

To reduce CH_4 emissions from wetland rice fields, recycling and addition of organic substrates rich in easily decomposable carbon should be minimized rather than maximized; but organic matter is an important soil energy source and soil conditioner. Sustained fertility of soils requires the maintenance of soil organic matter. For many resource-poor farmers, organic substrates are still the main or may be the only nutrient source. In India, crop residues are often removed from the field and used as fodder and fuel. In Southeast Asia, excess straw is often burned because recycling requires additional labor, space, and time. Long straw hampers land preparation, especially puddling and harrowing. Incorporation of fresh rice straw also causes net immobilization of nitrogen (Nagarajah et al. 1989). Burning of straw, especially the preferred burning at low temperatures, reduces nutrient losses but produces large amounts of aerosols and CH_4. In general, the use of organic amendments decline as chemical fertilizers become available and responsive rice cultivars are grown. Environmentally sound technologies must consider both maintaining and even increasing soil fertility and mitigating CH_4 emissions. Aerobic composting or prefermentation of crop residues is highly dependent on socioeconomics.

Rice Cultivars

Up to 90% of CH_4 released from rice fields to the atmosphere is emitted through rice plants (Seiler 1984; Holzapfel-Pschorn et al. 1986). Well-developed intracellular air spaces (aerenchyma) in leaf blades, leaf sheath, culm, and roots provide an efficient gas exchange between the atmosphere and the anaerobic soil. In older roots, the aerenchyma comprises about 50% of its cross section. The main gas exchange between root aerenchyma and soil seems to occur through fissures in the exodermis caused by outgrowing side roots. Within the rice plant, cell layers separating root and shoot aerenchyma determine the diffusion coefficient (Butterbach-Bahl 1993). Cutting shoots above the flood water, or part of roots, does not affect CH_4 diffusion through rice plants within 1–2 days. The CH_4 concentration in the growth medium and the number of tillers per plant are positively correlated with CH_4 emission rates (Mario et al. 1991). Older tillers within a single plant hill show higher CH_4 fluxes (Kimura 1992). Positive correlations are found between root biomass and CH_4 production and between aboveground biomass (Sass et al. 1990) or grain yield (Neue and Roger 1993) and CH_4 emissions. In general, yield of rice grain is a function of the production of plant biomass. The harvest index (ratio of grain yield to total dry weight) is about 0.3 for traditional tall rice cultivars and 0.5 for improved, short cultivars (Yoshida 1981). Rice roots become the major carbon source for CH_4 production at later growth stages because of root exudation and senescence of roots. Cultivar differences in root exudation have been reported (Ladha et al. 1986).

Oxygen diffusion from rice roots, and abundant CH_4-oxidizing bacteria present in the rhizosphere, provide a high potential for CH_4 oxidation. Up to 90% of the CH_4 produced is apparently oxidized in the rhizosphere and the oxidized soil-floodwater interface (Holzapfel-Pschorn et al. 1985; Schütz et al. 1989b; Frenzel et al. 1992). Large cultivar differences in root oxidation power (Neue and Roger 1993) and in emission rates (Parashar et al. 1990) have been reported. The pronounced diversity of morphological and metabolical traits of rice cultivars provides the possibility to develop high-yielding cultivars with lower potential for CH_4 emissions.

Cultural Practices and Crop Diversification

Mechanical disturbances of flooded soils caused by cultural practices (e.g., land preparation, transplanting, weeding, fertilization, and harvest) increase ebullition of soil-entrapped CH_4 (Neue et al. 1994). Less cultural disturbances of reduced soils and shorter flooding periods in direct-seeded rice, especially dry-seeded rice, should lower CH_4 emissions. The adoption of direct seeding (wet and dry seeding) instead of transplanting is highly dependent on the ability to manage water regimes. Often ricefields cannot be drained in the wet season, and in rainfed areas, farmers are reluctant to drain flooded fields because of uncertain rainfalls.

Crop diversification is a feasible option to reduce total CH_4 emissions in line with economic benefits (Neue et al. 1991). In rice-growing areas with year-round irrigation, production can be shifted by adopting a rice-upland cropping system (i.e., sequential cropping of one upland crop before or after one or two crops of rice). In China, the area cropped to a rice-wheat system has increased by about 40% since 1950 and occupies about 10% of the arable land (Huke et al. 1993). The average yield is 5.5 t/ha rough rice and 4 t/ha wheat grain, compared with 9 t/ha for a double rice crop. Other common upland crops that are widely grown in rotation with rice are mungbean, soybean, corn, and vegetables. Crop intensification and diversification has become possible because of investments in irrigation systems and the breeding of early maturing photoperiod-insensitive rice cultivars that mature up to 60 days earlier than traditional cultivars. Direct seeded rice can even be harvested 15–30 days earlier than when the same variety is transplanted. However, replacing wetland rice with upland crops is hardly possible in the wet season. Wetland rice is mostly grown in the wet season because fields are flooded. Harvesting water in rice fields reduces soil erosion and regulates water balance in watersheds.

Conclusion

Reducing the uncertainties of regional and global CH_4 emission rates and predicting CH_4 emission trends from rice agriculture requires mechanistic modeling of CH_4 fluxes in combination with a geographic information system of controlling factors. The major source is evidently irrigated rice, and CH_4 emission will increase as rice production increases with current technologies. Promising mitigation candidates are water management, organic amendments, fertilization, cultural practices, and rice cultivars. The current understanding of the processes controlling CH_4 fluxes and rice growth is sufficient to develop mitigation technologies. Information is lacking on application domains and the socioeconomic feasibilities of the various technical opportunities to reliably predict mitigation potentials and minimize possible trade-offs. Most rice farmers are poor and most rice-growing countries need more rice to feed growing populations. Governments will promote, and rice farmers will adapt, mitigation technologies only if rice production and productivity increases.

Acknowledgment. Although the research described in this chapter has been funded in part by the US Environmental Protection Agency under cooperative agreement number 817425 to the International Rice Research Institute, it has not been subject to the Agency's review and therefore does not necessarily reflect the views of the Agency, and no official endorsement should be inferred.

Ultraviolet-B Radiation

Measurement and Modeling of Ultraviolet-B Irradiance

F.L. Booker[1] and E.L. Fiscus[2]

Decreases in stratospheric ozone, which have been linked to emissions of manufactured halocarbons, may increase levels of solar ultraviolet-B (280–320 mm) (UV-B) radiation at ground level. The possible harmful effects of increased solar UV-B radiation has heightened interest in measuring and modeling this radiation. The relationship, however, between total column ozone and ground level UV-B irradiance is complex. Air mass, vertical ozone distribution, clouds, haze, and air pollutants all affect the transmission of solar UV-B radiation through the atmosphere (Madronich 1993). Surface albedo also influences the diffuse radiation flux. Furthermore, latitudinal and seasonal variations in surface UV-B irradiance because of stratospheric ozone circulation patterns and solar elevation far exceed the increases in surface UV-B irradiance that might accompany stratospheric ozone depletion in middle latitudes (Frederick 1993). The measurement and modeling of solar UV-B irradiance thus needs to be comprehensive to detect trends and to predict ground level UV-B irradiance under changing atmospheric conditions. In other situations, however, relatively low-cost broadband meters and simplified computer models can accommodate many research needs in agriculture, ecology, and climatology.

Instrumentation to Measure UV-B

Accurate measurements of solar UV-B are technically difficult because solar irradiance decreases five orders of magnitude from 320 to 290 nm, and the region of most interest is the shorter, low intensity wavelengths. Instrumentation must have good wavelength accuracy, sensitivity, high suppression of stray light, and stability to provide reliable data. A variety of instruments are available that meet these requirements.

[1] Department of Botany, Box 7632 North Carolina State University, Raleigh, NC 27695, USA
[2] United States Department of Agriculture, Agricultural Research Service, Air Quality, Plant Growth and Development Research Unit, and Department of Crop Science, North Carolina State University, Raleigh, NC 27695, USA

S. Peng et al. (Eds.) Climate Change and Rice

Ground-based measurements of UV-B irradiance can be obtained with: high-resolution scanning spectroradiometers that provide full spectral information, optical filter radiometer instruments that measure irradiance at selected narrow wavebands, and broadband meters that provide a weighted measure of UV-B irradiance over the UV waveband. Scanning spectroradiometers typically use mechanically driven diffraction gratings to disperse incident light, which is then detected by a photomultiplier tube. Suppression of stray light is achieved with a double monochromator or a single monochromator and optical filters (Kerr and McElroy 1993). Filter radiometers rely on optical interference filters to block unwanted radiation; whereas, the passband is detected by either a photomultiplier tube (Blumthaler 1993) or a vacuum photodiode (Roy et al. 1989). A filter radiometer can provide spectral information by using a number of narrowband filters that measure specific wavelengths in rapid succession.

Alternatives to the more expensive instruments are illustrated by a broadband instrument such as the Robertson-Berger (R-B) meter (DeLuisi et al. 1992). This instrument uses a magnesium tungstate phosphor in combination with optical filters. Phosphorescence induced by UV radiation is detected by an inexpensive photodiode. The combination of filters and phosphor gives the instrument a spectral response function indicative of human skin erythema.

The best approach for measuring UV-B irradiance depends on the goals of the planned research. General survey instruments suitable for short-term monitoring in agricultural or ecological research will likely be unsuitable for detecting trends in ground level solar UV-B radiation or in detailed modeling applications. In addition, plant researchers using UV-B radiation enhancement techniques are faced with special requirements because the spectral distribution of the sun differs from the UV-B lamps typically used to simulate decreases in stratospheric ozone. If a special weighting function, such as the generalized plant action spectrum (Caldwell 1971), is to be used to interpret the data, an instrument is required that will deliver a well-defined output regardless of the distinct source or proportions of the total irradiance emanating from each source.

To measure trends in UV-B irradiance at the surface of the earth that are attributable to changes in column ozone, high resolution, scanning spectroradiometers provide the most informative data. Spectral irradiance measurements can detect the characteristic wavelength-dependent pattern inherent in such changes (Frederick 1993). For example, measurements obtained with a Brewer spectroradiometer showed an increase in UV-B intensity near 300 nm but not at 324 nm from 1989 to 1993 in Toronto, Canada, that was consistent with measurements of decreased stratospheric ozone over the same period (Kerr and McElroy 1993). Comprehensive measurements of spectral irradiance are being made in the United States, Scandinavia, Europe, New Zealand, Australia, and polar regions using a variety of instruments, although no global, long-term UV monitoring network yet exists (Correll et al. 1992; Wester 1992; Madronich 1993). In general, the data show consistency with measured ozone amounts (Frederick 1993). The accumulating data from these spectral measurements may be valuable for identifying trends and for determining the physical mechanisms (e.g., ozone, aerosols, and cloudiness) responsible for such changes (Madronich 1993).

Optical filter radiometers capable of obtaining high-resolution measurements at several fixed, diagnostic wavelengths may be a less complex and cost-effective method to monitor trends in ground level UV-B than scanning spectroradiometers. Filter radiometers are available from Yankee Environmental Systems Inc (Turners Falls, MA) and International Light Inc (Newburyport, MA). In Sweden, custom-built filter radiometers are used to measure solar irradiance at about 306 and 360 nm to obtain estimates of erythemal UV-B, UV-A, and total column ozone (Wester 1992). Narrow-band irradiance measurements at 307 nm are useful because calculations of solar UV influx by latitude and season have shown that erythemally weighted radiation varies in nearly the same way as radiation at 307 nm (Johnson et al. 1976). A 4-nm half-bandwidth interference filter centered at 303 nm or a 4-nm half-bandwidth low pass filter centered at 305 nm will theoretically indicate UV-B irradiance weighted by the Caldwell (1971) generalized plant-action spectrum normalized to 300 nm. Simulated filters convolved with 168 actual spectroradiometer scans from several locations and times confirm an acceptable linearity between transmitted radiation and plant-action weighted irradiance whether the source is the sun, UVB-313 lamps (Q-Panel Co, Cleveland, OH), or a combination of the two.

Broadband meters have been widely used to monitor trends in solar UV-B radiation, although their efficacy in this role is controversial. A network of R-B meters was established in the United States in the early 1970s, with later additions in Australia and Europe, to measure solar erythemal radiation on a continuous basis. Evaluations of long-term measurements obtained with R-B meters, however, have shown both a decline in solar UV-B irradiances at eight R-B network stations from 1974 to 1985 (Scotto et al. 1984) and an increase in clear-sky solar UV-B irradiance at a high-altitude site in the Swiss Alps since 1981 (Blumthaler 1993). Both reports have been criticized because they reported trends in solar UV-B radiation that were in the opposite direction or larger than that expected from the changes in column ozone measured over the same period (Frederick 1993). Various factors (e.g., instrument drift, improper calibrations, changes in air pollutant levels, and large variability in the data) have been offered and in some cases refuted as explanations (DeLuisi et al. 1992; Blumthaler 1993; Frederick 1993; Madronich 1993). For example, recent evaluations of several R-B meters used in the network found that the instruments and spectral responses were basically stable, and that the calibration procedures and temperature drift did not account for the downward trend in UV-B radiation levels observed by Scotto et al. (1988) (DeLuisi et al. 1992; Kennedy and Sharp 1992). A comparison between clear-sky R-B data and computed R-B measurements based on Dobson ozone measurements at two sites in the United States (Bismark, ND and Tallahassee, FL) showed good agreement during the summer months (Frederick and Weatherhead 1992). However, the analysis revealed a downward trend in R-B data during the winter that was not supported by Dobson ozone data and could not be explained.

Although broadband R-B meters may not be adequate for unambiguously detecting the effects of stratospheric ozone depletion, the meter is useful for many applications in agriculture, forestry, ecology, and climatology to provide ground-based measurements of UV-B radiation. Its use in this role is particularly important in UV-B enhancement studies conducted in the field because accurate estimates of UV-B exposure and the corresponding simulated ozone depletion can be determined. In our studies, an R-B meter was used to monitor solar UV-B radiation continuously and a portable erythemal meter with a detachable sensor (Model 2D, Solar Light Co Inc, Philadelphia, PA) was used daily to measure and adjust the supplemental UV-B irradiance provided by lamp banks (Booker et al. 1992a; Miller et al. 1994). Modified versions of the R-B meter with improved spectral response, temperature stability, and data acquisition systems are available from Solar Light Co Inc (Philadelphia, PA), Yankee Environmental Systems Inc (Turners Falls, MA), VITAL Technologies (Ontario, Canada) and others. A broadband meter called the YMT sensor, similar in design to the R-B meter, has a spectral-response function that approximates the generalized plant action spectrum, which is desirable for monitoring solar UV-B radiation and controlling UV-B supplementation systems in plant research projects (Yu et al. 1991).

Ultraviolet-B Models

Modeling solar UV irradiance can be helpful for understanding the transmission of solar UV-B radiation through the atmosphere and for deducing ground level UV-B irradiance from measurements of column ozone and other pertinent atmospheric and surface factors. In addition, UV-B supplementation experiments require a model that relates treatment exposures to the simulated stratospheric ozone depletion. A number of radiative-transfer models have been developed that vary in approach and complexity (Björn 1989; Frederick and Snell 1990; Madronich 1993). A model developed and revised by Green (1983) has been widely used (Rundel 1986; Correll et al. 1992; Blumthaler 1993). Björn and Murphy (1985) and Björn (1989) encoded the revised algorithms, modified the reflection coefficient routines, added several weighting functions, and included an additional feature that allowed internal generation of column ozone values for the northern hemisphere. More recently, we expanded the Björn and Murphy (1985) and Björn (1989) codes to increase the flexibility and usefulness of the program (Fiscus and Booker 1993). Routines to allow yearly and daily irradiance calculations were added. The model also allows the user to input special weighting functions and annual files of relevant environmental data (e.g., humidity, barometric pressure, aerosols, and most importantly, column ozone data obtained from the TOMS instrument flown on the Nimbus-7 satellite (McPeters et al. 1993) or from any other source).

Comparisons between measured and computed irradiances showed that the revised Green-Björn and Murphy model predicted solar spectral irradiance in the UV-B region with accuracy (Björn and Murphy 1985; Björn 1989). Good agreement between the Green-Björn and Murphy model and other UV-B models has also been observed (Björn 1989). In addition, output from the Green-Björn and Murphy model weighted by the spectral response function of the R-B meter was highly correlated with data from the R-B network (Björn 1989).

We have found that output from the Green-Björn and Murphy model weighted by the spectral response function of the Model 2D meter sensor (Solar Light Co), which was designed to simulate an erythemal action spectrum (Parrish et al. 1982), fit the clear-sky boundary of our R-B data very well. This is understandable because spectra weighted by the Model 2D sensor and R-B meter spectral response functions are similar in shape. The daily counts from the R-B meter were converted to plant-action spectrum-weighted irradiances normalized to 300 nm ($UV\text{-}B_{BE}$) (Caldwell 1971). The conversion was done as previously described (Fiscus et al. 1994) by using the ratio of the Model 2D sensor-weighted irradiance to the $UV\text{-}B_{BE}$ irradiance calculated from the model. This ratio was computed for each day of the year using TOMS ozone column data for that day. The validity of this approach was confirmed by comparisons between the derived $UV\text{-}B_{BE}$ irradiances and calibrations of our R-B meter done with a scanning spectroradiometer on 5 days in 1990-1991 (Booker et al. 1992a). The comparison showed that the mean (± SD) calculated calibration (2.16 ± 0.10 R-B counts = 1 J/m^2 $UV\text{-}B_{BE}$) agreed with the mean spectroradiometer calibration (2.10 ± 0.14 R-B counts = 1 J/m^2 $UV\text{-}B_{BE}$) within 3%.

We found that $UV\text{-}B_{BE}$ irradiance computed by the model when TOMS column ozone data were input compared well with measurements of solar $UV\text{-}B_{BE}$ radiation obtained with our R-B meter over several years (Fig.1). The model accurately tracked the outer (clear sky) envelope of the data. The daily R-B data, however, are highly variable, which has been attributed mainly to cloudiness (Frederick and Snell 1990). To further explore the ability of the model to compute accurately daily irradiance at ground level, we examined several ways to compensate for the source of variability in the data set.

Ultraviolet-B Model Validation

The effect of fractional cloud cover on solar UV-B influx has been approximated by:

$$F_s/F_0 = (1 - 0.056\ C)\ , \qquad (1)$$

where F_s is the radiation flux at the surface, F_0 is the radiation flux for clear skies, and C is the average cloud cover in tenths of the sky (Johnson et al. 1976). Similar approaches have been used in other model calculations (Björn 1989; Madronich 1993).

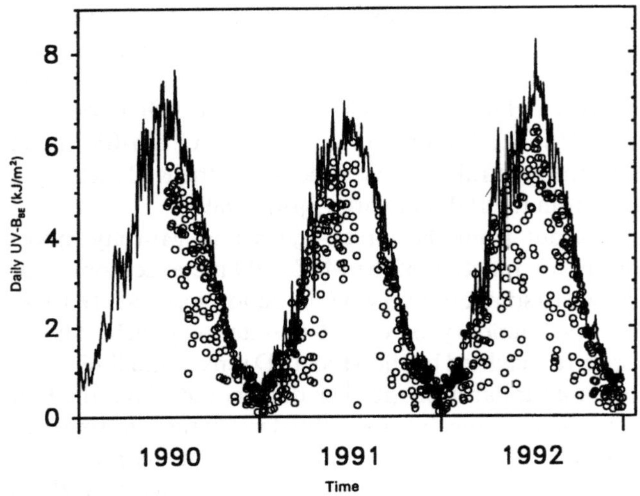

Fig. 1. Daily measured and modeled plant-action spectrum weighted UV-B (UV-B_{BE}) irradiance from 1990 to 1992 at Raleigh, NC (35.75°N, 78.67°W). Measurements were obtained with an R-B meter (O) and converted to UV-B_{BE} irradiance as described in the text. User input data for the revised Green-Björn and Murphy model (—) were: ozone column – TOMS data; environment – rural; ground cover – green farmland; barometric pressure 1020 mb; relative humidity 0.50; aerosols 0; action spectrum – generalized plant-action spectrum normalized to 300 nm

We used estimates of average cloud cover (in tenths) made daily by the National Weather Service at a local airport to calculate F_0 for each day in 1992 using the daily UV-B_{BE} irradiance measured by our R-B meter as F_s. The UV-B_{BE} data adjusted for cloud cover were then compared with model calculations (Fig. 2). The average fit between the UV-B_{BE} data adjusted for clouds and the model was improved, although the data remained highly variable (Table 1). Adjusting the irradiance on the basis of cloud cover increased the values of measured UV-B_{BE} irradiance data in 1992 from an annual mean ratio of 0.71 ± 0.24 (± SD) of clear-sky calculations to 1.10 ± 0.30 of the clear-sky maximum. Using various combinations of cloud-cover estimates made at 3-h intervals each day at the airport did not noticeably improve the fit. Estimates of average cloud cover provided only a rough approximation of the attenuation of solar UV-B irradiance by clouds. Additional comparisons between average cloud cover and minutes of sunshine recorded daily at the airport did not suggest that estimates of solar UV-B radiation at ground level would be substantially improved by incorporating sunshine data into attenuation computations.

Measurements of visible radiation from 400 nm to 700 nm (photosynthetically active radiation, PAR) might provide an independent index of factors other than ozone that attenuate solar UV-B irradiance (e.g., clouds and aerosols). Minimal wavelength dependence in the transmitted radiation would be expected to arise from passage through clouds (Madronich 1993), and one study showed

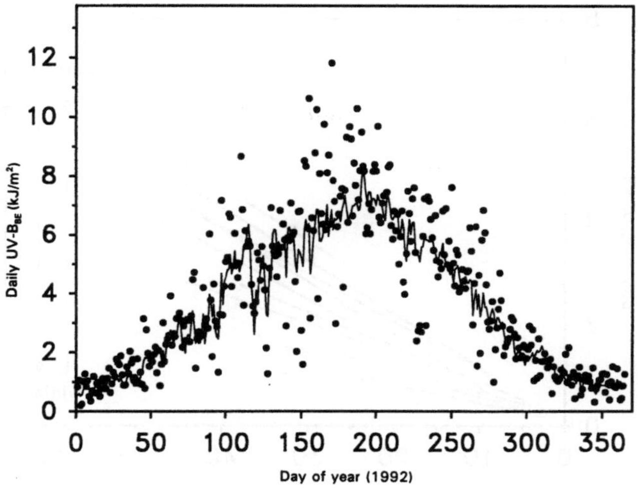

Fig. 2. Daily measured UV-B_{BE} irradiance adjusted using average daily cloud cover data (SC) and modeled UV-B_{BE} at Raleigh, NC. User input data for model calculations were the same as those described in Fig. 1 (— Green-Björn and Murphy; • SC adjusted R-B data). Note the change in scale on the y axis compared with Fig. 1

Table 1. Average ratios (± SD) between measured and modeled UV-B_{BE} irradiances for 1992 at Raleigh, NC (35.75°N, 78.67°W). Measurements were obtained with an R-B meter and converted to UV-B_{BE} irradiance as described in the text. User input data for the revised Green-Björn and Murphy model were: ozone column – TOMS data, environment – rural, ground cover – green farmland, barometric pressure 1020 mb, relative humidity 0.50, aerosols 0, and action spectrum – generalized plant-action spectrum normalized to 300 nm

Comparison	Ratio±SD
Measured UV-B_{BE} / modeled UV-B_{BE}	0.71 ± 0.24
UV-B_{BE} adjusted for cloud cover / modeled UV-B_{BE}	1.10 ± 0.30
UV-B_{BE} adjusted with PAR / modeled UV-B_{BE}	1.03 ± 0.09

that global irradiance was only slightly more attenuated than erythemal irradiance as cloudiness increased (Blumthaler 1993). The PAR instruments are also inexpensive and widely available; therefore, this seemed to be a reasonable approach.

The first step was to establish an empirical relationship between PAR and UV-B_{BE}. A linear model could be fit to the relationship between daily UV-B_{BE} and daily PAR for each month of the year in 1992, but the slopes (dUV-B_{BE}/dPAR) of the lines changed over the year (Fig. 3). The slopes increased during the spring, remained fairly constant from April through September, and then decreased again. This pattern was attributed to seasonal changes in column ozone and solar elevation that attenuated UV-B_{BE} irradiance relatively more than PAR (Blumthaler 1993). Because the seasonal changes were large, we decided to use

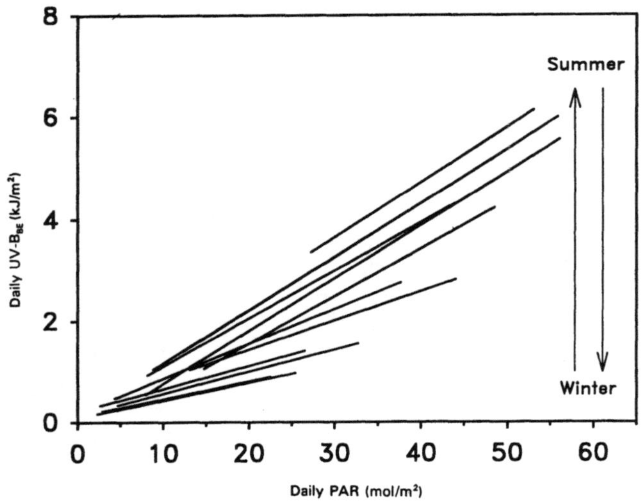

Fig. 3. Linear regression models of the relationship between measured UV-B_{BE} irradiance and PAR for each month of 1992 (Raleigh, NC)

the slopes of the linear models for each month in our calculations. In addition, a derivation of PAR attenuation was obtained to serve as an index factor. Maximum daily PAR (PAR_{MAX}) was approximated by plotting daily PAR, drawing an outer envelope to the data by inspection, and modeling the envelope with a polynomial. The PAR_{MAX} could then be estimated for each day. The use of PAR measurements to estimate attenuation of solar UV-B_{BE} irradiance was expressed as:

$$\text{UV-B}_{BE\,MAX} = \text{UV-B}_{BE} + k\,(PAR_{MAX} - PAR)\,, \tag{2}$$

where UV-$B_{BE\,MAX}$ is the UV-B_{BE} irradiance for clear skies, UV-B_{BE} is the measured UV-B_{BE} irradiance at the surface, k is the slope of the linear model for UV-B_{BE} versus PAR for the appropriate month, and PAR_{MAX} is the calculated maximum daily PAR.

The fit between the UV-B_{BE} irradiance data adjusted by this approach and the model was significantly improved (Fig. 4, Table 1). There was a short interval from day 180 to day 220 where the adjusted data were below the model calculations, probably because of an underestimate of the PAR_{MAX} boundary line in this region. Nonetheless, there was good overall agreement between the data and model calculations. The incorporation of concurrently measured visible radiation data, as an index of solar UV-B_{BE} attenuation, increased the annual mean ratio of the measured-modeled data to 1.03 ± 0.09 of the clear-sky calculations. Compared with UV-B_{BE} data adjusted with average daily cloud cover, UV-B_{BE} data adjusted with PAR measurements approximated model calculations more closely and with much less variability.

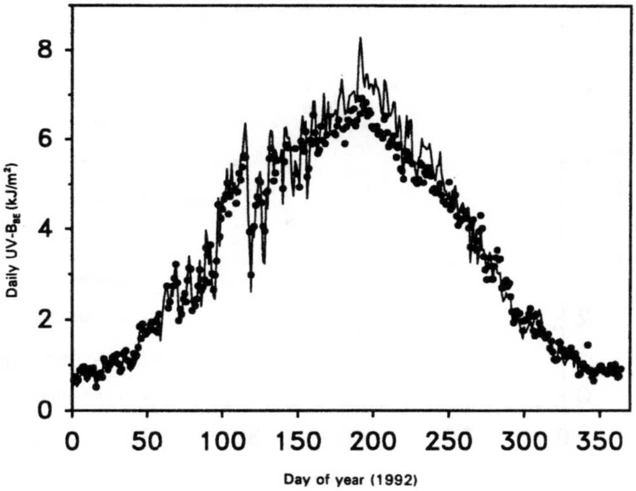

Fig. 4. Daily measured UV-B_{BE} irradiance adjusted using daily PAR data and modeled UV-B_{BE} irradiance for 1992 at Raleigh, NC. User input data for model calcuations were the same as those described for Fig. 1 (— Green-Björn and Murphy; • PAR adjusted R-B data)

The reason why daily estimates of average cloud cover failed to serve as a good indicator of attenuation was apparent from a plot of average cloud cover versus PAR/PAR_{MAX} (Fig. 5). The data are extremely variable, especially for high values of cloud cover. An index of solar UV-B_{BE} irradiance attenuation based on PAR was far more resolved and informative.

This analysis provided additional evidence that the estimates of clear sky solar UV-B_{BE} irradiance computed by the revised Green-Björn and Murphy model are reasonably accurate. It is important to establish the validity of the model because it allows the user to interchange measured irradiance data among weighting functions. For example, irradiance can be monitored with a broadband meter with one spectral response function and then the model can be used to convert the data to another weighting function. Data can be gathered with an erythemal meter and then converted to values weighted by the generalized plant-action spectrum. Estimates of clear sky irradiance computed by the model also allow for more accurate determinations of the fraction of solar UV-B irradiance attenuated by clouds and aerosols. This is important to consider when conducting and evaluating UV-B enhancement studies.

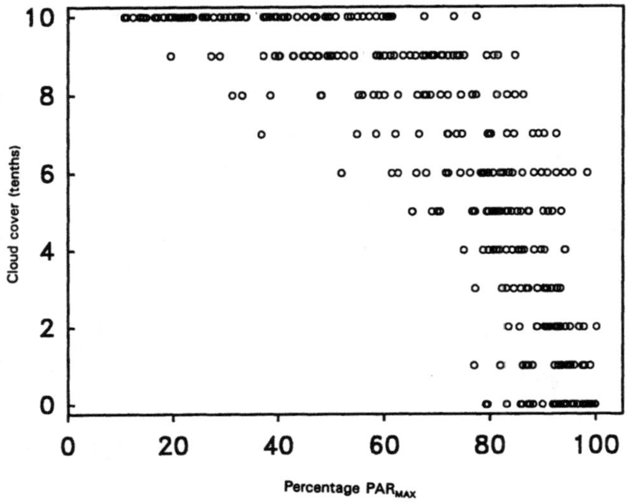

Fig. 5. Daily average cloud cover (in tenths) versus percentage of maximum PAR (Raleigh, NC, 1992)

Conclusion

A variety of instruments are available to measure solar UV-B irradiance, although their suitability depends on research goals. High resolution, scanning spectroradiometers provide the most informative data for detecting trends in solar UV-B irradiance. The United States Department of Agriculture plans to establish a monitoring network in the United States to detect trends in UV-B irradiance. High-resolution spectroradiometers will be used in conjunction with a network of broadband or multiband filter meters (with adequate sensitivity at wavelengths of about 300 nm) to detect trends of 5% per decade in UV-B irradiance at ground level (Gibson 1992). Specifications for these research-grade instruments are so stringent that existing instruments must be modified to meet the requirements. At the other end of the scale, however, relatively low-cost broadband meters and simplified computer models can accommodate many research needs in agriculture and ecology. We have further shown that concurrent measurements of visible radiation can provide valuable ancillary information when solar UV-B radiation is monitored and modeled. Monitoring solar UV-B radiation and developing models continue to be essential activities to provide the data and tools needed to assess the impact of stratospheric ozone depletion and global climate change on biological systems.

Acknowledgments. The TOMS data used in our study were provided by the Ozone Processing Team of NASA, Goddard Space Flight Center, Greenbelt, MD. We would also like to thank Dr Donald Krizek, USDA, Climate Stress Laboratory, Beltsville, MD for providing the R-B meter used in this study. Researchers interested in using the Green-

Björn and Murphy model can obtain a copy of our UVB program from E.L. Fiscus, USDA-ARS, 1509 Varsity Drive, Raleigh, NC 27606, USA.

The use of trade names in this publication implies neither endorsement by the United States Department of Agriculture or the North Carolina Agricultural Research Service, nor criticism of similar ones not mentioned.

Effects of Ultraviolet-B Irradiation on Spore Viability, Sporulation, and Mutation of the Rice-Blast Fungus

H. LEUNG, D. CHRISTIAN, P. LOOMIS and N. BANDIAN

How increases in UV-B radiation will affect terrestrial plant life and agricultural productivity has been the focus of many investigations (for reviews see Caldwell et al. 1989; Tevini and Teramura 1989). Rice production occurs over a broad geographic range where the level of UV-B radiation varies widely (Bachelet et al. 1991). In the tropics, daily UV-B radiance levels have been estimated to be as high as 11 kJ/m^2 (Bachelet et al. 1991). Because of high level of UV-B radiation in the tropics, an enhancement of UV-B resulting from stratospheric ozone depletion may be particularly damaging to tropical ecosystems. In anticipation of enhanced UV-B radiation in tropical rice-growing regions, Barnes et al. (unpubl. data) evaluated the response of 22 cultivars of rice exposed to enhanced UV-B irradiation under greenhouse conditions that simulated a 5% reduction in stratospheric ozone. They found that the enhanced UV-B treatment significantly reduced total plant dry matter, shoot height, leaf blade length, and total leaf area. Variation in sensititvity to UV-B was found among the cultivars tested and some high-yielding cultivars grown in tropical lowlands were found to be particularly sensitive to UV-B irradiation. Little is known, however, about the effects of UV-B on other biotic factors that directly affect rice production.

Several studies have been made on the impact of UV-B irradiation on plant diseases and the results suggest that the effects of UV-B depend on the pathogen and crop species under study. Carns et al. (1978) studied several diseases caused by fungi and found that UV-B irradiation either reduced viability of fungal spores and disease severity, or had no effect. In contrast, Biggs and Webb (1986) found that infection of field-grown wheat by the leaf-rust fungus (*Puccinia recondita*) increased significantly in a susceptible cultivar (Red Hart) but not in the resistant cultivar (Florida 301) under enhanced UV-B irradiation. Similarly, Orth et al. (1990) found that the severity of cucumber scab (caused by *Cladosporium cucumerinum*), was increased on a susceptible cultivar but not on a resistant cultivar. Finckh et al (M. Finckh, pers. comm.) investigated the susceptibility of rice cultivars to blast by exposing the plants to enhanced UV-B prior to inoculation. They found that the effects of preinoculation UV-B exposure were generally small; however, significant increases in lesion number per plant were observed in two cultivar-isolate combinations. No experiments have yet been

Department of Plant Pathology, Washington State University, Pullman, WA 99164-6430, USA

S. Peng et al. (Eds.) Climate Change and Rice

done to assess the direct effect of UV-B irradiation on the blast fungus and on host-pathogen interactions.

Rice-blast disease, caused by the fungus *Magnaporthe grisea* (anamorph *Pyricularia grisea*), is one of most damaging diseases of rice, and it is as widely distributed as the rice crop itself (Ou 1985). Control of rice blast throughout the tropics has relied mostly on the use of resistant cultivars. However, the success of this strategy is closely tied to the adaptive ability of the pathogen to host resistance. Climatic factors that affect the parasitic fitness of the pathogen could, therefore, have a direct impact on disease control. The blast fungus is particularly well suited as an organism for investigation of UV-B effects because sophisticated genetic and molecular tools are available for the manipulation of this fungus (Valent and Chumley 1991). An examination of the effects of UV-B radiation on a global disease such as blast, might provide insights into the impact of enhanced UV-B radiation on plant diseases in general.

To delineate the effects of UV-B irradiation on rice blast, we developed an experimental system to examine the biological responses of the fungus to UV-B. The blast fungus is sensitive to UV-B as measured by spore viability and sporulation. An increase in mutation rate at a pigment gene was observed at a nonlethal irradiance level, which suggests that enhanced UV-B may also contribute to the genetic variability of the pathogen population.

The Experimental System

Because the biological response of a target organism is highly dependent on the quality control of UV-B irradiation, we will briefly describe the experimental system developed to assess UV-B effects on the rice-blast fungus. Our objective was to set up an exposure regime that gives an irradiance range similar to that in the tropics. Based on the Green model (Green et al. 1980), Bachelet et al. (1991) estimated that the biologically effective levels of daily UV-B radiation on a clear day in Jakarta, Indonesia (February) and in Manila, Philippines (August) are 11 kJ/m^2 and 9 kJ/m^2, respectively (assuming a 7-h day). Under cloudy skies, daily UV-B radiance may decline to 4–7 kJ/m^2. The level of radiance may vary between dry and wet seasons.

The lamp frame used to irradiate the fungal cultures consisted of four UV-B lamps (UVB-313, Q-pannel, Cleveland, OH), and the distance between the lamps and the target plane could be readily adjusted. The UV-B irradiance was measured using a spectroradiometer (Model Optronic 742, Optronic Laboratories, FL). Calibration and operation of the spectroradiometer were described by Barnes et al. (1990). Irradiance was weighted using the generalized plant-action specturm (Caldwell 1971) and normalized at 300 nm to yield a measure of biologically effective UV-B irradiation. In addition, a hand-held UV-B/UV-A intensity meter (Model 3D, Solar Light Inc, PA) was used routinely to monitor the UV-B irradiance.

The transmission of UV-B through cellulose acetate decreased rapidly in the first 2 h and then stabilized at approximately 400 mW/m^2 after 24 h (Fig. 1A,B). The total irradiance can be calculated by integrating the area under the irradiance curve. Taking into account the decline in UV-B transmission over the first 4 h, the average hourly irradiance of UV-B (30 cm below the lamps, over the first 8-h period) was estimated to be 2.8 kJ/m^2. Therefore, a culture exposed for 4 h would receive approximately 11.2 kJ/m^2 of irradiance, which is equivalent to peak levels assumed in tropical Asia.

Representative irradiance measurements were made over a 16-month period (Fig. 2). The initial UV-B irradiance measurements, made at 3-month intervals, were similar (803 mW/m^2 in September 1992 and 846 mW/m^2 in December 1992). However, measurements in March 1993 showed a significant increase in irradiance (1151-1279 mW/m^2). A similar variation in UV-B irradiance was observed in the lamp system at IRRI, Philippines. The variation was subsequently deter-

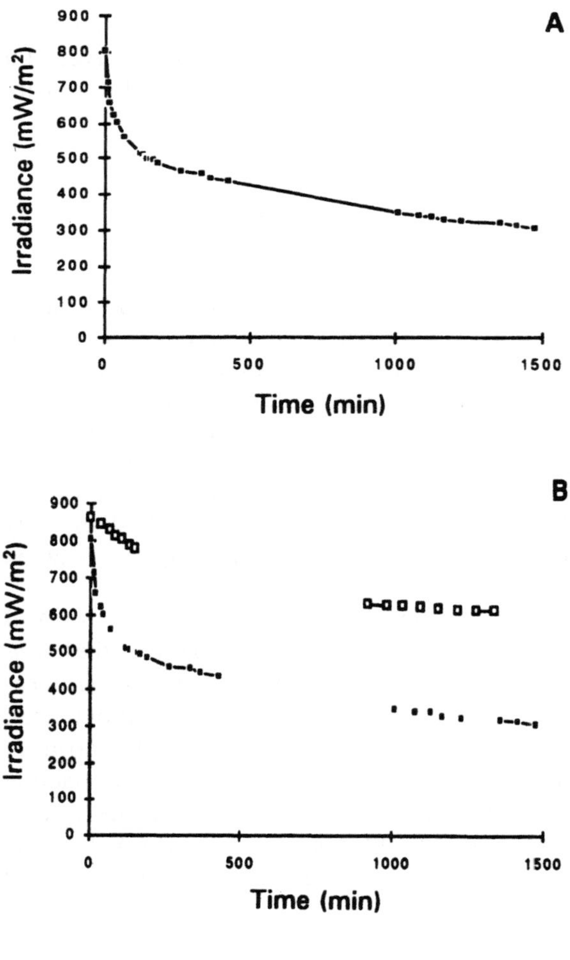

Fig. 1.A Aging of cellulose acetate wrapped around the lamps. The decline in transmission of UV-B is most rapid during the first 8 h. Irradiance stabilizes at approximately 400 mW/m^2 within the first 24 h. **B** Comparison of aging of cellulose acetate. Aging of cellulose acetate is much faster when it is wrapped around the UV-B lamp (■) than when it is 30 cm below the lamp (□). All exposure experiments are done by covering petri plates with cellulose acetate or Mylar D 30 cm below the lamps. The *Upper curve* is used to calculate UV-B irradiance

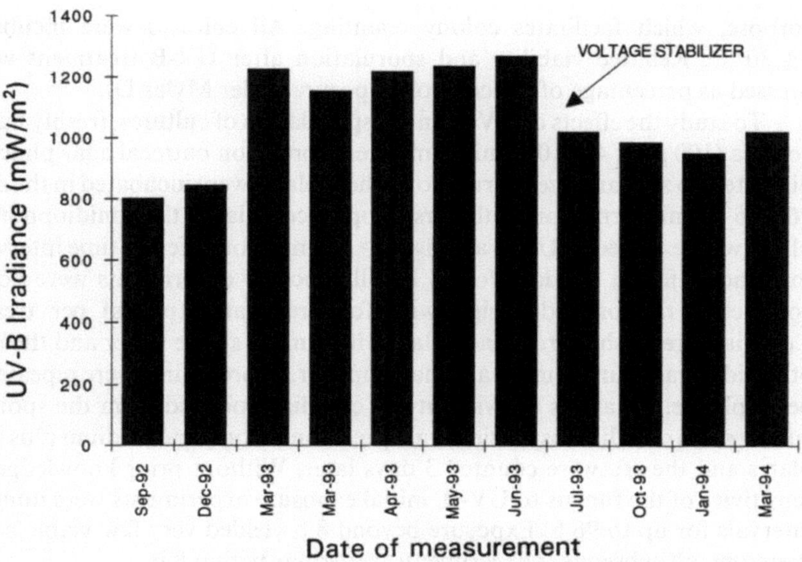

Fig. 2. Measurements of UV-B irradiance between September 1992 and March 1994. The rise in irradiance observed in March 1993 was apparently due to variation in the power source. The use of voltage stabilizers (implementation indicated by *arrow*) reduced fluctuations in UV-B irradiance

mined to be caused by voltage fluctuations in the power source. In June 1993, voltage stabilizers (citadel power line conditioner, Best Power Technology Inc, WI) were installed to provide stable voltage to the UV-B lamps. Since then, UV-B irradiance has been maintained within a range of 900-1000 mW/m^2.

Effects of Ultraviolet-B on Viability and Sporulation

Two *M.grisea* isolates (Guy 11 and Po6-6) were used to establish the range of UV-B irradiation levels appropriate for measuring biological effects on *M.grisea*. Guy 11 (mating type MAT-2) is a sexually fertile, rice-infecting strain (Leung et al. 1988). Strain Po6-6 is a rice-infecting strain virulent to most semidwarf cultivars grown in the Philippines and is commonly used in greenhouse tests at IRRI. All culture conditions and manipulations of the fungus were described by Leung and Taga (1988).

Two parameters (spore viability and sporulation) were used to determine the sensitivity of the fungal isolates to UV-B. Spore viability was defined as the number of colony-forming units (cfu) after 100 μl of freshly harvested conidial suspension (about 2×10^4 conidia/ml) was plated on complete medium plus 2.5% sorbose. The fungus grows as tight colonies on medium supplemented with sorbose, which facilitates colony counting. All cultures were incubated at 28–30 °C. Relative viability and sporulation after UV-B treatment were expressed as percentage of the control (exposure under Mylar D).

To study the effects of UV-B on the sporulation of cultures, freshly harvested conidia (100 μl of 4×10^4 conidia/ml) were spread on oatmeal agar plates (5-cm diameter) to synchronize sporulation. These plates were incubated in the dark for 96 h to permit formation of the first crop of conidia on the conidiophores. The plates were exposed to UV-B and light (> 320 nm) for different time intervals and then incubated in the dark for 24 h. All exposure experiments were done in a completely randomized design with four replicates (plates) per treatment. Conidia were washed from each plate with 1 ml of sterile water and the number of conidia was counted using a hemacytometer. Spore counts were repeated twice per replicate. To assess the viability of conidia produced from the sporulating cultures, the conidial suspension was spread onto complete medium plus sorbose plates and the cfu were counted 3 days later. Without prior knowledge of the sensitivity of the fungus to UV-B, initial exposure experiments were done at 4-h intervals for up to 96 h. Exposure beyond 8 h yielded very few viable colonies; therefore, all subsequent experiments were done within 8 h.

The UV-B treatments were done by covering the plates with cellulose acetate (transmits UV-B) and the corresponding control plates were covered with Mylar D (blocks wavelengths <320 nm). This procedure prevented desiccation and contamination of the exposed plates as well. The decay of cellulose acetate placed 30 cm below the lamp was much slower than when it was wrapped around the lamps (Fig. 1B). Because of the decline in UV-B transmission by cellulose acetate under irradiation, fresh cellulose acetate and Mylar sheets were used for each exposure experiment (Adamse and Britz 1992).

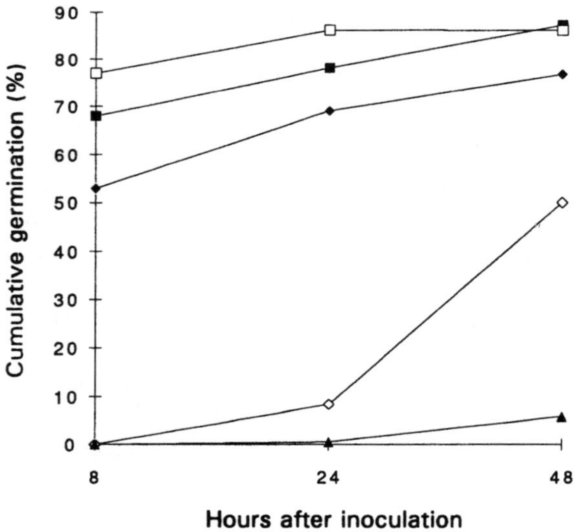

Fig. 3. Effects of UV-B irradiation on cumulative germination as scored on three water-agar plates inoculated with conidia of Guy 11. Spore germination is inhibited by 2-h UV-B exposure ($p<0.01$) but the final germination rate 24–48 h later is not significantly different from the control ($p>0.1$). Germination is severely suppressed by 4–6 h of UV-B irradiation: ■ dark; □ light 2 h; ◊ UV-B 4 h; ▲ UV-B 6 h

Monitoring of spore germination of Guy 11 at different exposure times showed that germination was affected by UV-B irradiation (Fig. 3). Under no UV-B irradiation, over 70% of the conidia germinated 8 h after plating. A 2-h UV-B exposure (5.6 kJ/m^2) significantly inhibited germination for the first 8 h ($p<0.01$), but the final germination percentage (24–48 h later) was not affected. After 4 h of UV-B exposure, no germination was observed until 24 h after plating. A similar pattern of inhibitory effect of UV-B on spore germination has been reported for *Cladosporium cucumerinum* (Owens and Krizek 1980).

Spore viability was significantly reduced in Guy 11 after 3 h of exposure to UV-B (approximately 8.4 kJ/m^2) (Fig. 4A). At 4 h, survival of Guy 11 dropped below 10%. Similar to the test of spore viability, the negative effect of UV-B is evident in sporulating cultures after 4 h of exposure (Fig. 4B). The reduction in sporulation appeared to stabilize beyond 4 h of exposure. Because a significant increase in sporulation was observed under light treatment at all exposure times (4, 12, 24, and 48 h of exposure, $p<0.05$), the leveling of sporulation might be caused by a balance between the inhibitory effect of UV-B and the stimulatory effect of irradiance by longer wavelengths transmitted by Mylar D. Most spores recovered after UV-B exposure were morphologically defective and their viability was significantly reduced. Strain Po6-6 was considerably more tolerant to UV-B than Guy 11 with 70% survival after 4 h of exposure (Fig. 4C). Approximately 6-7 h of exposure was required to cause 90% lethality for Po6-6. Because greater than 50% lethality was observed after 4 h of UV-B exposure, we can infer that damage caused by UV-B must have occurred before germination.

From these experiments, we conclude that UV-B has a significant effect on spore viability and sporulation of the rice-blast fungus. The effects are measurable within the UV-B irradiance level assumed in the tropics. Although it is not realistic to assume the fungus to be under direct and continuous sunlight in the field, it is significant that biological effects can be measured under a relatively short exposure regime. This experimental system allows us to investigate the more subtle effects of UV-B within a short time frame. The results of Guy 11 and Po6-6 also suggest intraspecific variation in sensitivity to UV-B irradiation. It is necessary to examine a broader collection of isolates to develop a general model of UV-B effects on the fungus.

Intraspecific Variation in Ultraviolet-B Sensitivity

Isolates from two disease-screening sites in the Cavinti and IRRI Blast Nursery at Los Baños, Philippines, were analyzed using DNA fingerprinting to determine the genetic diversity of the pathogen population (Chen 1993). Information from this study allows us to identify isolates representing distinct genetic lineages from the Philippines. These isolates were examined for their sensitivity to UV-B irradiation using the described experimental system.

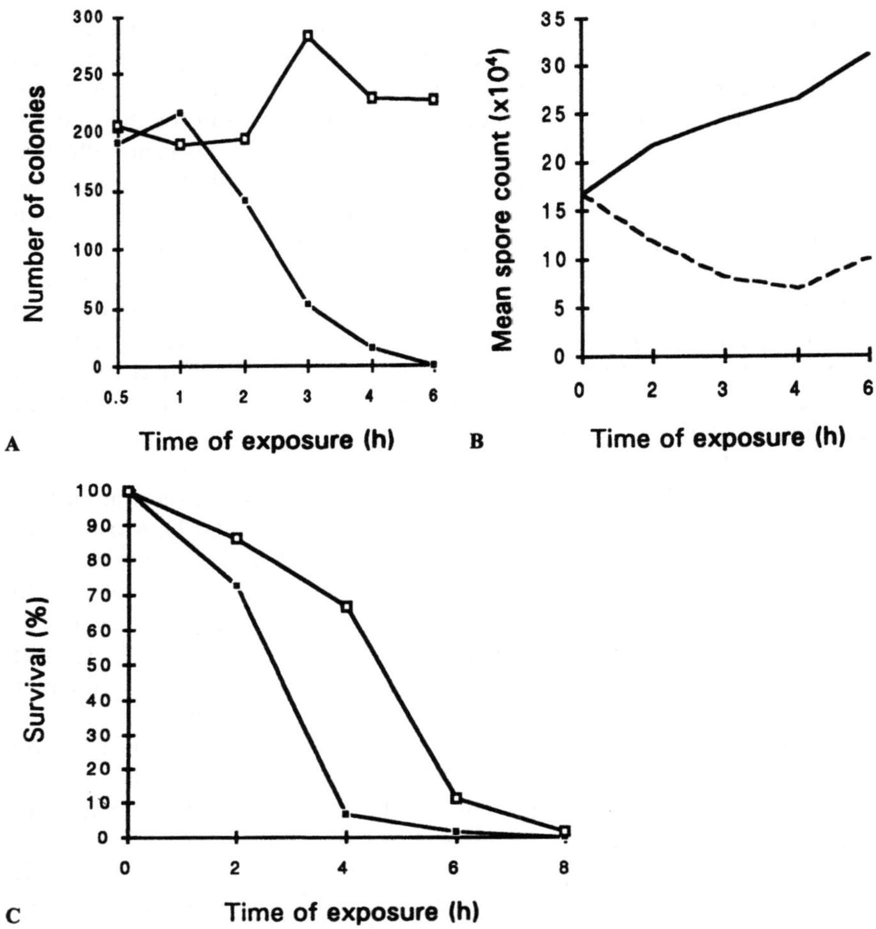

Fig. 4. A. Effect of UV-B irradiation on viability of Guy11 spores (■ UV-B, □ light). **B** Inhibitory effect of UV-B on sporulation of Guy11. Four-day-old plates were exposed for the specified time, incubated for 24 h in the dark, and then the spores were harvested by washing the plates with 1 ml of water. Spores were counted using a hemacytometer (- - - UV-B; - - light). **C** Relative sensitivity of Guy11 and Po6-6 to UV-B [Po6-6 (□) is significantly more tolerant to UV-B than Guy11 (■) at all exposure times]

A total of 22 isolates from 15 clonal lineages were evaluated for spore viability and sporulation following 2, 4, 6, and 8 h of UV-B exposure. All exposure experiments were done using a completely randomized design with four replicates. Analysis of variance showed significant differences in UV-B sensitivity among the Philippine isolates at 2, 4, 6, and 8 h of exposure ($p<0.001$). Figure 5 illustrates the variation in UV-B sensitivity among Philippine isolates after 4 h of UV-B exposure. Overall, sporulation was suppressed by UV-B exposure; however, because of the inherent variability of sporulation capacity of the field

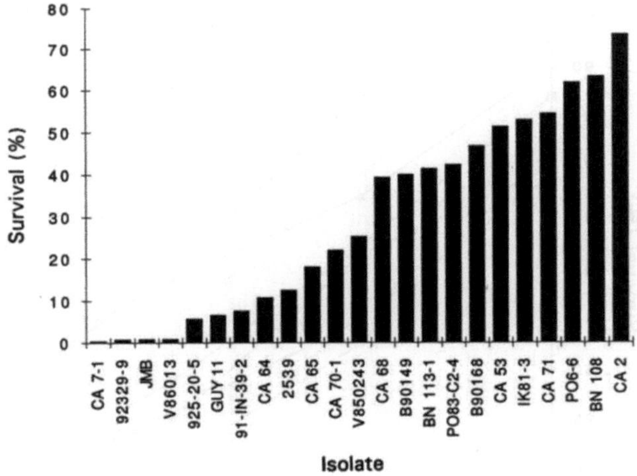

Fig. 5. Spore viability of isolates from the Philippines after 4 h of UV-B exposure (11.2 kJ/m^2). Viability of each isolate is expressed as percentage of the no UV-B control. The standard error is 8.7%

isolates, we were unable to discern differential responses in sporulation to UV-B treatment.

The survey of representative isolates from the Philippines indicates that there is substantial variation in UV-B tolerance among isolates. At 4 h of exposure (11 kJ/m^2), spore lethality can range from 15% to 95% (mean = 40 ± 9%). Strain 2539, a fertile laboratory isolate, showed intermediate tolerance to UV-B at 4 h of exposure (60% lethality). Sexual crosses between 2539 and isolates of different UV-B sensitivity were made and ascospore progeny are now available to investigate the inheritance of UV-B sensitivity in the fungus.

Because much information is available on the effects of UV-C (<280 nm) on microorganisms, it would be informative to test whether sensitivity to UV-B and UV-C is correlated. Six Philippine field isolates that are significantly different in their UV-B sensitivity were exposed to a germicidal lamp (254 nm) for 4,8,12 and 16 min (which corresponded to 0.19, 0.29, and 0.38 kJ/m^2, respectively). Figure 6 shows the kill curves for these six isolates under UV-C irradiation. No significant difference in UV-C sensitivity was detected among these six isolates at any exposure time, and no correlation was apparent between the survival after UV-C and UV-B exposure. These preliminary results suggest that the mechanism(s) of tolerance to UV-B and UV-C in *M.grisea* is (are) qualitatively different.

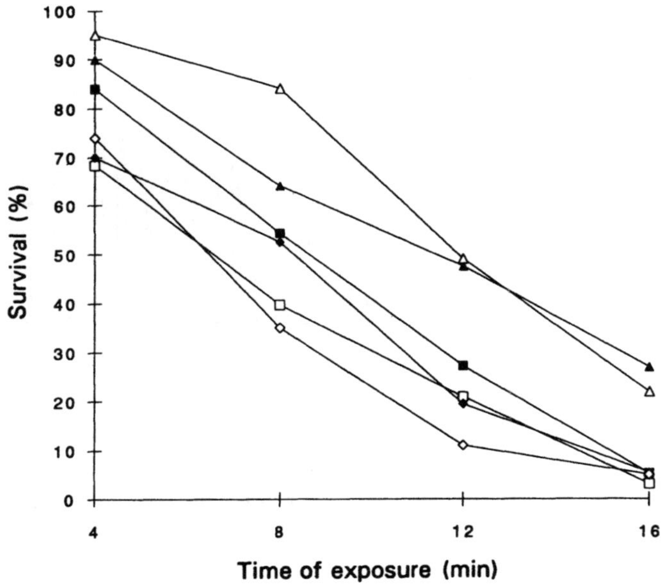

Fig. 6. Kill curves of six Philippine isolates under UV-C (254 nm) irradiation. The percentage survival is an average of two experiments for each isolate. No significant difference in UV-C sensitivity among the six isolates was observed at each exposure time ($p>0.1$): ■ CA 7-1, □ 92329-9, ♦ V86013, ◊ Po6-6, ▲ CA 2-1, and Δ CA 6)

Ultraviolet-B Effects on Genetic Variability

Although adverse effects on spore viability and sporulation can be measured under direct exposure to UV-B, it is unlikely that a significant portion of the fungal population is under continuous exposure in nature. Shading by the canopy and oblique leaf angle will reduce direct exposure of the fungus to intense sunlight. Furthermore, sporulation of the fungus occurs mostly during the night and early morning under periods of high humidity. It is reasonable to assume that only a small proportion of the pathogen population is directly affected by the adverse effects of UV-B, and that the impact on the total population of the pathogen could be relatively minor. However, a long-lasting effect on the pathogen population could result from a brief exposure to UV-B, if such exposure enhances the rate of mutation. Genetic changes caused by UV-B could be maintained and amplified even if only a small subpopulation of the pathogen was affected.

To test whether UV-B plays a significant role in generating variability in the pathogen population, a buff pigment loci was used as a marker to monitor the mutagenic effect of UV-B. The buff marker offers a number of advantages as a mutagen sensor. Buff colonies can be readily distinguished among the grayish-black colonies of the wild-type strain and the buff phenotype is known to be caused by a mutation of a single gene (Chumley and Valent 1990). Previous genetic studies with Guy 11 indicate that spontaneous mutations at the *BUF* locus occur at low frequency ($<10^{-4}$); whereas buff mutation occurs in about 0.3% of UV-C-treated survivors (Leung, unpubl. data).

We examined the frequency of buff mutants in survivors of Guy 11 after 2, 4, and 6 h of exposure to UV-B. Table 1 shows that there is an increase in mutation rate at the *BUF* locus under increased UV-B irradiation. A slight increase in the number of white colonies, which is also controlled by a single gene, was also observed after UV-B exposure, at which time no deleterious effect on the viability of Guy 11 was observed. This supports our hypothesis that UV-B can change the genetic makeup of the pathogen population.

Another mechanism generating genotypic variation (in contrast to genic variation produced by mutation) is sexual or parasexual recombination. Because the sexual stage of the fungus has not been found in nature, it is generally believed that the genetic variability observed in the pathogen population is of somatic origin. Parasexual recombination in *M. grisea* has been demonstrated by forced heterokaryon formation between auxotrophs (Genovesi and Magill 1976; Crawford et al. 1986), but it is not known whether such an event can be enhanced by environmental stimuli. Two dominant selectable markers, one conditioning hygromycin B resistance (Leung et al. 1990) and the other conditioning resistance to phosphinothricin (Leung et al. unpubl. results) have been introduced into Guy 11 and other strains by DNA-mediated transformation. In our preliminary experiments, conidia from the two marked strains were mixed and spread onto complete medium agar. After 3 days of cocultivation, the culture was overlaid with agar containing both antibiotics. Preliminary results suggest that a low frequency of recombinants with resistance to both antibiotics can be recovered in the control experiment (no UV-B exposure). These genetically marked strains are being used to investigate whether parasexual recombination is stimulated under UV-B irradiation.

Table 1. Mutagenic effect of UV-B on *Magnaporthe grisea* strain Guy 11 as measured by the occurrence of buff colonies.

Treatment[a]	UV-B irradiance (kJ/m^2)	Survival (%)	No. of colonies screened	Buff colonies No.	Buff colonies (%)	White colonies No.	White colonies (%)
Dark	0	100	10000	0	0	0	0
Light (2 h)	0	100	9000	0	0	1	0.01
Light (4 h)	0	100	9800	0	0	0	0
UV-B (2 h)	5.6	90	18000	7	0.04	1	0.01
UV-B (4 h)	11.2	30	10000	14	0.14	0	0
UV-B (6 h)	16.8	20	1200	3	0.25	2	0.16

[a] For light treatment, plates were covered with Mylar D. For UV-B treatment, plates were covered with cellulose acetate.

Conclusion

We have established an experimental system to evaluate the effects of UV-B on the blast fungus. Spore viability and sporulation in the blast fungus are adversely affected by daily UV-B levels assumed in the tropics (6–11 kJ/m^2). A wide range of sensitivity was observed among field isolates from the Philippines. However, given the large size of the pathogen population, it is not clear whether the adverse effects of UV-B on viability and sporulation will significantly affect the epidemiology of the disease. We have also shown that UV-B can enhance mutation in a pigment gene at a nonlethal irradiance level. We hypothesize that the mutagenic, rather than lethal, effects of UV-B present the greatest impact on the pathogen population. To generalize the mutagenic effects of UV-B on the blast fungus, mutation rates of additional marker genes under UV-B irradiation must be tested. Furthermore, photoreactivation by long wavelengths, which may moderate the mutagenic effect of UV-B, will also need to be investigated. In addition, we have constructed a lamp frame in a growth chamber to examine the overall effect of UV-B on the host-pathogen interaction.

Experiments are being conducted to determine whether preinoculation and postinoculation UV-B exposure have a direct effect on disease development.

Acknowledgments. This work is supported by United States Environmental Protection Agency (US EPA) Cooperative Agreement No. CR-817426. It has, however, not been subject to the Agency's review and therefore does not necessarily reflect the views of the Agency, and no official endorsement should be inferred. The spectroradiometer and accessories are on loan from US EPA, Environmental Research Laboratory, Corvallis. We thank Steve Holman of Mantech Environmental Technology for his assistance in using the spectroradiometer, and Dahu Chen, Menchu Bernardo, and Rebecca Nelson of IRRI for providing the Philippine isolates and the lineage data. The contribution of M. Bonman and P. Barnes to the initiation of the project and their helpful suggestions are gratefully acknowledged.

Effects of Ultraviolet-B Radiation on Growth of Rice Plants in the Field

I. Nouchi and K. Kobayashi

Anthropogenic chlorofluorocarbons in the atmosphere may cause the depletion of the stratospheric ozone layer and a subsequent increase in UV-B radiation (280–320 nm) reaching the surface of the earth (NASA 1988). A 1% decrease in the ozone layer would result in an approximately 2% increase in biologically effective UV-B (UV-B_{BE}) radiation at temperate latitudes (Chemical and Engineering News 1986). It is important to assess adverse effects of enhanced UV-B on terrestrial ecosystems. Many studies concerning the effects on plants have been conducted over the past two decades in the United States and western Europe (Teramura 1983; Tevini and Teramura 1989). In Japan, however, attempts to assess the effects of enhanced UV-B on plants are limited (Takeuchi et al. 1989, 1993 Nouchi 1991, 1993; Suge et al. 1991; Kumagai and Sato 1992; Sato and Kumagai 1993).

Numerous studies have shown that UV-B radiation can adversely affect plant growth and a multitude of physiological and biochemical processes in plants (Teramura 1983; Bornman 1989; Tevini and Teramura 1989; Runeckles and Krupa 1994). The UV-B radiation inhibited some reactions in the photosystem II reaction center, reduced activities of ribulose-1,5-bisphosphate carboxylase and ATP-synthase (Strid and Porra 1992), disrupted thylakoid membrane proteins and membrane structure (Chow et al. 1992), and altered pigment composition (Strid and Porra 1992) and nucleic acids (Pang and Hays 1991; Jordan et al. 1992). In addition to the depression in photosynthesis, plant growth characteristics (e.g., plant height, leaf area, land dry matter production) are depressed in some species by enhanced UV-B regimes (Teramura 1983; Tevini and Teramura 1989).

However, the majority of this information has been obtained in growth chambers and greenhouses where higher plants tended to be considerably more sensitive to UV-B radiation than in the field (Caldwell et al. 1989). The reduced visible irradiance in greenhouses may contribute to increased sensitivity, because visible irradiance can enable photo-repair processes to protect against otherwise accumulating injuries in the plant tissue. In addition, greenhouse-grown plants have lower levels of UV-absorbing pigments and much thinner cuticles (both of

National Institute of Agro-Environmental Sciences, 3-1-1, Kannondai, Tsukuba, Ibaraki 305, Japan

S. Peng et al. (Eds.) Climate Change and Rice

which act as barriers to UV-B) compared with field-grown plants (Caldwell et al. 1983a; Steinmuller and Tevini 1986). Therefore, UV-B irradiation experiments under greenhouse conditions may not provide reliable estimates of the effects of predicted UV-B increases, and UV-B studies in the field are needed.

Attempts to assess the effects of UV-B enhancement on field-grown plants are still limited (Lydon et al. 1986; Teramura and Murali 1986; Barnes et al. 1988; Sinclair et al. 1990; Teramura et al. 1990; Booker et al. 1992b; Miller et al. 1994) and the results are contradictory. Sinclair et al. (1990) reported the results of a field study conducted on six soybean cultivars grown under simulated 16% ozone depletion. The results showed that UV-B radiation had little effect on production of dry matter and on final seed yield. Miller et al. (1994) also reported enhanced UV-B (in excess of doubling of ambient UV-B_{BE}) did not suppress growth or seed yield of Essex, Coker 6955, or S53-34 soybean in a 3-year field UV-B study. These contradict the conclusion by Teramura et al. (1990) from a 6-year field study on the UV-B sensitive Essex cultivar of soybean, which showed a 19–25% reduction in seed yield when grown under a simulated 25% ozone depletion. In their study, however, there was no adverse effect on seed yield except once when grown under a simulated 16% ozone depletion. Although these experiments were conducted in the field, the UV-B irradiation systems have limited the diurnal UV-B supplementation to a square-wave pattern with a constant lamp output throughout the daily treatment period. Such a constant lamp output may cause unrealistically high UV-B enhancement under low solar UV-B irradiation (e.g., in the early morning or on a cloudy day). Therefore, modulation of supplemental UV-B irradiance in the field is essential to reflect the changes in solar irradiance.

Modulated UV-B supplementation systems have been developed by Caldwell et al. (1983b) and Yu et al. (1991). In the present study, we also developed a UV-B irradiation system with modulated lamp output, and used the system to irradiate 17 rice cultivars. Rice is one of the most important staple food grains, and it is important to determine if rice growth and yield are affected by enhanced UV-B radiation. Cultivars from different regions of the world were exposed to enhanced UV-B in the field. With this experiment, we hoped that the growth responses and the leaf pigments of the rice plants could be studied in more realistic UV-B regimes.

System for Supplemental Ultraviolet-B Radiation

A continuous proportional-control system similar to that of Yu et al. (1991) for supplemental UV-B radiation was designed to monitor solar UV-B radiation and deliver a desired supplemental UV-B irradiance by controlling the output of fluorescent sunlamps. This system consists of three major components: two UV-B sensors, the feedback control system, and eight lamp banks (Fig. 1). Eight 40-W fluorescent sunlamps (F40UVB, Philips Lighting Co, NJ), 120 cm in length

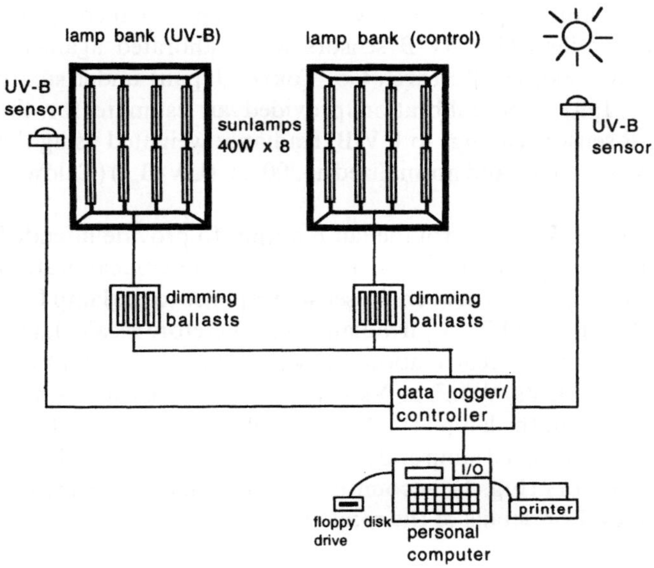

Fig. 1. Schematics of the continuous proportional-control system for supplemental UV-B irradiation in the field

and connected with weather-protected sockets, were positioned into four rows on aluminum frames. The distance between rows was 40, 35, and 40 cm starting from the edge row. The lamp bank was suspended above the plants on an iron frame using four wire ropes. Pulleys were used to raise or lower the lamp bank. An area of uniform supplemental UV-B irradiation under each lamp bank was 1.9 m × 1 m. Lamp banks were oriented to a north-south direction so that shadows of the lamps passed rapidly over the plant canopy.

Lamps were filtered either with a 0.13-mm cellulose diacetate film (Cadillac Plastics Co, Baltimore, OH) for supplemental UV-B radiation treatment, or with a 0.13-mm Mylar D plastic film (DuPont Co, Wilmington, DE) for a control treatment. The cellulose diacetate and Mylar D films effectively absorb radiation with wavelengths shorter than 290 nm and 320 nm, respectively. The light intensity of fluorescent sunlamps was regulated by changing the input voltage to dimming ballasts (Advance Transformer Co, Chicago, IL). The dimming ballasts were housed in a steel box equipped with a small cooling fan. The box was mounted remotely from the lamps to avoid shading on the plants. A UV-B sensor (Ms-210D, Eiko Seiki Co, Tokyo, Japan) was installed at plant height under a UV-B irradiating lamp bank to measure the mixed solar and lamp UV-B. The sensor was designed to resemble the DNA action spectrum. Another sensor of the same type was set in the open to measure ambient solar UV-B irradiance. Outputs of the two sensors were compared, and the lamp output was adjusted to attain the proportional UV-B enhancement. The lamps were turned off when UV-B was below a threshold level to avoid unstable and variable lamp output. The UV-B data acquisition and the lamp output control were run by a personal computer using a data logger-controller. The UV-B sensors were calibrated against a spectroradiometer (MSR-7000, OptResearch Co, Tokyo, Japan) under sunny conditions on 10 July 1992. The calibration provided an estimate for the conversion of the UV-B sensor readings to UV-B irradiance weighted using the generalized plant-action spectrum and normalized at 300 nm (UV-B_{BE}) (Caldwell 1971).

The conversions were used to control the lamp output to provide intended 50% enhancement in the UV-B_{BE} over the control. However, a later recalibration of the sensors showed a discrepancy between the sensor response to the lamp UV-B and that to the solar UV-B. The UV-B_{BE} irradiance was therefore recalculated, and was used in this paper. In the recalculation, the sensor output (Vs, mV) was related to the solar UV-B_{BE} (UVs, W/m^2) as Vs=449 UVs, and the other sensor output (V1, mV) was related to the lamp UV-B_{BE} (UV1, W/m^2) as V1=131 UV1. The UV-B_{BE} irradiance for the mixed lamp and solar UV-B was recalculated from the outputs of the two sensors (Fig. 1), accounting for the sensor responses to each of the UV-B sources. For details of the recalculation, refer to Nouchi and Kobayashi (1995).

The spectroradiometer was calibrated against a 40-W deuterium arc lamp (Model UV-40, Optronic Laboratories Inc, FL) and a regulated power supply (HEUM-51P-OR, Matsusada Precision Devices Co, Shiga, Japan). The heights of the fluorescent sunlamps were adjusted weekly to 35 cm above the top of the plant canopy throughout the experiment as the plants grew. The filters were replaced weekly and biweekly for the UV-B and control treatments, respectively, because the transmittance of the filters decreases with the time of irradiation.

Plants

The field UV-B irradiation experiment was performed from late August through late October at National Institute of Agro-Environmental Sciences, Tsukuba, Japan (36°01'N, 140°07'E, 25 m elevation). Seeds of 17 lowland rice cultivars from different geographical regions of the world were obtained from Dr. S. Maruyama, National Agriculture Research Center, Tsukuba. The seeds were pregerminated in tap water for 3 days at room temperature and sown into seedlings boxes on 6 July 1992. After 3 days of incubation, seedling boxes were moved to a greenhouse, and the plants were grown for about 1 month. The seedling were transplanted in 4-l plastic pots (200 cm^2 in surface) on August 6 with one plant per pot. Each pot contained an aluvial soil with 12 g of mineral fertilizer ($N:P_2O_5:K_2O$ = 0.08:0.08:0.08). The pots were flooded with water throughout the experiment. Exposure to UV-B irradiation treatments was initiated on 25 August and terminated on 22 October.

Using a randomized block design, ten plants of each cultivar were assigned to the UV-B treatments. Two or three pots of each cultivar were assigned to each of the two UV-B treatments with four replicates. To reduce spatial variability of UV-B irradiance, the pots were rotated within a replicate weekly.

Growth Measurements

Plant height and tiller number were measured on 21 August, 3 September, 17 September, and 22 October. At the end of the 2 months of UV-B irradiation, all the cultivars were harvested. Each plant was divided into leaf blades, leaf sheaths and culms, roots, dead leaves, and panicles. Leaf area was measured using a leaf area meter (AAC-400, Hayashi Denko Co Ltd, Tokyo, Japan), and the plant parts were dried at 70 °C for 72 h and weighed.

Measurements of Pigments

Eight leaf discs (1 cm in diameter) were sampled from fully expanded top leaves 1 month after the start of the UV-B irradiation. They were extracted with 25 ml of 99.5% ethanol for 2 days in the dark at 4 °C. The chlorophyll content was determined according to the method of Knudson et al. (1977). The concentrations of ethanol-extractable UV-absorbing compounds were taken from the same ethanol extract used for the chlorophyll analysis and were measured with a spectrophotometer (UV-1200, Shimadzu Co, Kyoto, Japan). The absorption spectra were recorded between 220 and 700 nm. The peak absorbance at 340 nm was arbitrarily used for comparative analysis of the ethanol extract.

Statistical Analysis

Statistical differences in the growth parameters and pigment content were determined by analysis of variances and treatment means were compared with the Scheffe's F-test at the $p = 0.05$ level.

Results

Temporal Variation in UV-B$_{BE}$ Irradiance

Before the irradiation experiment, shading by the lamp bank and supported framework was assessed. Daily integral solar UV-B$_{BE}$ irradiance was measured above and below the lamp bank with the UV-B sensors from 17 August to 24 August. The ratio of UV-B$_{BE}$ below the lamp bank to that above the lamp bank was in the range 0.72–0.77 (mean 0.75) (data not shown).

Diurnal changes in UV-B$_{BE}$ irradiance for ambient sunlight and the supplemental UV-B treatment are shown in Fig. 2 for 9 and 15 September and 10

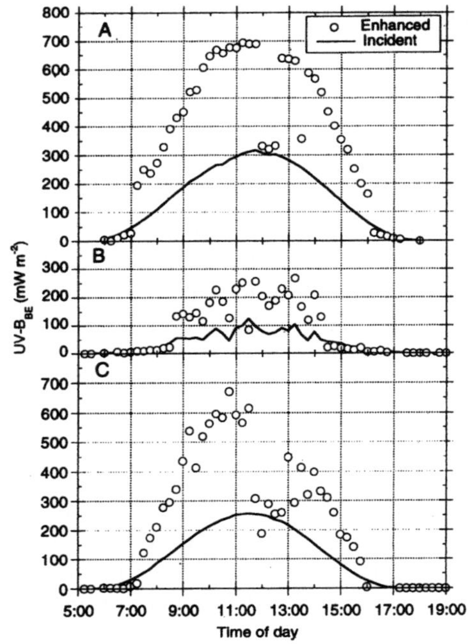

Fig. 2A–C. Diurnal changes in UV-B$_{BE}$ irradiance on clear and cloudy days above (ambient incident) and below (enhanced UV-B) the lamp bank at the plant height. **A** 9 September 1992. **B** 15 September 1992. **C** 10 October 1992

October 1992. A large deviation of UV-B$_{BE}$ in the supplemental UV-B treatment against the ambient sunlight occurred once a day at about noon on sunny days (9 September and 10 October). These deviations were caused by shadows cast by the lamps and supporting framework.

Figure 3A shows seasonal changes in daily UV-B$_{BE}$ irradiance in ambient sunlight and supplement UV-B treatment. The daily UV-B$_{BE}$ irradiance in ambient sunlight ranged from 0.3 to 8.3 kJ/m^2. The total daily integrated UV-B$_{BE}$ above (ambient incident) and below (enhanced UV-B) the lamp bank was 259.2 and 579.5 kJ/m^2, respectively. Because solar UV-B$_{BE}$ irradiance below the lamp bank decreased by 25% compared with the ambient incident, the ratio of the UV-B$_{BE}$ irradiance in the supplemental UV-B treatment to that of the control was 2.67 on average from 25 August to 22 October.

Effects of Ultraviolet-B on Rice Growth

Enhanced UV-B did not induce any symptoms of foliar injury on leaves of all the 17 rice cultivars. Furthermore, no significant effects of a 2-month irradiation of enhanced UV-B were observed on any growth parameters of all the tested rice cultivars at harvest (Table 1). In addition, no significant effect of UV-B radiation was observed on plant height and tiller number at periodical observations during the growing season (data not shown).

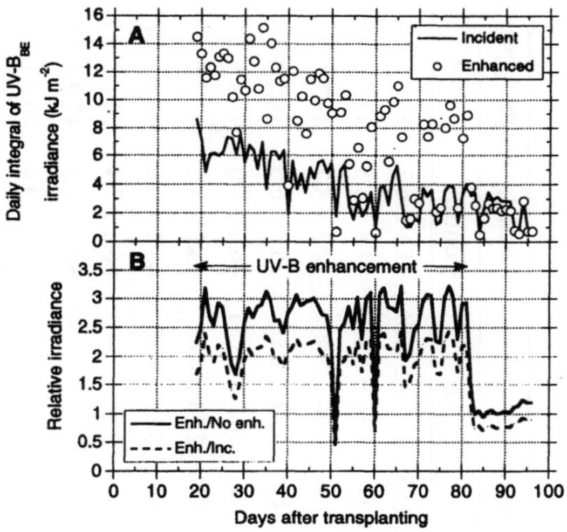

Fig. 3A,B. Seasonal changes in daily integral of UV-B_{BE} irradiance for ambient incident (above the lamp bank), enhanced UV-B (below the lamp bank), and control (below the lamp bank). **A** Daily integrals of UV-B_{BE} irradiance for ambient incident and enhanced UV-B. **B** Relative daily integrals of the UV-B_{BE} irradiance for the enhanced UV-B treatment to the ambient incident and to the control

Pigment Content

Total chlorophyll content in the top leaf of 17 cultivars after 1 month of UV-B treatment tended to increase for all but one cultivar. However, a significant increase in chlorophyll content was observed only for Ketan, Jamuna, and Lemont (Table 2). Similarly, UV-absorbing compounds tended to increase after 1 month of UV-B treatment for all cultivars except Lemont and Jamuna, but a significant increase in UV-absorbing compounds was observed only for Akenohoshi, IR 26, and Banten.

Discussion

Modulated systems for supplemental UV-B radiation experiments have been designed and built by Caldwell et al. (1983b) and Yu et al. (1991). Because the system of Yu et al. (1991) is the most practical to operate and most maintenance-free of the two field UV-B modulation systems, we constructed a similar continuous proportional-control system for supplemental UV-B radiation in the field. The results of the 2-month irradiation experiment showed that our UV-B modulation system operated over the long term under field conditions with almost no maintenance except for exchange of filters and adjustment of lamp bank distance above the plant canopy to accommodate plant growth.

Under the lamp banks, solar UV-B_{BE} irradiance was on average 25% less than ambient because of the shading by the lamp bank and supporting framework. The loss of solar UV-B_{BE} below lamp banks in our system was comparable to that in other studies. Caldwell et al. (1983b) reported an average shading of the

Table 1. Effects of UV-B on plant height, tiller, panicle dry weight, and total dry weight of rice[a]

Cultivars	Country of origin	Plant height (cm)		Tiller number		Panicle (g/plant)		Total dry weight (g/plant)	
		Control	UV-B	Control	UV-B	Control	UV-B	Control	UV-B
Akenohoshi	Japan	84.5 ± 4.9	85.0 ± 2.8	19.3 ± 2.3	20.0 ± 2.3	10.39 ± 3.98	10.75 ± 2.18	41.6 ± 6.3	41.0 ± 4.4
Akihikari	Japan	85.0 ± 4.1	85.4 ± 1.8	17.0 ± 3.0	18.9 ± 2.0	8.71 ± 2.24	7.16 ± 1.95	37.8 ± 5.5	37.7 ± 3.9
Koshihikari	Japan	96.1 ± 3.2	93.6 ± 4.9	18.9 ± 3.6	20.9 ± 3.7	10.25 ± 2.35	9.96 ± 2.90	38.2 ± 3.7	39.3 ± 3.7
Nipponbare	Japan	84.1 ± 5.7	81.8 ± 4.8	20.1 ± 5.1	24.0 ± 3.9	8.61 ± 2.47	9.61 ± 2.81	35.8 ± 6.6	38.2 ± 3.4
Sasanishiki	Japan	92.1 ± 4.5	93.1 ± 4.1	27.7 ± 4.9	25.5 ± 4.9	7.79 ± 2.14	10.23 ± 3.84	46.6 ± 6.0	43.4 ± 5.6
Aoniwai	China	77.5 ± 4.4	79.4 ± 4.8	26.2 ± 2.5	28.4 ± 2.0	2.56 ± 0.72	2.83 ± 0.67	43.6 ± 4.2	44.9 ± 4.7
Nankin 11	China	75.6 ± 3.4	74.3 ± 4.6	25.1 ± 4.1	26.7 ± 3.5	3.37 ± 0.75	3.87 ± 1.17	45.1 ± 6.0	47.3 ± 10.3
IR 26	Philip	70.0 ± 6.8	70.6 ± 6.0	39.7 ± 3.8	37.5 ± 7.1	0	0	36.2 ± 7.4	34.8 ± 5.1
Banten	Indon	93.4 ± 6.7	92.5 ± 4.4	23.8 ± 3.5	23.9 ± 3.2	0	0	50.4 ± 7.5	47.9 ± 6.1
Ketan	Indon	99.1 ± 7.5	98.0 ± 2.6	18.6 ± 4.5	18.7 ± 2.8	0	0	47.5 ± 9.4	46.0 ± 4.0
Dular	India	105.2 ± 7.1	107.4 ± 7.2	19.4 ± 2.7	20.3 ± 3.7	7.24 ± 1.25	7.78 ± 2.29	66.9 ± 8.8	69.3 ± 11.0
Jamuna	India	69.1 ± 5.0	67.1 ± 3.1	33.5 ± 3.0	32.5 ± 3.5	0.88 ± 0.31	0.59 ± 0.47	42.4 ± 6.4	39.8 ± 4.0
Kele	India	96.6 ± 6.0	92.4 ± 7.3	38.9 ± 3.4	43.5 ± 6.3	2.72 ± 0.51	2.88 ± 0.59	65.5 ± 9.9	66.5 ± 8.9
Bomba	Spain	105.9 ± 4.6	101.2 ± 4.1	19.2 ± 3.9	22.1 ± 2.7	1.26 ± 0.77	0.96 ± 0.42	52.3 ± 7.0	50.8 ± 10.7
Raffaelo	Italy	90.4 ± 4.3	88.3 ± 4.6	19.8 ± 3.0	19.3 ± 2.6	1.34 ± 0.59	0.81 ± 0.43	46.6 ± 5.8	40.8 ± 9.2
Lemont	USA	77.3 ± 3.3	76.8 ± 3.0	17.6 ± 4.8	18.0 ± 2.5	0.04 ± 0.09	0.10 ± 0.23	36.0 ± 4.4	38.4 ± 5.2
CP 231	USA	99.9 ± 7.7	99.9 ± 6.0	15.1 ± 3.5	15.4 ± 3.8	1.10 ± 0.37	1.15 ± 0.33	44.5 ± 9.0	46.0 ± 3.3

[a] Each value is the mean and standard deviation of ten plants; Philip=Philippines, Indon=Indonesia.

Table 2. Effects of UV-B on chlorophyll and UV-absorbing compounds in rice leaves[a]

Cultivars	Country of origin	Chlorophyll (μg/cm^2)		A340	
		Control	UV-B	Control	UV-B
Akenohoshi	Japan	32.7 ± 2.1	36.3 ± 5.4	0.956 ± 0.105	**1.313 ± 0.041
Akihikari	Japan	32.5 ± 5.4	32.5 ± 5.4	0.863 ± 0.048	1.068 ± 0.154
Koshihikari	Japan	29.0 ± 0.5	29.7 ± 2.7	0.763 ± 0.041	0.99 ± 0.117
Nipponbare	Japan	30.7 ± 7.5	33.3 ± 0.5	0.807 ± 0.135	0.845 ± 0.030
Sasanishiki	Japan	23.1 ± 3.8	27.6 ± 2.3	0.737 ± 0.164	0.909 ± 0.164
Aoniwai	China	40.1 ± 3.2	41.4 ± 8.7	1.005 ± 0.207	1.193 ± 0.381
Nankin 11	China	28.5 ± 4.8	36.4 ± 3.6	1.108 ± 0.103	1.315 ± 0.345
IR 26	Philippines	34.6 ± 3.0	43.1 ± 8.5	0.662 ± 0.095	**0.971 ± 0.015
Banten	Indonesia	26.5 ± 6.0	29.8 ± 3.0	1.046 ± 0.085	*1.571 ± 0.190
Ketan	Indonesia	26.3 ± 3.5	*37.1 ± 5.6	1.006 ± 0.451	1.688 ± 0.131
Dular	India	32.5 ± 8.6	36.8 ± 2.7	0.936 ± 0.015	1.195 ± 0.239
Jamuna	India	27.6 ± 2.7	*38.9 ± 4.2	1.077 ± 0.344	0.975 ± 0.131
Kele	India	22.8 ± 4.8	31.9 ± 5.5	0.893 ± 0.244	0.913 ± 0.061
Bomba	Spain	33.3 ± 5.7	32.5 ± 2.8	0.931 ± 0.049	1.187 ± 0.193
Raffaelo	Italy	30.5 ± 2.2	35.5 ± 6.1	0.987 ± 0.348	1.167 ± 0.506
Lemont	USA	28.2 ± 5.3	*43.0 ± 1.8	1.226 ± 0.121	1.081 ± 0.215
CP 231	USA	38.7 ± 2.9	45.4 ± 9.4	1.251 ± 0.071	1.488 ± 0.410

[a] Each value is the mean and standard deviation of three plants. UV-absorbing compound content is expressed absorbance of leaf extract (6.28 cm^2) in ethanol solvent (25 ml) at 340 nm. The difference in pigment content between UV-B treatments and the control are significantly diferent at the 5% level (*) and 1% level (**) as determined by Scheffe's F-test.

daily UV-B_{BE} irradiance of 20%. Booker et al. (1992a) reported that solar UV-B_{BE} irradiance inside an open-top chamber was on average 24% less than ambient. Daily solar UV-B_{BE} irradiance at Tsukuba was 7.0 kJ/m^2 during a cloudless day near the summer solstice. Our measurement of global solar UV-B_{BE} irradiance with the spectroradiometer at Tsukuba under sunny conditions was almost the same as that calculated by the model of Green et al. (1980).

Rice is less sensitive to UV-B than the very sensitive cucumber and soybean under greenhouse and controlled environment conditions (Krupa and Kickert 1989), although few studies have examined the effects of UV-B radiation on rice (Biggs and Kossuth 1978; Biggs et al. 1984; Teramura et al. 1991; Dai et al. 1992; Barnes et al. 1993). Teramura et al. (1991) grew 16 rice cultivars from 7 different geographical regions of the world in a greenhouse for 12 weeks under conditions simulating 20% ozone depletion. Their results show a significant decrease in total biomass with increased UV-B radiation for about one-third of all cultivars. In addition, Dai et al. (1992) reported that plant height, leaf area, and dry weight of some rice cultivars (IR30, IR45, and IR74) were reduced by 4 weeks of UV-B irradiation in a phytotron. Teramura et al. (1991) and Dai et al. (1992) also showed a wide range of sensitivity to UV-B among rice cultivars. On the other hand, Biggs et al. (1984) reported statistically insignificant losses in grain yield for one cultivar in a field simulating 8% ozone depletion for Florida. The present sutdy showed no effects on the growth of 17 rice cultivars from 8 different gerographical regions of the world after enhanced UV-B irradiation in the field for about 8 weeks. Heading date, plant height, tiller number, panicle dry weight, and dry weight of whole plants were unaffected by the enhanced UV-B irradiation. Plants are apparently more sensitive to a given UV-B dose under greenhouse rather than field conditions (Caldwell et al. 1989); therefore, our field UV-B irradiation system can provide more reliable estimates of potential impacts on growth and yield associated with anticipated increases in solar UV-B radiation. Therefore, a result showing no effect of increased UV-B on any growth parameter suggests that increased UV-B has less influence on rice than might have been anticipated.

A decrease in chlorophyll content has been shown in several sensitive species or cultivars in reponse to UV-B radiation in a growth chamber study, but a chlorophyll decrease is usually observed only under enhanced UV-B and low visible light (Tevini and Teramura 1989). In our field experiment, chlorophyll content on a leaf-area basis generally tended to increase for almost all the tested cultivars. A similar increase in chlorophyll was reported by Dai et al. (1992), who found that total chlorophyll of IR30 and IR40 rice cultivars increased significantly after 4 weeks of UV-B treatment in a phytotron. The increase in chlorophyll content on a leaf-area basis could be accounted for by other responses to UV-B radiation such as increased leaf thickness, which would conceivably increase chlorophyll content on area basis. However, we have no data for increased leaf thickness in chlorophyll analysis at 1 month of irradiation. The UV-absorbing compound also tended to increase with enhanced UV-B in almost all the cultivars. The UV-absorbing compounds accumulate in leaves of most higher plants when irradiated with UV-B. These compounds, including flavonoids, have been suggested as protection against the damaging effects of UV-B radiation. Similar results of increased UV-absorbing compounds in rice leaves were reported by Teramura et al. (1991) and Dai et al. (1992), although some rice cultivars showed a significant decrease in total biomass and photosynthesis. It is unclear, therefore, whether the UV-absorbing compounds are fully effective in protecting rice growth from UV-B irradiation in the field.

The present study indicated that exposure to a 100% increase in UV-B_{BE} irradiance against ambient solar UV-B_{BE} and a 167% increase against control UV-B_{BE}, had no effects on growth of the 17 rice cultivars. However, because this study was conducted from late summer to midautumn, and the plants received less UV-B_{BE} than in a normal season for rice growth, the effects of UV-B on rice might have been quite limited. Studies in a normal rice-growing season are now under way with the UV-B irradiation system in the field.

Acknowledgments. We wish to thank Drs. M.M. Caldwell and S.D. Flint, Utah University, Drs. J.E. Miller, F.L. Booker, E.L. Fiscus, and Mr. R.B. Philbak, USDA-ARC,

North Carolina University, Drs. A.H. Teramura and H.E. Reed, University of Maryland, and Dr. D.M. Olzyk, US EPA Corvallis Environmental Research Laboratory, for valuable suggestions in constructing a field UV-B modulation system. We also thank the staffs of the experimental farm for their help in setting the UV irradiation system and measuring plant biomass. The present research was supported by the Global Environment Research Program Budget, which is set aside by the Japan Environment Agency to promote comprehensive research on the Global Environment.

Genetic Study of Resistance to Inhibitory Effects of Ultraviolet Radiation in Rice

T. KUMAGAI and T. SATO

There have been many reports on the effects of UV-B radiation on growth, development, and pigment biosynthesis in crop plants (Teramura 1983; Tevini and Teramura 1989). Investigations that involve the whole plants and the complete life cycle conducted in the phytotron or greenhouse are required. Studies conducted in the field are invaluable because they help anticipate the potential changes in productivity of crops as a result of imaginary environmental changes. They are also indispensable in the search for genetic resources that may survive in, or may adapt to, the environmental changes. However, in spite of the global importance of cereals as staple food crops, few studies have examined the effects of UV-B radiation on cereals. There is a noteworthy study on soybeans, in which Murali and Teramura (1986b) and Teramura et al. (1990) examined the potential for alterations in the yield and seed quality of soybeans grown for 6 years in the field. They reported the necessity for multiyear experiments and the need to increase the understanding of the interaction between UV-B radiation and other environmental stresses to assess the potential consequences of stratospheric ozone depletion.

With respect to rice plants, Teramura et al. (1991) found in greenhouses that, among 16 rice cultivars from seven different geographical regions, Kurukaruppan (Sri Lanka), Himali (Nepal), and Tetep (Vietnam) cultivars were more tolerant to the effects of elevated UV-B radition, and suggested that geographical location might influence sensitivity to UV-B radiation. Recently, Dai et al. (1992) examined the effects of UV-B radiation on the growth of four lowland rice cultivars (IR line) in a phytotron to evaluate the use of morphological and physiological parameters to identify sensitive and less-sensitive genotypes in future screening. They found that the distinct responses and relative ease in measurement of stomatal opening and ion leakage made these parameters suitable indices to select rice cultivars less sensitive to UV-B radiation after 2 weeks of UV-B treatment. Based on the relative change in total biomass production between UV-B irradiated and control plants, they also found that IR 74 was the most sensitive and IR 64 the least sensitive genotype.

We found that elevated UV radiation (containing a large amount of UV-B with a small amount of UV-C) inhibited the development of leaves and tillers, the

Institute of Genetic Ecology, Tohoku University, Sendai 980, Japan

S. Peng et al. (Eds.) Climate Change and Rice

increase in biomass production, the elongation of plant height, the photosynthetic rate, and the chlorophyll content in rice plants in a phytotron (Kumagai and Sato 1992). In our experiments, unfiltered UV-B lamps were used instead of high quality UV-B radiation, and simulation of stratospheric ozone depletion was not taken into account. However, we consider that cultivars showing resistance to the inhibitory effects of UV radiation containing a large amount of UV-B with a small amount of UV-C would also exhibit higher tolerance to high quality UV-B radiation because various biological phenomena that have action peaks in the UV-C region also use the UV-A and UV-B regions of the spectrum to a lesser degree. For example, the action for cyclobutyl pyrimidine dimer induction in DNA in intact alfalfa seedlings peaks about 280 nm but extends to 365 nm (Quaite et al. 1992). On the basis of these findings, we examined cultivar differences in the resistance to injuries caused by UV radiation among 198 rice cultivars belonging to five Asian rice ecotypes (aus, aman, boro, bulu, and tjereh) from the Bengal region and Indonesia and to Japanese lowland and upland rice groups (Sato and Kumagai 1993). Various cultivars that had different sensitivities to the effects of UV radiation were involved in the same ecotype and the same group. The Japanese lowland rice group and the boro ecotype were more resistant. Among Japanese lowland rice cultivars, Sasanishiki exhibited more resistance to UV radiation, whereas Norin 1 showed less resistance, although these two cultivars are closely related.

Investigations of the inheritance of resistance to injuries caused by UV radiation are important in the search for genetic resources resistant to UV radiation and in the development of cultivars resistant to UV radiation. However, few studies have been conducted.

Effects of Elevated Ultraviolet Radiation on Growth and Development of Rice Cultivars

The effect of UV radiation on the increase in plant height of four Japanese rice cultivars (Norin 20, Zosan 1, Sasanishiki, and Norin 1) was first examined in a phytotron (Kumagai and Sato 1992). When these cultivars were grown under irradiation with visible light supplemented with or without UV radiation (containing a large amount of UV-B with a small amount of UV-C) (Fig. 1), the increase in plant height of each rice cultivar was inhibited by supplementary UV radiation. Browning in Norin 20, Norin 1, and Zosan 1 appeared 3 days after the transfer of the plants to the UV irradiated conditions. Browning gradually spread from older to younger leaves as the duration of cultivation was prolonged. About 1.5 months later, Norin 1 was severely damaged. Sasanishiki was most resistant to the damaging effect of UV radiation, whereas Norin 1 was the least resistant. We analyzed more precisely the mechanism whereby UV radiation affected biomass production and photosynthesis in Norin 1 and Sasanishiki (Fig. 2). The photoperiod consisted of 12 h of light and 12 h of dark, and the temperature was

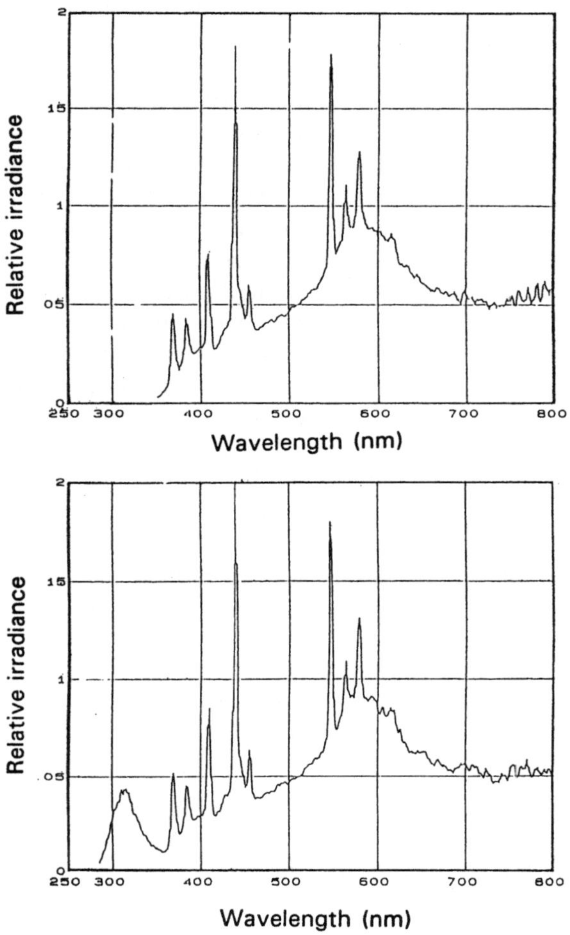

Fig. 1. Spectral distribution of the white light in growth cabinets with (*lower*) and without (*upper*) supplementary UV radiation. White light was provided by metal halide lamps (Toshiba, DR 400/T(L)), tungsten lamps (Toshiba, RF 220 V 200 W), and flurorescent lamps (Toshiba, FR80HWA). The UV light supplemented to white light was provided by fluorescent tubes (Toshiba, fl20S-E). The ratio of the irradiance in the UV-B and UV-C regions was 1:0.17. When necessary, UV radiation was filtered through stainless-steel net filters to reduce irradiance. Spectral distribution was measured by a radiospectrometer with a multichannel detector (Hamamatsu Photonics Co) and a CT-10C diffracting monochrometer (Japan Spectroscopic Co). Each fluence rate of PAR (photosynthetically active radiation) and UV region was measured by Data Logger (Li-cor Co, L1-1000) and SD104 UV sensor (Macom Co)

maintained at 27 °C during the day and 17 °C at night. The degree of inhibition was based on the relation: (value of the control-value of sample)/value of control × 100.

The value of the control was derived from plants grown under visible radiation without supplementary UV radiation. Each fluence rate per second of PAR (μmol/m^2) was 112 for the visible radiation without supplementary UV radiation, and 117 for the visible radiation with supplementary UV radiation. The irradiance of UV radiation was varied at four different levels, control (C), low (L), medium (M), and high (H) levels. The fluence rate (W/m^2) in the UV region was 0.022 for the control, 0.133 for the low level, 0.294 for the medium level, and 0.573 for the high level. Both biomass production and photosynthetic activity in Norin 1 decreased with the increase in UV radiation: the degree of inhibition of biomass production in the plants grown for 2 weeks under the

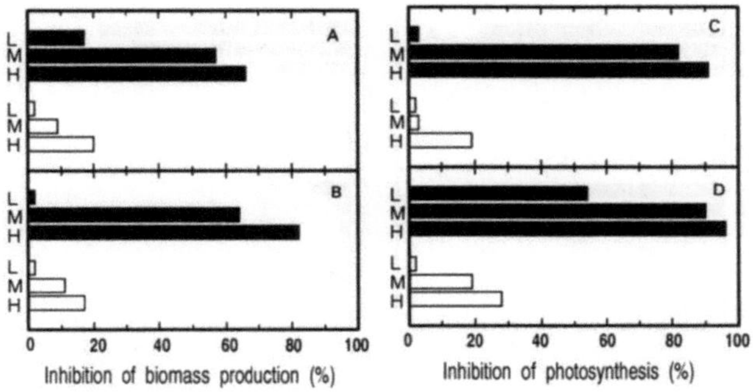

Fig. 2A–D. Inhibition of biomass production and photosynthesis in Sasanishiki and Norin 1 by increases in UV radiation supplemented to visible light. Norin 1 (*black columns*) and Sasanishiki (*Open columns*) were grown for 2 (**A, C**) or 4 weeks (**B, D**) in low (*L*), medium (*M*), and high (*H*) levels of UV radiation supplemented to visible radiation. (After Kumagai and Sato 1992)

medium and high levels of UV radiation was 58% and 64%, respectively. The degree of inhibition of photosynthesis determined in the fifth leaf of the plants grown for 2 weeks under the high levels of UV radiation was about 90%. A reduction was observed in biomass production and photosynthesis determined in the seventh leaf of plants grown for 4 weeks under the high levels of UV radiation: the degree of inhibition of biomass production amounted to almost 80% and that of photosynthesis exceeded 90%. The extent of biomass production and photosynthesis in Sasanishiki also decreased with the increase in UV radiation. However, the degree of inhibition of biomass production and photosynthesis in this cultivar was much lower than that of Norin 1.

The growth components [e.g., plant height, leaf formation, fresh weight (FW) and dry weight (DW) of biomass, and total chlorophyll content] in Norin 1 and Sasanishiki grown under the low levels of visible radiation also decreased with an increase in UV radiation (Kumagai and Sato 1992). When both cultivars were grown under the high levels of visible radiation the increase in UV radiation also resulted in a reduction of each growth component (see Fig. 3, which shows only the data concerning the effects of UV radiation on fresh weight and chlorophyll content). However, the degree of reduction of all the parameters decreased compared with the low irradiance level of visible radiation. This indicated that the parameters in both cultivars grown under low levels of visible radiation were more inhibited by UV radiation than those of cultivars grown under the high levels of visible radiation. The degree of inhibition of all parameters of Norin 1 was higher than those of Sasanishiki, irrespective of the amount of visible radiation applied.

According to rice field experiments (unpubl. data), tiller formation in Norin 1 was markedly inhibited by elevated UV radiation in the early stage of growth compared with that of Sasanishiki. This was not the case for plant height. With an increase in growing time, the degree of reduction in tiller formation was gradually

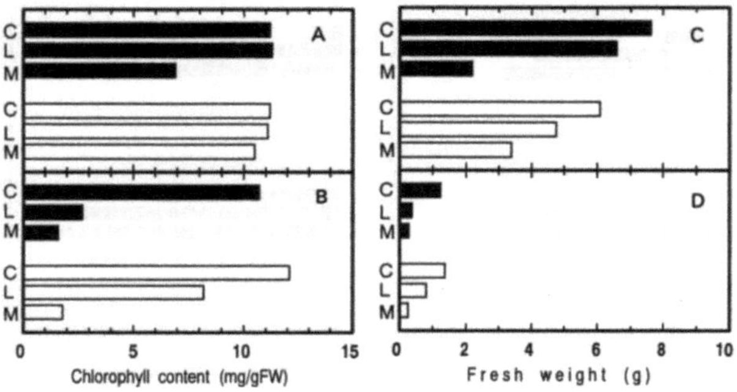

Fig. 3A–D. Effects of the level of UV radiation supplemented with a low or high level of visible radiation on the growth of Norin 1 and Sasanishiki. Norin 1 (*black columns*) and Sasanishiki (*Open columns*) were grown for 4 weeks under a high (**A, C**) or low (**B, D**) levels of visible radiation supplemented with control (*C*), low (*L*) and medium (*M*) levels of UV radiation. (After Kumagai and Sato 1992)

recovered. The number of sterile seeds increased in both cultivars grown under elevated UV radiation: the percentage of sterility was about 10% higher than in the nonelevated UV irradiated control. Grain size became a little smaller. About 35% yield reduction was observed. The protein content of grain increased by about 10%, indicating the lowering of rice quality.

It is evident that there were cultivar differences among rice cultivars in the degree of resistance to UV radiation, and that Sasanishiki was the most resistant to UV radiation among the tested cultivars. Results also suggested that the resistance to the effect of UV radiation was related either to a lower sensitivity to UV radiation or to a greater ability to recover from the injuries caused by UV radiation through the exposure to visible radiation. According to the data base of Hokuriku Agricultural experiment Station in Japan (1977, 1980), Norin 20 is an early maturing variety that requires approximately 79 days for heading, whereas the others require about 110 days. The plant height of this cultivar is as short as 70 cm; it has a large number of panicles per plant but a small number of grains per panicle. Zosan 1 is a tall variety (102 cm) with long panicles, a small number of panicles per plant, but a large number of grains per panicle. The plant height of Sasanishiki, a leading variety in the Miyage prefecture in Japan, is 77 cm, and the size of the panicle and number of grains per panicle are intermediate between those of Norin 20 and Zosan 1.The plant height of Norin 1 is 89 cm, and its panicles and grains are similalr to those of Sasanishiki. Our results indicate that the degree of resistance to UV radiation is independent of the plant shpae.

It is possible that the rice cultivars that have more resistance to UV radiation could be selected by examining the effect of various levels of UV radiation supplemented to visible radiation in a phytotron. Sato and Kumagai (1993) examined cultivar differences in resistance to the inhibitory effects of UV radiation among five Asian rice ecotypes (aus, aman, boro, bulu, and tjereh) and the Japanese lowland (JLR) and upland rice groups (JUR). The aus, aman, and boro ecotypes, from the Bengal region, and the tjereh ecotype, from Indonesia, are indica rices. The bulu ecotype, from Indonesia, is a tropical japonica, and the Japanese lowland and upland rice groups are temperate japonica rices.

Each fluence rate (W/m²) in the UV region was 0.01 for the control, 0.069 for the low level, 0.138 for the medium level, and 0.219 for the high level. When experimental plants were grown under the low irradiance level, a wide range in the frequency distribution was observed in every ecotype and group, but no remarkable differences among ecotypes and groups were observed in the pattern of the distribution or in the mean of the ratio of the FW of the sample and control (open columns in Fig. 4). Different sensitivities to UV radiation were observed in the same ecotype and in the same group: one type was promoted by UV radiation (i.e., resistance); whereas, the other type was inhibited by UV radiation (i.e., sensitive). The proportion of cultivars resistant to the low irradiance level in individual ecotypes or groups differed: 79% for boro, 69% for aus, 68% for aman, 57% for bulu, and 58% for tjereh. As for the Japanese rice cultivars, 75% of the JLR and 54% of the JUR were resistant. With an increase in the irradiance level of UV light to the medium level, the pattern of the frequency distribution of the ratio of the FW of the sample and control significantly shifted toward the lower values and the mean decreased markedly (black columns in Fig. 4). However, it

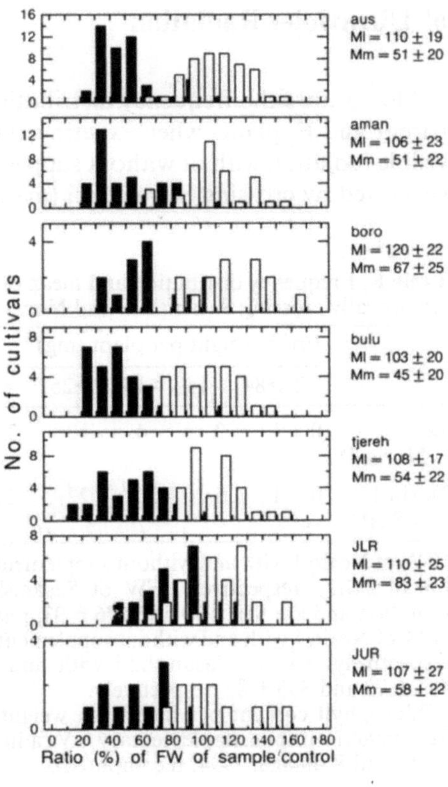

Fig. 4. Frequency distribution of the ratio (%) of FW of the sample to that of the control. *Open* and *Black columns* show the results of plants grown under the low and medium levels of supplementary UV radiation, respectively. *Ml* and *Mm* are mean ± SD of ratio of FW in plants grown for 20 days under the low and medium UV radiation, respectively. (Sato and Kumagai 1993)

should be noted that JLR maintained a mean of 83%, which was the highest value, and the boro ecotype showed 67%, whereas the others showed a mean below 58%. Furthermore, the proportion of cultivars resistant to the medium irradiance level was 21% for the JLR and 14% for the boro ecotype; whereas the others were below 4%. With an increase in the irradiance level of UV light to the high level, the pattern of the frequency distribution of the ratio of the FW of the sample and control in all ecotypes and groups was similar to that seen in plants grown under the medium irradiance level. However, the mean of each ecotype decreased somewhat (data not shown). Even in cultivars grown under the high irradiance level, two cultivars of JLR and one cultivar of each of boro, bulu, and tjereh exhibited resistance to the UV radiation. Similar results were observed for the effects of elevated UV radiation on the increase in plant height and chlorophyll content in the third leaf. Overall, it was clear that the JLR group and the boro ecotype were more resistant to the inhibitory effects of UV radiation: Sasanishiki belonged to the most resistant group and Norin 1 belonged to the less resistant one. The data suggested that the differences in resistance to UV radiation were not related to the geographical origins of the rice cultivars examined.

Genetic Analysis of Resistance to the Inhibitory Effects of Ultraviolet Radiation

Table 1 shows the frequency distribution of fresh weight per plant in both the parent and F_2 plants when experimental plants were grown for 20 days under visible radiation with or without supplemental UV radiation. The F_2 plants were generated by crossing Sasanishiki (male) with Norin 1 (female) and vice versa.

Table 1. Frequency distibution and mean of fresh weight (FW) in F_2 plants generated by reciprocally crossing Sasanishiki and Norin 1

	Fresh weight per plant (mg)											
	>2	>84	>140	>196	>252	>308	>364	>420	>476	>532	>588	>614
Sas × Norin 1[a]	0	1	0	4	12	36	26	25	8	4	1	0
Norin 1 × Sas[b,c]	0	1	5	20	22	21	18	7	2	1	2	0

[a] FW of Norin 1 with and without supplementary UV radiation are 331 ± 48 and 743 ± 68 (mean ± SD), respectively. FW of Sasanishiki with and without supplementary UV radiation and are 556 ± 79 and 676 ± 93, respectively.
[b] FW of Norin 1 with and without supplementary UV radiation are 248 ± 45 and 686 ± 99, respectively. FW of Sasanishiki with and without supplementary UV radiation are 485 ± 72 and 565 ± 82, respectively.
[c] Chlorophyll content per unit fresh weight of Sasanishiki × Norin 1 and Norin 1 × Sasanishiki without supplementary UV radiation are 728 ± 68 and 613 ± 75, respectively. (Sato and Kumagai 1994, recomputed).

Experimental plants were cultured for 2 weeks at 26–14 °C in a greenhouse. At the three-leaf stage, the plants were transferred to large growth cabinets (Koitotron, Type KG, Koito Co, Tokyo, Japan), and were further cultured for 20 days in visible radiation with or without supplementary UV radiation. The photoperiod consisted of 12 h of light and 12 h of dark, and the temperature was maintained at 25 °C during the day and 17 °C at night. Each fluence rate of PAR per second (μmol/m²) and UV region (313 ± 30 nm; W/m²) was 112 and 0.023 for the visible radiation without supplementary UV radiation, and 117 and 0.294 for the visible radiation with supplementary UV radiation. The increase in fresh weight of Norin 1 was more inhibited by supplemental UV radiation compared with that of Sasanishiki. The reduction in the mean fresh weight of Norin 1 reached about 65%, whereas the reduction in Sasanishiki was only about 15%. The value of the control was derived from plants grown in visbile radiation without supplemental UV radiation. When grown in visible radiation without supplemental UV radiation, the mean fresh weight in F_2 plants was similar to that of each parent. However, when the two different types of F_2 plants derived from reciprocal crosses were grown in visible radiation with supplemental UV radiation, the frequency distributions of the fresh weight of each F_2 plant extended beyond the range of the individual parents, and the means of fresh weight were close to those of Norin 1. Table 2 shows the frequency distribution of chlorophyll content per unit fresh weight in both the parent and F_2 plants grown for 20 days under visible radiation with or without supplemental UV radiation. Results similar to those of fresh weight were obtained. Namely, the means of chlorophyll content per unit fresh weight in F_2 plants grown without supplemental UV radiation were similar to those of their parents. Under elevated UV radiation, the frequency distributions of the two types of F_2 plants extended beyond those of their parents and the means of chlorophyll content were close to those of Norin 1. Maternal inheritance was not involved. Furthermore, it suggests that resistance to the inhibitory effects of UV radiation might be controlled by recessive genes.

Inheritance of chlorophyll content per unit fresh weight was further examined in F_3 lines. Eighty-two F_3 lines, generated by self-fertilization of F_2 plants

Table 2. Frequency distribution of chlorophyll content per unit fresh weight (CHL) in F_2 plants generated by reciprocally crossing Sasanishiki and Norin 1

	Chlorophyll content per unit fresh weight (mg/g FW)								
	>0.22	>0.66	>1.10	>1.54	>1.98	>2.42	>2.86	>3.30	>3.74
Sasanishiki × Norin 1[a]	0	8	27	57	30	33	12	0	1
Norin 1 × Sasanishiki[a]	3	13	38	39	35	19	8	0	0

[a] CHL of Norin 1 with and without supplementary UV radiation were 1.41 ± 0.32 and 4.48 ± 0.58 (mean ± SD), respectively. CHL of Sasanishiki with and without supplementary UV radiation were 2.92 ± 0.39 and 3.82 ± 0.61, respectively. CHL of Sasanishiki × Norin 1 and Norin 1 × Sasanishiki without supplementary UV radiation were 3.73 ± 0.41 and 4.26 ± 0.36, respectively. (Sato and Kumagai 1994, recomputed)

Table 3. Frequency distribution of the degree of browning observed in F_3 plants generated by self-fertilizing F_2 plants of Sasanishiki and Norin 1. (Sato and Kumagai 1994)

	Degree of browning					No. of plants
	1	2	3	4	5	
F_3 Plants	2	60	341	1210	1542	3155
Sasanishiki	16	29	30	7	0	82
Norin 1	0	0	0	13	75	88

from the cross Sasanishiki and Norin 1, were used for genetic analysis of the inhibitory effects of UV radiation. Each line consisted of 30 individuals. There was a negative correlation between the degree of browning and the mean of chlorophyll content per unit fresh weight of plants. The mean of chlorophyll content per unit fresh weight (mg/g FW) at each grade were 3.27 at grade 1, 2.79 at grade 2, 2.34 at grade 3, 1.69 at grade 4, and 0.95 at grade 5. Therefore, we considered that the degree of browning observed could be used as an index of chlorophyll content. The degree of browning observed in the fourth leaf was scored according to the following grades: grade 1 no bowning, grade 2 less than 20% of area of the fourth leaf browned, grade 3 from 20% to 50% of the area of the fourth leaf browned, grade 4 from 50 to 80% of the area of the fourth leaf browned, and grade 5 more than 80% of the area of the fourth leaf browned. Table 3 shows the frequency distribution of the degree of browning of Sasanishiki, Norin 1, and F_3 plants when they were grown for 20 days in visible radition with supplemental UV radiation. Sasanishiki exhibited scores ranging from 1 to 4, whereas Norin 1 showed scores of only 4 and 5. Based on these observations, the individual F_3 plants showing grade 1 to 3 were regarded as being resistant and those showing grade 4 and 5 were regraded as being sensitive. The phenotypes of F_3 lines were classified into three types corresponding to the genotypes of F_2 plants: resistant homozygotes, segregated heterozygotes and sensitive homozygotes. All lines consisting of individual plants exhibiting grade 1–3 and grade 4–5 were designated as resistant and sensitive homozygotes, respectively. The lines containing the characteristics of those two groups were designated as segregated heterozygotes. The segregation ratio of resistant homozygous lines, segregated heterozygous lines, and sensitive homozygous lines was 1:65:16. These values did not fit the segregation ratio of 1:2:1. Therefore, the resistance to the inhibitory effects of UV radiation in the Japanese lowland rice cultivars examined here could be controlled by recessive polygenes.

The results indicated that Sasanishiki, which was bred to be resistant to cool summer damage, must have acquired resistance from some progenitor during the breeding. However, it is not clear why Sasanishiki has strong resistance to UV radiation, and it may be impossible to trace the process of breeding. Consequently, it is ;indispensable that the relationship between the genetic mechanism and physiological mechanism of the resistance to the inhibitory effect of UV radiation be resolved.

Effect of Enhanced Ultraviolet-B Radiation on Growth and Production of Rice Under Greenhouse and Field Conditions

Q. DAI, S. PENG, A.Q. CHAVEZ, and B.S. VERGARA

The potential damage caused by the decrease in stratospheric ozone and the resulting increase in UV-B radiation has received widespread recognition. In higher plants, reductions in leaf area, plant height, and biomass production have been reported in a number of UV-B-sensitive species (Teramura 1983). Effects on physiological and biochemical process have also been studied (Tevini and Teramura 1989), including photosynthetic capacity (Ziska et al. 1993), respiration (Teramura et al. 1980), root activity (Dai et al. 1992), ion absorption (Murali and Teramura 1985), superoxide dismutase (SOD) and catalase activities (Krizek et al. 1993), and malondialdehyde (MDA) and polyamine contents (Kramer et al. 1991). In addition to the direct effects on crops, enhanced UV-B also influences crop production indirectly by altering competitive relationships with weeds (Gold and Caldwell 1983). Barnes et al. (1988), working on wheat and wild oat, demonstrated that UV-B-induced shifts in the balance of competition between these species were associated with differential effects of UV-B on shoot morphology, which led to changes in relative leaf display in mixed canopies.

To date, most information regarding the impact of UV-B on plants is on temperate species. Little is known of the effects of enhanced UV-B on rice, which provides about 3 billion people with approximately 70% of their calorie requirements (De Datta 1981). The levels of ambient UV-B radiation in tropical and high-altitude regions are greater than in other areas (Bachelet et al. 1991); therefore, it is important to test the sensitivity of rice to elevated UV-B in the tropics. Several experiments have examined the influence of enhanced UV-B on rice (Teramura et al. 1991; Dai et al. 1992; Barnes et al. 1993; He et al. 1993; Nouchi et al. 1993). Teramura et al. (1991), using 16 rice cultivars from seven different geographical regions, reported that more than one-third of the cultivars significantly reduced total biomass production under 15.7 kJ/m^2 of daily UV-B radiation. Dai et al. (1992) found that UV-B-sensitive rice cultivars significantly decreased their leaf area, dry weight, and dry matter production. All these experiments were conducted under controlled environments using UV-B-emitting lamps in a growth chamber or greenhouse. Although the results from greenhouse and growth chamber studies can help determine the relative sensitivity of different cultivars to UV-B, the quantitative assessment of yield reduction

APPA Division, International Rice Research Institute, P.O. Box 933, 1099 Manila, Philippines

S. Peng et al. (Eds.) Climate Change and Rice

in UV-B treated rice crop would require additional field experimentation. Little information is available with regard to field experiments. To evaluate the response of rice (growth and yield production) to elevated UV-B radiation under a natural environment, several field experiments were conducted.

Materials and Methods

Greenhouse Experiments

Seeds of IR74, a lowland rice cultivar widely planted in Asia, were pregerminated at 30 °C for 24 h and then sown in 1-l plastic pots containing 1.7 kg of Maahas clay soil (Andaqueptic Haplaquoll) fertilized with 0.4 gN, 0.04 g P, and 0.25 g K per kilogram of soil. Plants were grown in a greenhouse for 10 days and then subjected to UV-B for 3 weeks in an unshaded, temperature- and humidity-controlled greenhouse (day-night temperature 27/21 °C, RH 70%) at IRRI. Photosynthetic photon flux (PPF 400–700 mm) at the top of the rice canopy at solar noon on a clear-sky day in the greenhouse, as measured with a quantum sensor (Li-Cor Model Li-185b, Lincoln, NE), was aproximately 940 μmol m^{-2} s^{-1}. The UV-B setup and measurements were the same as reported by Dai et al. (1992). Irradiance of UV-A was determined at 1-nm intervals from 321 nm to 400 nm and summed every 10 nm. The ratio of UV-A to UV-B was the quotient of total UV-A over biologically effective UV-B.

A randomized complete block (RCB) design with four replications was used. The sample pots were placed within a uniform radiation area determined by a UVX radiometer with a UVX 31 sensor (San Gabriel, CA). Irradiation varied by less than 10% within the sample area. The average biologically effective UV-B radiation (UV-B_{BE}) in the UV-B treated plots, when weighted according to the general plant-action spectrum of Caldwell (1971) and normalized to unity at 300 mm, was 13.0 kJ/m^2 per day. Control plants received no UV-B irradiance because of the natural characteristic of the glass roof. After 3 weeks of UV-B treatment, six plants (two pots) from each replication were sampled for growth analysis. Plant height, tiller number, leaf area, leaf dry weight, and sheath dry weight were determined. Plant dry weight was the sum of leaf and sheath dry weight.

Data were analyzed by ANOVA. The difference between treated and corresponding control (p<0.05) were analyzed by LSD test. Greenhouse experiments were conducted twice.

Field Experiments

The UV-B exposure facilities in the IRRI Experimental Farm consisted of 24 sets of lamp frames that use a ballast system. Each lamp frame consisted of 12 preaged (100 h) UV-emitting fluorescent lamps (UV-B 313, Q-Panel, Cleveland, OH) spaced 30 cm apart. The lamps were kept 60 cm above the canopy of the rice plant throughout the experiment and were enclosed by either presolarized 0.13-mm thick cellulose diacetate (transmission down to 290 nm) for the UV-B treatment or 0.13-nm thick clear polyester film (optically equivalent to Mylar D, absorbing almost all radiation below 320 nm) for the control treatment. The plastic films or the UV-B and treatments were replaced with new ones weekly. The UV-B irradiance was measured with a double-monochromator spectroradiometer (Optronic Model 752, Orlando, FL). Wavelength alignment of the instrument was checked at 312.9 nm by a low-pressure mercury-vapor lamp. The instrument was calibrated for absolute responsivity against a 200 W tungsten-halogen standard lamp (Optronic Model OL 752–10, Orlando, FL). The average biologically effective UV-B radiation (UV-B_{BE}) in the UV-B treated plots was 13.0 kJ/m^2 per day (as in the greenhouse experiments). Control plants received ambient UV-B irradiance. The effective sampling area was determined using a UVX radiometer with a UVX 31 sensor (San Gabriel, CA). The UV-B irradiation varied by less than 10% within the sampling area. The UV-A (321–400 nm) and the ratio of UV-A to UV-B were measured and calculated as in the greenhouse studies.

The first field experiment was conducted at the IRRI farm during the 1992 dry season. Two lowland rice cultivars (IR64, a less sensitive cultivar, and IR74, a sensitive cultivar, identified from previous greenhouse experiment, Dai et al. 1992) were used. A split-plot design with UV-B treatment as the main plot and cultivar as the subplot was employed. Each treatment had five replicates. Pregerminated seeds were sown in seedling trays in the greenhouse for 15 days and then transplanted in the field at 20 × 20 cm spacing. In the wet season of 1992 and the dry and wet seasons of 1993, IR72 (less sensitive) and IR74 were used in the field studies conducted in the same site with 15 × 15 cm plant spacing. This time, the experiment design was factorial in RCB with six replications.

Basal fertilizer rates for dry seasons were at 60–40–40 kg (N-P-K) per hectare, and 30–30–30 kg (N-P-K) per hectare for wet seasons. Nitrogen (60 and 30 kg/ha for dry and wet seasons, respectively) was top dressed at middle tillering and panicle initiation (PI). At the flowering stage, 40 kg N/ha was top dressed for the dry season experiments. Normal water and crop-protection management was implemented.

At the middle tillering, panicle initiation, and flowering stages, 12 hills from each replication were sampled. Plant height, tiller number, leaf area index (LAI), specific leaf weight (SLW), and bioass production (leaf and stem dry weight) were measured. The contents of UV-B-absorbing compound (flavonoid) in the youngest fully expanded leaves were also determined at these growth stages using the method of Murali and Teramura (1986a).

Photosynthetic rate (A) and transpiration rate (E) were measured over a 20–s period with a portable photosynthesis system (LI-6200, LI-COR, Lincoln, NE). Stomatal conductance (g) to water vapor was calculated based on measured T, vapor pressure deficit, and a predetermined boundary layer conductance (Anon. 1987) at various growth stages. At maturity, plant height, spikelet filling precentage, 1000-grain weight, panicle number per m^2, and grain yield were determined from 36-hill samples from each replication. The data were analyzed by SAS (1987) and differences (p<0.05) between UV-B treated and control plants were determined by LSD test.

Results

In the greenhouse experiments, enhanced UV-B radiation had a significant effect on plant height, tiller number, leaf area, leaf dry weight, sheath dry weight, and plant dry weight, but there was no significant change in tiller number during the wet season of 1992 (Table 1). The results of the field experiments were completely different from the greenhouse studies. At the middle tillering (Table 2), panicle initiation (Table 3), and flowering stages (Table 4), enhanced UV-B had no significant effect on plant height, tiller number, leaf area, LAI, SLW, leaf dry weight, stem dry weight, and plant dry weight in two cultivars tested for three consecutive seasons. However, amounts of UV-B-absorbing compound (presumably flavonoids) generally differed significantly between UV-B treated plants and control. In the 1992 wet season, UV-B-treated IR72 had 24–29% more UV-B-absorbing compound than the control, whereas UV-B-treated IR74 had 5–14% higher levels of UV-B-absorbing compound than the control. In general, production of UV-B-absorbing compound was greater at middle tillering and panicle initiation than at flowering stage.

At maturity, enhanced UV-B had no significant effects on plant height, spikelet filling percentage, 1000-grain weight, panicle number per m^2, and grain yield in both IR72 and IR74 for four seasons, except that IR74 had reduced grain yield (12%) in the 1992 dry season (although this difference was not statistically significant). Variation in rice yields among seasons was observed. The grain yields in the dry seasons were greater than that in wet seasons (Table 5).

Table 1. Effect of enhanced daily UV-B radiation (13.0 kJ/m^2) on plant height (PH), tiller number (TN), leaf area (LA), leaf dry weight (LDW), sheath dry weight (SDW), and plant dry weight (PDW) of IR74 in the greenhouse. Each value is the average of 24 plants[a]

UV-B treatment	PH (cm)	TN	LA (cm^2)	LDW (g/plant)	SDW (g/plant)	PDW (g/plant)
6 March 1992 (dry season)						
Treated	55.3	8.8	260.9	0.6774	0.4525	1.1299
Control	64.1	10.3	310.2	0.8794	0.6184	1.4978
Change (%)	–13.6*	–14.6*	–15.9*	–23.0**	–26.8**	–24.6**
8 September 1992 (wet season)						
Treated	53.0	5.7	163.8	0.3490	0.2405	0.5895
Control	60.5	6.0	189.8	0.4077	0.2989	0.7066
Change (%)	–12.4*	–5.6	–13.7*	–14.4*	–19.5**	–16.6*

[a] The* and** respresent significance at p<0.05 and p<0.01, respectively.

Table 2. Effect of enhanced UV-B radiation on the growth of rice plants at the middle tillering stage under field conditions. Each value is the average of six replicates[a]

UV-B		PH (cm)	Tillers/m² (no.)	LAI	SLW (mg/cm²)	Dry weight (g/m²)			Flav. absorb. (305 nm)
						Leaf	Stem	Total	
Wet season 1992									
IR72	Treated	53.3	1137	3.21	2.63	85.5	65.0	150.5	0.228
	Control	54.5	1101	3.01	2.74	84.5	64.5	149.0	0.182
	Change (%)	–2	3	7	–4	1	1	1	25**
IR74	Treated	46.7	1275	2.83	2.64	74.0	55.7	129.7	0.240
	Control	48.7	1222	2.87	2.62	76.5	55.0	131.6	0.210
	Change (%)	–4	4	–1	1	–3	1	–1	14*
Dry season 1993									
IR72	Treated	43.7	1034	3.20	2.63	84.1	69.7	153.7	0.211
	Control	44.2	992	3.30	2.59	85.3	71.6	156.9	0.191
	Change (%)	0	4	–3	2	–1	–3	–2	10*
IR74	Treated	45.3	1177	4.51	2.59	117.1	96.0	213.1	0.164
	Control	46.3	1198	4.59	2.53	115.9	94.8	210.6	0.154
	Change (%)	–2	–2	–2	2	1	1	1	6
Wet season 1993									
IR72	Treated	53.9	599	2.64	2.97	78.4	92.3	170.7	0.124
	Control	54.7	599	2.74	3.00	82.2	95.0	177.7	0.102
	Change (%)	–1	0	–4	–1	–5	–3	–4	22**
IR74	Treated	50.4	661	3.04	2.76	83.4	106.9	190.3	0.121
	Control	50.2	611	2.96	2.58	76.0	103.3	179.3	0.105
	Change (%)	0	8	3	7	10	3	6	15**

[a] PH plant height, LAI leaf area index, and SLW specific leaf weight. The * and ** represent significance at $p < 0.05$ and $p < 0.01$, respectively.

The photosynthetic rate, stomatal conductance, transpiration rate, and internal CO_2 concentration in IR74 were not affected by enhanced UV-B radiation (Table 6). Irradiance at 290–303 nm in the field was smaller than in the greenhouse, whereas, irradiance at 304–320 nm was greater in the field than in the greenhouse, even when the $UV\text{-}B_{BE}$ levels were the same in the greenhouse and field (Table 7). The sum of UV-A (321–400 nm) in the field was 7.55 times greater than in the greenhouse. Irradiance at the 351–380 nm in the field was about 12 times greater than in the greenhouse. The ratios of UV-A to UV-B in the field and in the greenhouse were 11.8 and 2.5, respectively. The ratios of UV-A to $UV\text{-}B_{BE}$ in the field and greenhouse were 61.7 and 8.2, respectively.

Discussion

Enhanced UV-B radiation significantly reduced many components of growth and morphology of IR74 in the greenhouse, although the magnitude of the response differed among the parameters. Plant height, leaf area, and plant dry

Table 3. Effect of enhanced UV-B radiation on the growth of rice plants at the panicle initiation stage under field conditions. Each value is the average of six replicates[a]

UV-B		PH (cm)	Tillers/m² (no.)	LAI	SLW (mg/cm²)	Dry weight (g/m²)			Flav. absorb. (305 nm)
						Leaf	Stem	Total	
Wet season 1992									
IR72	Treated	80.8	961	6.79	3.30	231.2	258.1	489.3	0.094
	Control	83.0	986	7.27	3.28	244.6	269.4	514.0	0.073
	Change (%)	−3	−3	−7	1	−5	−4	−5	29**
IR74	Treated	79.6	1246	7.61	3.02	236.8	251.0	487.7	0.098
	Control	81.0	1227	7.07	3.21	232.6	245.7	478.2	0.089
	Change (%)	−2	2	8	−6	2	2	2	10*
Dry season 1993									
IR72	Treated	73.6	929	6.90	3.80	261.8	400.8	683.3	0.107
	Control	76.5	935	6.90	3.86	266.6	396.5	686.2	0.094
	Change (%)	−4	−1	0	−2	−2	1	0	14**
IR74	Treated	70.1	986	7.80	3.64	285.2	386.4	690.6	0.131
	Control	69.6	939	7.80	3.58	279.9	380.2	678.8	0.126
	Change (%)	1	5	0	2	2	2	2	4
Wet season 1993									
IR72	Treated	77.4	577	4.55	3.64	165.0	245.0	418.5	0.071
	Control	75.1	541	4.14	3.67	150.9	246.4	406.6	0.066
	Change (%)	3	7	10	−1	9	0	3	8
IR74	Treated	73.2	524	4.22	3.62	152.5	286.2	453.6	0.079
	Control	76.5	493	4.48	3.63	162.0	295.9	472.5	0.071
	Change (%)	−4	6	−6	0	−6	−3	−4	11*

[a] PH plant height, LAI leaf area index, and SLW specific leaf weight. The * and ** represent significance at $p < 0.05$ and $p < 0.01$, respectively.

weight were more sensitive to UV-B than tiller number (Table 1). This agrees with previous findings (Teramura et al. 1991; Dai et al. 1992; Barnes et al. 1993).

Most information concerning the effects of enhanced UV-B radiation on plants are from growth chamber or greenhouse experiments. Few studies have been conducted under field conditions. It is not possible to directly extrapolate the grain-yield responses to UV-B radiation in the field from greenhouse experiments because plants respond differently to UV-B in each environment (Teramura and Murali 1986; Tevini and Teramura 1989). Conflicting results on the effects of UV-B on crop productivity have been obtained. The results of a 6-year field experiment conducted in Maryland, US, demonstrated that enhanced UV-B radiation affected soybean yields using two cultivars (Teramura et al. 1990). Essex, a UV-B sensitive cultivar, had reduced grain yield of 19–25% for 4 years under simulated conditions of 25% ozone depletion, whereas no significant changes in yield were recorded for 2 years of prolonged drought stress. However, neither photosynthesis nor growth were affected by supplemental UV-B radiation in Williams, a UV-B-insensitive cultivar. In another field study with six soybean varieties simulating a 32% enhancement of UV-B irradiance, no influence on seed yield was detected (Sinclair et al. 1990). Ziska et al. (1992), in a field experiment working on cassava, found no significant changes in total biomass,

Table 4. Effect of enhanced UV-B radiation on the growth of rice plants at the flowering stage under field conditions. Each value is the average of six replicates[a]

UV-B		PH (cm)	Tillers/m^2 (no.)	LAI	SLW (mg/cm^2)	Dry weight (g/m^2) Leaf	Stem	Pan.	Flav. absorb. (305 nm)
Wet season 1992									
IR72	Treated	100.9	561	6.20	4.09	266.9	513.6	115.1	0.103
	Control	101.6	600	6.46	4.08	276.8	526.9	111.4	0.083
	Change (%)	–1	–7	–4	0	–4	–3	3	24**
IR74	Treated	108.3	535	6.99	4.27	314.5	577.5	98.8	0.077
	Control	109.5	517	7.05	4.19	314.9	569.5	93.6	0.073
	Change (%)	–1	4	–1	2	0	1	6	5
Dry season 1993									
IR72	Treated	89.9	706	7.30	4.89	375.0	611.2	166.0	0.095
	Control	91.8	685	7.70	4.89	356.2	614.0	159.9	0.096
	Change (%)	–2	3	–5	0	5	0	4	–1
IR74	Treated	104.9	727	10.20	4.31	439.6	714.9	152.5	0.111
	Control	105.2	703	9.80	4.51	440.4	734.5	156.6	0.116
	Change (%)	0	3	4	–4	0	–3	–3	–4
Wet season 1993									
IR72	Treated	87.6	402	4.44	4.42	196.6	458.8	105.0	0.067
	Control	88.9	416	4.62	4.47	206.0	459.4	107.7	0.066
	Change (%)	–1	–3	–4	–1	–5	0	–3	2
IR74	Treated	98.8	411	5.28	4.47	236.1	424.6	110.6	0.076
	Control	99.4	392	5.31	4.53	239.6	502.7	104.9	0.073
	Change (%)	–1	5	–1	–1	–1	4	5	4

[a] PH plant height, LAI leaf area index, SLW specific leaf weight, Pan panicle. The ** represents significance at $p < 0.01$.

Table 5. Response of rice yield and yield components to enhanced daily UV-B radiation (13 kJ/m^2) under field conditions for four consecutive seasons

		UV-B	Plant height (cm)	Filled spikelet (%)	1000-grain weight (g)	Panicles/m^2	Yield (t/ha)[a]
DS 1992[b]	IR64	Treated	100.8	89	24.1	278	7.0
		Control	101.5	89	24.7	272	7.2
	IR74	Treated	92.7	84	24.2	283	6.3
		Control	92.1	86	24.3	314	7.2
WS 1992	IR72	Treated	103.5	70	22.7	427	5.7
		Control	104.5	64	22.3	473	5.2
	IR74	Treated	92.8	67	22.2	519	4.8
		Control	94.4	67	22.4	480	4.5
DS 1993	IR72	Treated	97.8	78	21.6	589	7.7
		Control	98.3	78	21.3	590	7.6
	IR74	Treated	94.8	84	20.2	648	8.5
		Control	94.4	84	20.3	626	8.4
WS1993	IR72	Treated	91.0	71	25.1	374	4.2
		Control	91.3	65	25.4	358	4.3
	IR74	Treated	87.3	68	23.4	399	4.4
		Control	89.4	79	23.1	379	4.7

[a] Based on 14% moisture content.
[b] DS dry season, WS wet season.

Table 6. Response of photosynthetic rate (A), stomatal conductance (g), transpiration rate (E), and internal CO_2 concentration (*Ci*) of IR74 to enhanced daily UV-B radiation (13 kJ/m²) in the field[a]

	Date measured (DAT[b])	Observations (n)	UV-B	A (μmol CO_2 m^{-2} s^{-1})	g (mol CO_2 m^{-2} s^{-1})	E (mmol H_2O m^{-2} s^{-1})	C_i (ppm)
DS 1992	70	15	Treated	23.1	1.57	16.8	287
			Control	22.9	1.54	15.6	283
WS 1992	26	10	Treated	24.0	1.28	19.7	294
			Control	23.3	1.54	18.1	283
	48	8	Treated	25.3	2.11	17.1	293
			Control	24.1	1.77	15.4	295
	63	5	Treated	20.5	1.54*	16.8	276
			Control	20.8	1.26	16.5	275
DS 1993	45	16	Treated	18.5	1.24	15.5	285
			Control	18.3	1.27	15.6	290

[a] The * represents significance at $p < 0.05$.
[b] Days after transplanting.

plant height, leaf dry weight, and stem biomass production under simulated conditions of 15% ozone depletion. However, the biomass of root (harvest organ of cassava) and the root-shoot ratio was decreased significantly under UV-B treatment. In a field study in Germany using filtered lamps to simulate 10% and 25% ozone reductions, no UV-B effects were demonstrated in three cabbage cultivars, in lettuce, or in rapeseed. Biggs and associates grew ten crop species in Gainesville, Florida (29°N) and found yield reductions between 5 and 90% in half of them, among them crops like wheat (–5%), potato (–21%), and squash (–90%), whereas rice, peanut, and corn were unaffected (Biggs and Kossuth 1978; Biggs et al. 1984). Because flavonoid contents increased in most plants, UV-B might have an effect on food quality; however, this was not specifically tested (Dumpert and Knacker 1985).

Results of four consecutive seasons of field studies conducted at IRRI indicate that enhanced UV-B radiation (about 20% ozone depletion) had no effect on rice growth parameters and grain yield of IR72 and IR74 under field conditions (Tables 2–5). No effects on photosynthetic rate, stomatal conductance, transpiration rate, and internal CO_2 concentration were detected (Table 6). However, IR74 showed a 25% reduction in dry matter in the greenhouse experiment under UV-B treatment for 3 weeks at the same level of simulated UV-B (Table 1).

This discrepancy may be partly caused by the different background UV-B levels in the field and greenhouse studies. The UV-B-treated plants in both greenhouse and field experiments were exposed to the same simulated level of UV-B_{BE}, whereas the control plants received no UV-B in the greenhouse but received ambient UV-B radiation in the field. Therefore the degree of UV-B enhancement for UV-B-treated plants over control plants in the greenhouse was

Table 7. Irradiance of UV-B (290–320 nm) and UV-A (321–400 nm) spectra in both greenhouse and field under UV-B-emitting lamps

nm	Field (F) ($W\ cm^{-2}\ nm^{-1}$)	Greenhouse (G) ($W\ cm^{-2}\ nm^{-1}$)	F/G
290	1.2921E-08	1.3758E-08	0.94
291	2.9589E-08	3.5161E-08	0.84
292	7.3596E-08	9.1777E-08	0.80
293	1.7969E-07	2.3249E-07	0.77
294	3.9872E-07	5.3079E-07	0.75
295	8.0667E-07	1.0819E-06	0.75
296	1.3854E-06	1.8648E-06	0.74
297	2.1552E-06	2.8850E-06	0.75
298	3.0373E-06	3.9939E-06	0.76
299	3.9698E-06	5.1186E-06	0.78
300	4.9047E-06	6.1499E-06	0.80
301	6.0071E-06	7.2500E-06	0.83
302	7.2510E-06	8.2795E-06	0.88
303	8.7668E-06	9.1868E-06	0.95
304	9.9747E-06	9.5923E-06	1.04
305	1.1151E-05	1.0007E-05	1.11
306	1.2413E-05	1.0444E-05	1.19
307	1.3802E-05	1.0871E-05	1.27
308	1.5420E-05	1.1194E-05	1.38
309	1.6672E-05	1.1584E-05	1.44
310	1.7747E-05	1.1809E-05	1.50
311	2.0517E-05	1.2172E-05	1.69
312	2.2827E-05	1.2793E-05	1.78
313	2.5007E-05	1.4120E-05	1.77
314	2.6084E-05	1.3614E-05	1.92
315	2.6107E-05	1.2560E-05	2.08
316	2.6794E-05	1.2434E-05	2.15
317	2.8455E-05	1.2256E-05	2.32
318	3.0951E-05	1.2440E-05	2.49
319	3.1063E-05	1.2313E-05	2.52
320	3.2662E-05	1.2260E-05	2.66
Sum	4.0663E-04	2.4918E-04	1.63
$UV\text{-}B_{BE}$ [a]	7.7640E-05	7.7655E-05	1.00
321–330	4.0660E-04	1.1203E-04	3.63
331–340	4.7655E-04	7.9232E-05	6.01
341–350	5.1538E-04	5.3503E-05	9.63
351–360	5.7249E-04	4.5714E-05	12.52
361–370	6.9312E-04	6.2777E-05	11.04
371–380	6.9078E-04	5.0496E-05	13.68
381–390	6.4754E-04	6.7640E-05	9.57
391–400	7.8844E-04	1.6303E-04	4.84
Sum	4.7909E-03	6.3442E-04	7.55

[a] Weighted according to the general plant-action spectrum of Caldwell (1971) and normalized to unity at 300 nm.

greater than in the field. The irradiance of certain UV-B spectra, especially 290–303 nm, in the greenhouse was also greater than in the field (Table 7). This portion of UV-B has more damaging effects on biological system than longer wavelength (Tyrrell 1986). Furthermore, the ratio of UV-A to $UV\text{-}B_{BE}$ in the field was 7.5

times greater than in the greenhouse (Table 7). Higher amounts of UV-A radiation can ameliorate the damaging effect induced by UV-B through the operation of photoreactivation and DNA repair (Tyrrell 1986; Middleton and Teramura 1993). Finally, visible solar radiation is much greater in outdoor experiments. Studies have shown that plants grown under high PAR for a period of time became less sensitive to UV-B (Warner and Caldwell 1983; Mirecki and Teramura 1984). Protective mechanisms may be less developed under controlled and lower conditions of visible light, especially the DNA repair. There is widespread DNA repair in the photoenzymatic mechanism for removal of pyrimidine dimers (the most common DNA photoproduct by radiation in the 250–300 nm range). If the normal cell was exposed to an artifical source of UV-B, lacking the longer wavelength radiation present in daylight, it would appear more sensitive. Because daylight contains a strong complement of the effective radiation, DNA repair should be strongly stimulated during natural irradiance (Rupert 1986). The repair or protective mechanisms stimulated by UV-A or high visible solar radiation must be confirmed for rice plants.

Conclusion

Enhanced daily UV-B radiation (13.0 kJ/m^2) had a significant effect on the growth and dry matter production of rice plants in greenhouse experiments. However, the significant effect was not evident in field experiments that simulated the same UV-B level. The reason for this discrepancy may be the higher amount of UV-A in the field and the greater amount of irradiance in the 290–303 nm region in the greenhouse.

Acknowledgments. The research described in this chapter was funded by United States Environmental Protection Agency, under agreement number CR 817426 to the International Rice Research Institute. It has, however not been subject to review by the Agency and therefore does not necessarily reflect the views of the Agency, and no official endorsement should be inferred.

Estimation of Biologically Effective Ultraviolet-B Radiation in the Tropics Using Action Spectra of Higher Plants

Y.-P. Cen

The potential effect on the biosphere of increased ultraviolet radiation is cause for concern. To estimate and compare the radiation stress on plants under various conditions, both the energy-distribution spectra of the radiation and a weighting function to describe the relative importance of various wavelength components are needed. The biologically effective UV-B radiation (UV-B_{BE}) as defined by Caldwell (1971) expresses the amount of UV radiation as a function of the biological effectiveness of the radiation. With UV-B_{BE}, it is easier to compare results from various sources of UV-B irradiation in laboratory and field studies using both artificial and solar UV-B sources. The generalized plant UV-B action spectrum (Caldwell 1968, 1971) is widely used as weighting function to simplify comparisons. It is also recommended that the action spectrum be normalized to 1.0 at 300 nm for comparison with other action spectra.

Several action spectra on higher plants published by Caldwell (1971), Negash and Björn (1986), Steinmetz and Wellmann (1986), Quaite et al. (1992), and the newly constructed action spectrum for the enhancement of ultraweak luminescence by UV radiation in leaves of *Brassica napus* by Cen and Björn (1994) were used to calculate UV-B_{BE} for simulated daylights in Ecuador (0°N, 81°W) and IRRI (14.1°N, 121.2°E). A simulated daylight in a temperate area (Sweden, 55.7°N, 13.4°E) was also included for comparison. The objectives of this study were to compare UV-B radiation and UV-B_{BE} in the tropics with temperate regions to determine the variation in estimated UV-B_{BE} using different action spectra of higher plants under normal and decreased stratospheric ozone levels.

Solar Ultraviolet Radiation in the Tropics

The stratospheric ozone levels are naturally lowest in equatorial regions where most ozone is produced (Stolarski 1988; McFarland and Kaye 1992). The daylight UV spectra is influenced by many factors (e.g., solar angle, atmospheric

Plant Physiology, APPA Division, International Rice Research Institute, P.O. Box 933, 1099 Manila, Philippines

S. Peng et al. (Eds.) Climate Change and Rice

conditions, albedo, and altitude). Higher UV-B radiation was demonstrated in the tropics where the latitude is lower, the stratospheric ozone layer is thinner, and the solar zenith angle is smaller (Fig. 1). A computer program (Björn and Murphy 1985; Björn 1989) was used to simulate daylight under normal and decreased stratospheric ozone levels. The simulated daily UV-B spectra in Ecuador at the equator, Philippines in the tropics, and Sweden in the temperate zone showed that the total amount of UV irradiation was higher at the equator and in the tropics. The UV-B spectra at the equator and in the tropics were more toward the short wavelengths compared with the spectrum of the temperate zone (Fig. 2).

Action Spectra

An action spectrum is the specific response of a system to radiation as a function of wavelength (Caldwell 1971; Jagger 1985; Caldwell et al. 1986). It gives information about the biomolecule responsible for the photon absorption and the relationship between the chromophores and respective responses (Peak and Peak 1983; Ensminger 1993). The UV action spectrum can be used as a weighting function to evaluate biologically effective UV-B radiation (Caldwell 1968, 1971).

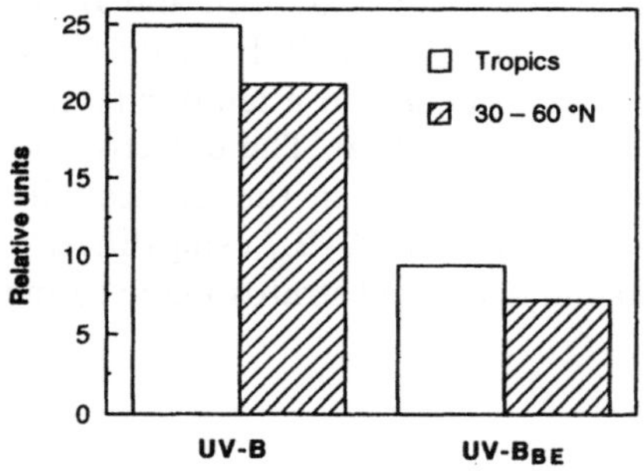

Fig. 1. The daily UV-B radiation and biologically effective UV-B (weighted by general action spectrum of Caldwell) radiation in tropics (0–20°N, 120°E) and other latitude areas (30–60°N, 120°E). The units of y-axis are Wh m^{-2} day^{-1} and kJ m^{-2} day^{-1} for UV-B and $UV\text{-}B_{BE}$, respectively. Daylight values were obtained using a computer program. (Björn and Murphy 1985; Björn 1989)

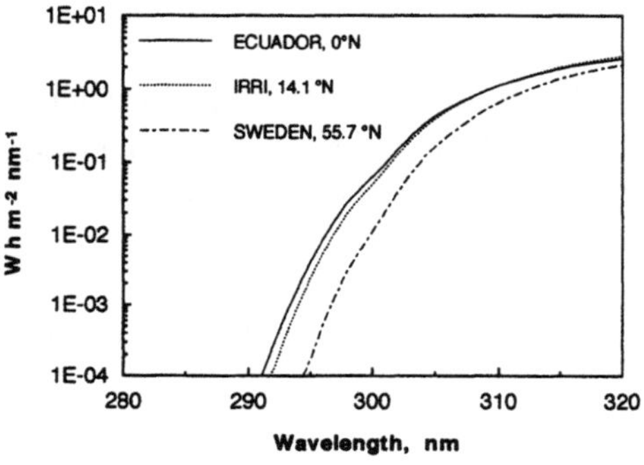

Fig. 2. The daily UV-B radiation on a clear sky day (August 5) in Ecuador (— 0°N), Philippines (••• 14.1°N), and Sweden (-•-• 55.7°N). Data were obtained by using a computer program. (Björn and Murphy 1985; Björn 1989)

However, a single action spectrum is inappropriate for evaluating different plant responses to UV-B radiation, and a separate action spectrum should be determined for a certain response for a particular plant species.

Ultraweak luminescence emitted from living organisms is positively correlated to the levels of free radicals in these organisms, and therefore the enhancement of ultraweak luminescence by UV radiation in plant tissues is also a measure of potential UV damage (Abeles et al. 1978; Abeles 1986; Campbell 1988; Hideg 1993). We have constructed two action spectra for the enhancement of ultraweak luminescence by UV radiation (270–340 nm) from both the adaxial and abaxial leaf surfaces of *Brassica napus* (Cen and Björn 1994). There was a similar spectral pattern for these two spectra except for wavelengths lower than 290 nm. Fairly steep slopes with increasing wavelength were seen in the shorter UV region (< 290 nm), compared with those for longer wavelengths (Cen and Björn 1994). The higher sensitivity of the abaxial leaf surface indicates the importance of the protective role of the adaxial epidermis against UV radiation (Cen and Bornman 1993). Different action spectra of higher plants normalized at 300 nm are presented in Fig. 3. (Caldwell 1971; Thimijan et al. 1978; Negash and Björn 1986; Steinmetz and Wellmann 1986; Quaite et al. 1992; Cen and Björn 1994). Different action spectra showed different wavelength sensitivities in the UV-B region.

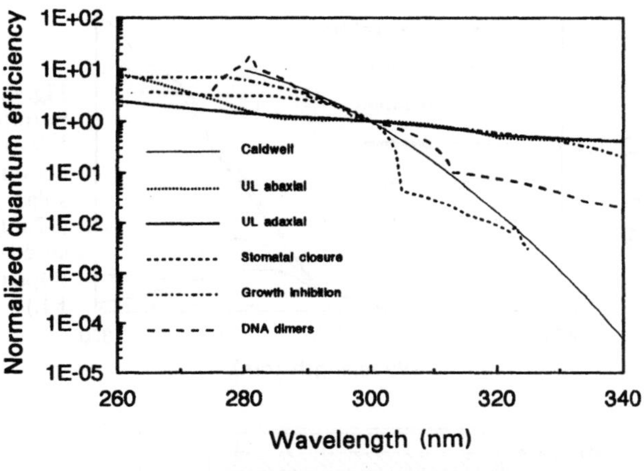

Fig. 3. Several action spectra for UV effects on higher plants. *Caldwell* Generalized plant-action spectrum (Caldwell 1971; Thimijan et al. 1978); *UL abaxial*; UL enhancement from the abaxial leaf surface (Cen and Björn 1993); *UL adaxial*; UL enhancement from the adaxial leaf surface (Cen and Björn 1994); *stomatal closure* (Negash and Björn 1986); growth inhibition, hypocotyl inhibition (Steinmetz and Wellmann 1986); DNA dimers (Quaite et al. 1992)

The Use of Action Spectra to Estimate UV-B_{BE}

The use of different action spectra as weighting functions resulted in different values of UV-B_{BE} for the same daylight UV-B radiation. The variations reached a factor of 11 among the action spectra used (Fig. 4). The increase in UV-B_{BE} in the tropics relative to temperate regions differed according to the action specta used, from 49% with hypocotyl inhibition (Steinmetz and Wellmann 1986) to 143% in stomata closure (Negash and Björn 1986). If stratospheric ozone depletion occurs, higher increases in daily UV-B_{BE} would be found in tropical than in temperate regions (Fig. 5). With 5% stratospheric ozone depletion, the daily UV-B_{BE} in the tropics would increase from 0.3 to 2.6 kJ/m^2 using different weighting functions compared with 0.1 to 2.0 kJ/m^2 in Sweden. The average increase in the daily UV-B_{BE} would be 1.5 kJ/m^2 Ecuador, 1.6 kJ/m^2 in the Philippines, and 1.1 kJ/m^2 in Sweden (Fig. 6). The maximum increase in the dialy UV-B_{BE} was weighted by hypocotyl inhibition (Steinmetz and Wellmann 1986); whereas, the minimum increase was weighted by stomatal closure (Negash and Björn 1986).

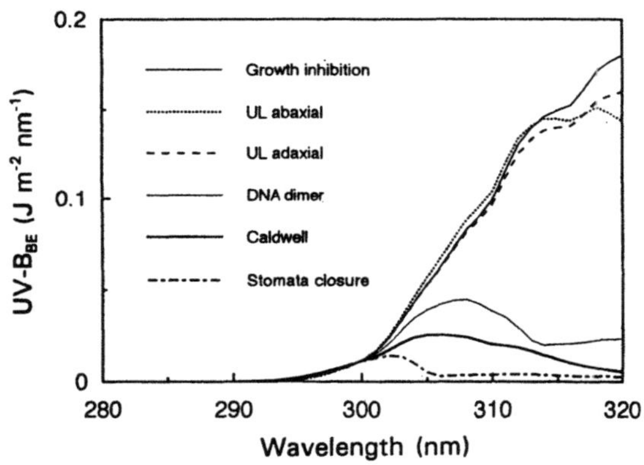

Fig. 4. The spectral energy distribution of biologically effective UV-B radiation weighted by the action spectra of higher plants for solar UV-B radiation at IRRI (12:00, 1 March). Daylight measured with a spectroradiometer. (Optronic Model 752, Orlando, FL)

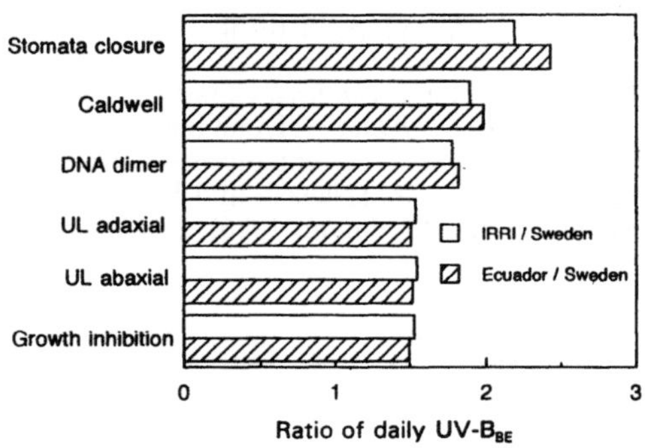

Fig. 5. The ratio of daily UV-B_{BE}, Ecuador over Sweden and Philippines over Sweden, weighted by different action spectra of higher plants. Daylight values were obtained using a computer program. (Björn and Murphy 1985; Björn 1989)

Fig. 6. The increase in daily UV-B_{BE} under 5% stratospheric ozone depletion at Ecuador, Philippines, and Sweden. The UV-B_{BE} values are weighted by different action spectra of higher plants. Daylight values were obtained using a computer program. (Björn and Murphy 1985; Björn 1989)

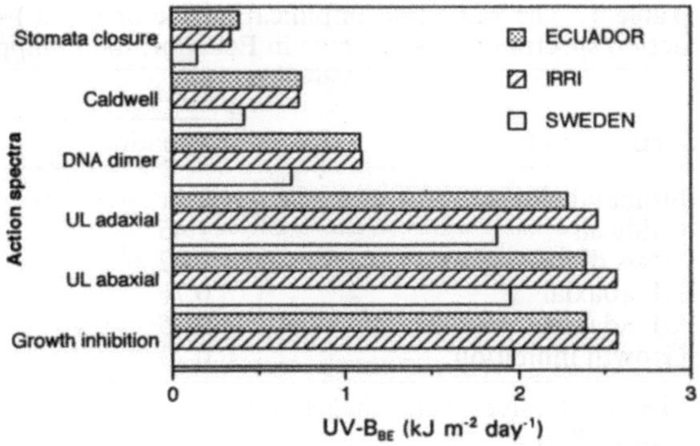

Fig. 7. The relative changes of solar UV-B radiation (◊) and UV-B_{BE} (◆, weighted by the generalized plant-action spectrum of Caldwell) because of ozone depletion at IRRI, Los Baños, Philippines (5 August). Daylight values were obtained using a computer program. (Björn and Murphy 1985; Björn 1989)

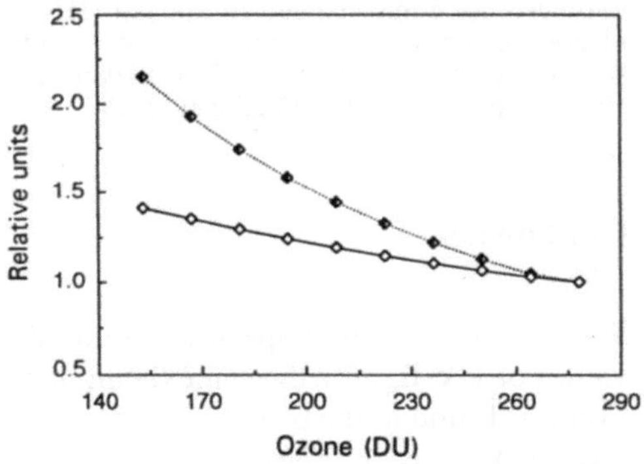

Sensitivity of UV-B_{BE} to Stratospheric Ozone Depletion

When all other conditions are the same, UV-B_{BE} shows a stronger dependence on ozone depletion than the unweighted solar UV-B radiation (Fig. 7). The sensitivity of UV-B_{BE} to ozone depletion can be described by the radiation amplification factor (RAF), which is the relative change of UV-B_{BE} resulting from a unit change in the stratospheric ozone layer (Caldwell 1981; Caldwell et al. 1989; Rundel 1983). The RAF varied from 0.73 to 2.11 in Ecuador, 0.77 to 2.10 in the Philippines, and 0.90 to 1.98 in Sweden according to the different action spectra of higher plants that were used (Table 1). The RAF values for different plant responses also differed between tropical and temperate areas. The RAF values for stomatal closure were higher in tropical than in temperate areas, whereas the other bioresponses showed lower values in tropical regions.

Table 1. The radiation amplification factor (RAF) calculated[a] according to different action spectra of higher plants in Ecuador, the Philippines, and Sweden

	RAF		
Action spectra[b]	Ecuador	Philippines	Sweden
Stomatal closure	2.11	2.10	1.98
Caldwell	1.45	1.48	1.60
DNA dimer	1.22	1.26	1.40
UL abaxial	0.76	0.80	0.94
UL adaxial	0.75	0.79	0.92
Growth inhibition	0.73	0.77	0.90

[a] Daylight values were calculated using a computer program (Björn and Murphy 1985; Björn 1989)
[b] Caldwell, generalized plant action spectrum (Caldwell 1971); UL abaxial, ultraweak luminescence enhancement on abaxial leaf surface (Cen and Björn 1994); UL adaxial, ultraweak luminescence enhancement on adaxial leaf surface (Cen and Björn 1994); stomatal closure (Negash and Björn 1986); growth inhibition (Steinmetz and Wellmann 1986); and DNA dimer (Quaite et al. 1992).

Conclusion

When different action spectra were used as weighting functions, the estimated values of UV-B_{BE} for the same daylight radiation differed. Higher UV-B radiation was found in the tropics where the latitude is lower, the stratospheric ozone layer is thinner, and the solar zenith angle is smaller. When all other conditions are the same, UV-B_{BE} would increase more because of ozone depletion in the tropics than in temperate areas. However, the sensitivity of UV-B_{BE} to ozone depletion as described by the RAF, varied according to the action spectra of higher plants that were used. This may lead to different conclusions about the significance of solar UV-B radiation increase caused by stratospheric ozone depletion.

The generalized plant-action spectrum of Caldwell (1968, 1971) is commonly used to estimate UV-B_{BE} when studying the responses of higher plants to UV-B radiation. Variations in estimated UV-B_{BE} using different action spectra of higher plants may cause confusion with regard to the response of higher plants to UV-B radiation. To compare results between different research groups, it would be helpful to report the biologically effective UV-B radiation and the energy distribution spectra of UV-B radiation. Because no single weighting function is ideal for all situations, available action spectra can serve only as a guide to estimate the biological effectiveness of UV-B radiation.

Carbon Dioxide and Temperature

Global Carbon Cycle

J. Goudriaan

The concentration of greenhouse gases in the atmosphere will probably double at a time scale of a few human generations. This increase may induce climatic changes that will exceed the adaptive capacity of nature and of human society. The intensification of the otherwise beneficial natural greenhouse effect causes concern because it may lead to temperature increases of the order of a few degrees centigrade.

The dominating greenhouse gas is not CO_2, but water vapor. However, CO_2 has a central role because it fluctuates less with climate than water vapor does. Water vapor is a passive companion that tends to enhance the warming effect of other greenhouse gases.

Outline of the Global Carbon Cycle

Atmospheric CO_2 increases at a rather steady rate, at least on a decade time scale. During the last 50 years its rate of increase has grown exponentially at a rate of about 2.4% per year. A single descriptive equation for the atmospheric concentration of CO_2 (C) as a function of year number (t) is:

$$C = 285 + 52\exp[0.024(t - 1980)]. \qquad (\mu l/l)\ (1)$$

This equation does a reasonable job in fitting the observed values. It assumes a preindustrial level of 285 μl/l and it gives a level of 337 μl/l for the reference year 1980. Each μl/l of CO_2 in the atmosphere corresponds with an amount of carbon of about 2.1 Pg.[1] The rate of carbon emission to the atmosphere by combustion of fossil fuel (F) is approximated by:

$$F = 4.8\exp[0.024(t - 1980)]. \qquad (Pg/year)\ (2)$$

The mean rate of increase of atmospheric carbon has been about 55% of the rate of emission of CO_2 from burning fossil fuels. The proportion of emitted CO_2

Department of Theoretical Production Ecology, Agricultural University, POB 430, 6700 AK Wageningen, The Netherlands

[1] (1 Pg = 10^{15} g = 1 Gt).

S. Peng et al. (Eds.) Climate Change and Rice

remaining in the atmosphere, called the remanent fraction, is not a natural constant. It depends on the activity and absorptive capacity of the sinks for CO_2, such as the ocean and the biosphere. Also, if the rate of emission into the atmosphere deviates from the exponential increase that it has followed so far, the remanent fraction will change. Equations (3) and (4) allow for this effect by distinguishing between actual atmospheric carbon (A) and the equilibrium atmospheric carbon (A_{eq}). This equilibrium level corresponds with a fraction of 20% of the accumulated emission. This value is predicted, using the chemical buffering of CO_2 in seawater, and the presumed potential of the biosphere to sequester carbon. The instantaneous fraction remaining in the atmosphere is estimated to be 64%, but the actual atmospheric concentration will approach the lower equilibrium level with a time lag of 120 years:

$$dA_{eq} / dt = 0.20F \qquad \text{(Pg/year) (3)}$$

$$dA / dt = 0.64F + (A_{eq} - A) / 120 . \qquad \text{(Pg/year) (4)}$$

This set of equations must be initialized at a CO_2 concentration C of 290 μl/l (both A and A_{eq} at 610 Pg of carbon) in the year 1800, to give the actual level of 337 μl/l in 1980. The initial CO_2 concentration C (290 μl/l), is a bit higher than the historically correct value of 280 μl/l, but its positive offset is a simple way to allow for the biotic emission from large-scale deforestation and reclamation during the 20th century. The recent deforestation rate is compensated by the active role of the biosphere.

Equations (2), (3), and (4), and the atmospheric carbon content/concentration ratio of 2.1 Pg/(μl/l) together reproduce the single descriptive Eq. (1), and permit the study of alternative scenarios for the time course of the rate of the combustion of fossil fuels (F). Over the past several thousand years, atmospheric CO_2 has been relatively stable at about 280 μl/l until the effects of the industrial revolution became noticeable. Further back in time, during the last glacial period, atmospheric CO_2 was as low as 200 μl/l. This fluctuation of about 200 μl/l during glacial periods and 280–300 μl/l during interglacial periods has probably occurred several times during the last million years, when glacial periods occurred about every 100 000 years. The causes of these glacial periods are still debated, and probably consist of many cooperating factors (Archer and Maier-Reimer 1994).

At different time scales, different processes are important. On a time scale shorter than a decade, seasonal fluctuations dominate. The seasonal effect of carbon uptake by plant growth is prominent in atmospheric CO_2 records that have been obtained during the past 30 years, such as the well-known record of Mauna Loa. Most remarkably, this seasonal fluctuation is almost absent in the Southern Hemisphere. The best explanation is that carbon uptake by algal growth in the oceans is buffered by the large carbonate content of sea water. The alternative explanation that carbon uptake in the ocean is much smaller than that on land has been disproved recently by simultaneous measurements of atmospheric oxygen and CO_2. Until recently, the corresponding seasonal cycle of atmospheric oxygen could not be measured because of the very high background concentration. It has now been shown that the oxygen concentration cycles above both the continents and the oceans. Oxygen is not buffered in sea water in the same way as CO_2; therefore, it follows the seasonal cycle of photosynthetic absorption and emission by plants.

The annual sum of daily net growth rate of plant biomass of an ecosystem is called Net Primary Productivity (NPP). The global NPP of terrestrial and marine plant biomass are of the same order of magnitude, each about 50 Pg carbon. The total flux of about 100 Pg is drawn from the atmosphere that contains a total of about 700 Pg C. Therefore, a relatively small disturbance of global NPP can have a large temporary effect on the observed rate of increase of atmospheric CO_2 (Taylor 1993). About one-seventh of the whole CO_2 content of the atmosphere is recycled each year. This is such a short period of time that it seems that any disturbance caused by injecting CO_2 from fossil fuels should be absorbed without difficulty. This impression is reinforced when the large ocean pool of carbon is considered: almost 40 000 Pg C. However, this is a false impression.

Limited Absorption Capacity for Carbon Dioxide

The largest carbon reservoir on earth is rock carbonate. This huge reservoir participates in the global carbon cycle only on a geological time scale. In a process of granite rock weathering, carbonate and SiO_2 are formed from silicate and atmospheric CO_2, thereby removing CO_2 from the atmosphere. Eventually, this process will absorb practically all CO_2 emitted from the burning fossil fuels, but it will only do so after tens of thousands of years (Berner and Lasaga 1989).

The next largest reservoir of carbon in the carbon cycle is dissolved bicarbonate in the oceans. This reservoir is active on a decadal time scale. In comparison with carbon in the oceans, CO_2 in freshwater of lakes and rivers can be totally ignored. The amount of CO_2 that can dissolve in water is proportional to the partial pressure of CO_2 in the air above it. This is true for both freshwater and seawater, but in seawater bicarbonate ions have a much higher concentration than dissolved CO_2 because of the high pH (about 8.1). In fact, a chemical equilibrium exists in the chain $CO_2 \leftrightarrow HCO_3^{-1} \leftrightarrow CO_3^{-2}$, which shifts to the right with increasing pH (Butler 1982). When the CO_2 concentration in the air increases, more CO_2 dissolves and the equilibrium shifts towards HCO_3^{-1}. The HCO_3^{-1} releases H^+ ions; therefore, the pH goes down, offering strong negative feedback to further absorption. The net result is that incremental absorption of CO_2 by seawater is not proportional to partial pressure of atmospheric CO_2 and total amount of carbon dissolved (about 25 g/m^3), but to only about one-tenth of this product. Therefore, the effective carbon pool of ocean carbon is only about 4000 Pg, and 36 000 Pg can be considered as being chemically bound.

This 4000 Pg is much more carbon than the biosphere contains. The second reason for limited absorptive capacity is the slowness of mixing of deep ocean water and surface water. The upper few hundred meters of the ocean are in good contact with the atmosphere, but the deeper layers are separated by a thermal inversion and the exchange rate with them is slow. The average residence time of water in the deep Pacific is several hundred years. The water of the Atlantic is recirculated at a much faster rate, in less than 100 years' time. The overall effect is that of the 4000 Pg chemically accessible carbon, only about one-tenth is immediately in contact with the atmosphere. The other nine-tenths will absorb its portion in due time, but only after hundreds of years.

In the long run (a 1000 years), redistribution of emitted CO_2 will occur between the atmospheric reservoir of 700 Pg and the oceanic effective reservoir of 4000 Pg. A fraction of about 700/(700 + 4000) or about 15% will remain in the atmosphere. On a decadal time scale, this remanent fraction is much larger because of a lack of time for mixing. A similar calculation gives a remanent fraction of about 700/(700 + 400) or about 64%. This value is very close to the observed remanent fraction and can serve as a guideline for scenario studies.

The Role of the Terrestrial Biosphere

Deforestation releases from 1 to 2 Pg of C per year to the atmosphere in the form of CO_2. When this source is added to the known emission rate from burning fossil fuel, atmospheric CO_2 should rise faster than it actually does. This discrepancy is the problem of the so-called missing carbon. The total annual NPP of the terrestrial biosphere is about 50 Pg C per year, but it is worthwhile to consider this flux and its definition in more detail. The NPP is defined as the annual sum of the daily net growth rates of plant dry matter (expressed in carbon). This definition means that respiration of the vegetation has been taken into account. The flux before subtraction of respiration is the Gross Primary Productivity (GPP), which can be considered as the sum of gross CO_2 assimilation rates. However, the calculation of this flux is even more imprecise than that of NPP, because it is based on rough estimates of respiration. Estimates of NPP are based on field observations of production rates of new leaves, twigs, branches and roots and are therefore somewhat more reliable. The order of magnitude of global NPP is 50 Pg C per year, whereas that of GPP is about twice as much (Fig. 1). The NPP is largely consumed by grazers, such as insects and various kinds of mammals, or decomposed by fungi and microbes. Live biomass is estimated to be about 500 Pg, mainly in the form of wood. After death it turns to litter and eventually to soil organic matter (about 1500 Pg of C). The magnitudes of the pools of organic matter in various forms and of the fluxes of formation and consumption are related by residence times and partitioning coefficients. The outflow from each state variable is calculated as amount divided by longevity; therefore, its equilibrium level is simply the product of longevity and inflow. Their outflows cascade down to litter, humus, and resistant soil carbon. From biomass to humus, and from humus to resistant soil carbon, a considerable fraction of carbon is lost by

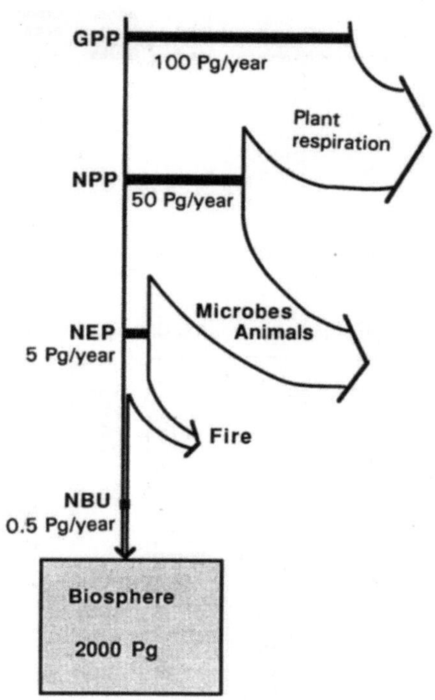

Fig. 1. Major terrestrial carbon fluxes, as consumed by the plants themselves, by heterotrophic organisms, and by fire. Net biospheric uptake is the remainder of large positive and negative fluxes

respiratory processes. The complement of this fraction returns to the atmosphere as respiratory CO_2. More complex approaches to decomposition of organic matter may be needed, but are not considered in this chapter.

Above-ground biomass consists of leaves, stemwood, roots, branches, and litter. Combination of the driving forces and residence times leads to a steady-state distribution of surface densities of carbon for a biosphere model that consists of six different ecosystem types (Fig. 2; Goudriaan 1990). In this figure, the width of each bar represents the area of the vegetation type, and the height of each bar the carbon mass per unit area (areal density). Therefore, the surface area of each bar is proportional to the total mass of carbon in the component that is represented. The high areal density of soil carbon in temperate forests and in grasslands compared with agricultural land and tropical forests is noteworthy.

Many ecosystems in the world, however, are not in a steady state. They accumulate dry matter during a number of years and are then disrupted by fire or other drastic events; therefore, they go through sawtooth-like life cycles. Fire is often a natural process, and necessary for rejuvenation of ecosystems. Locally, disruption by fire is a discontinuous event, but summed over the entire globe there is a continuous release of CO_2 by fire, which is practically compensated for by regrowth at the other places. The total carbon released by fire is of the order of 4–7 Pg/year. The largest portion in this flux is caused by periodic burning of tropical grasslands (Hall and Scurlock 1991). Although this flux is almost as large as the

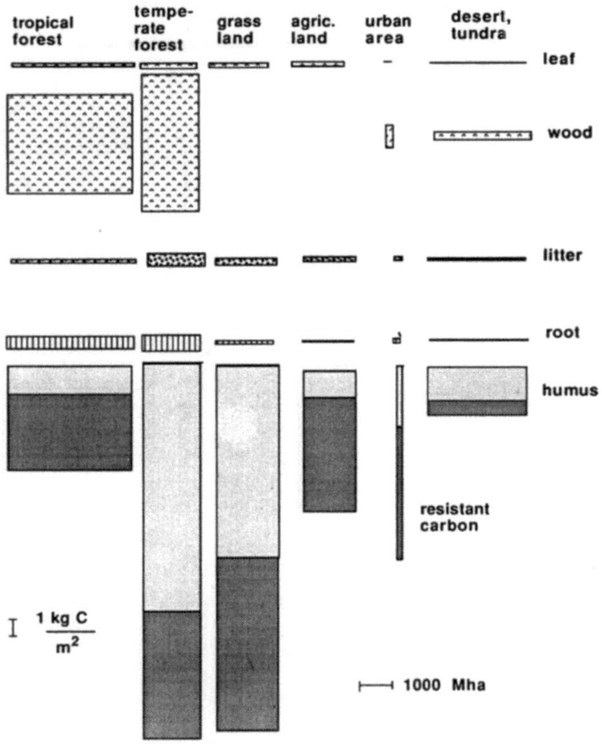

Fig. 2. Simulated carbon areal density (*heights of the columns*) and land area for each vegetation type (*widths of the columns*). The *areas of the columns* are proportional to the amounts of carbon in each compartment

rate of combustion of fossils fuels (about 6 Pg C/year in 1990), grassland burning does not contribute to the increase of atmospheric CO_2. Within 1 year, most grassland biomass turns to CO_2, if not by fire, then by decomposition. Moreover, fire residues such as charcoal are added to resistant soil carbon.

On a global scale, grasslands occupy about 20% of the vegetated land area and contribute about 23% to annual carbon flux and total carbon storage. This surprisingly large share in terms of carbon storage is entirely due to the large amount of organic matter that is maintained in grassland soils. This is caused by a combination of large partitioning of dry matter to below-ground parts and a relatively slow rate of decay of soil organic matter. When forest is converted into agriculture, the standing biomass is removed (often partly burned), and the disturbance of the soil will give rise to enhanced decomposition of organic matter in the soil and to a strong efflux of CO_2. In arable land, the disturbance is repeated annually by ploughing, and after a few decades the carbon content of the soil declines to a much lower level than originally present in the virgin soil. In grassland, however, the situation is different. The soil is much less disturbed, the crop is perennial, and the remnants of leaves, stolons, roots, stubble, and cattle manure accumulate and give rise to an organic matter level in the soil that is as high as in a forest soil.

In the Holdridge diagram (Holdridge 1967; Monserud and Leemans 1992) the term cool temperate steppe indicates the climatic zone where grasslands naturally occur. This zone is rather dry because annual potential evapotranspiration exceeds annual precipitation. These truly natural grasslands occur in the prairie zone of North America, the chernozem zone of the Ukraine, and in the pampas of Argentina. Their combined area is about 900 Mha and contains a total amount of soil carbon of 120 Pg (Post et al. 1985). At present, grasslands extend far beyond their natural range, and in fact the most productive grasslands are found in moist temperate forests zones, which have a larger moisture supply than the cool temperate steppe. These natural forests have long ago been replaced by agricultural and pasture lands.

The area of these anthropogenic grasslands is also about 900 Mha, but the carbon content of their soil is higher than that of natural grasslands, probably about 200 Pg.

Carbon Isotopes as a Tracer

Time courses of carbon isotopes can yield important information. The largest global tracer experiment ever conducted was unintentional, the hydrogen bomb experiments that peaked around 1962. Because of these experiments, the concentration of ^{14}C in the atmosphere approximately doubled. After the international ban in 1965, this peak gradually declined by dilution of ^{14}C in the ocean and biosphere (Fig. 3). The rate of this decline provides a clue to estimate the exchange rate. In early studies, the exchange rate with the biosphere was ignored, and the estimated rate was totally ascribed to the ocean. This yielded exchange rates in the order of 100 Pg C/year. In fact, part of the exchange is with the biosphere, and only the remainder is with the ocean. Therefore, the best method is to use a model that allows for both exchange rates, to impose the exchange rate of the biosphere (NPP) and then to find the exchange rate at the ocean surface by curve fitting. This method yields a lower rate of exchange with the ocean (50 Pg/year) than the one normally assumed, but it describes very well the measured decline rate of ^{14}C in the atmosphere.

A second unintentional global tracer experiment is the emission of fossil carbon. No ^{14}C remains in fossil fuel; therefore, the atmospheric ^{14}C content before 1960 gradually declined (Suess effect). Later, the hydrogen bomb experiments completely swamped this effect and made it impossible to follow. However, the other carbon isotope (^{13}C) was not disturbed by the nuclear experiments and remained available as a tracer. Naturally, about 1% of carbon is in the form of ^{13}C, but in several processes a slight discrimination occurs. The process of photosynthetic CO_2 assimilation in C_3 plants discriminates against the heavier

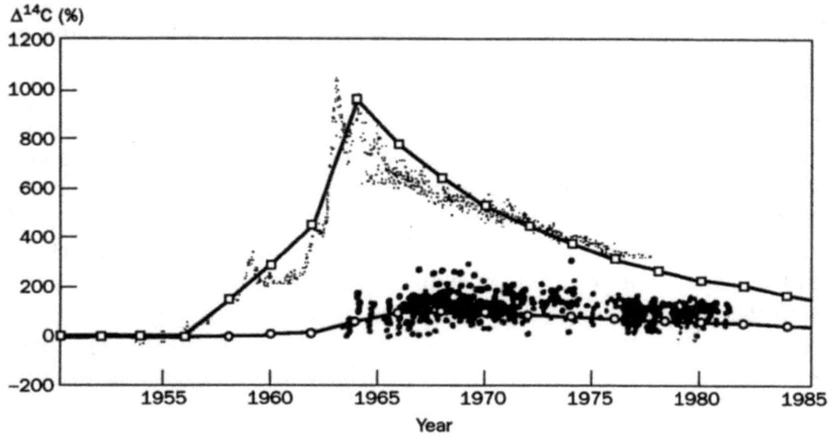

Fig. 3. Observed (Bolin 1986) and simulated time course of ^{14}C carbon in the atmosphere (*squares*) and in the surface waters of the ocean (*circles*). The model results (after fitting the atmosphere-ocean exchange rate) are given by the *solid lines*

^{13}C; therfore, the $^{13}C/^{12}C$ ratio in plant biomass is about 25 parts per thousand lower than in the atmospheric source. This has been the case since photosynthesis started; therefore, the depletion is equally present in fossil carbon. Additional discrimination in methanogenesis has further depleted the $^{13}C/^{12}C$ ratio in reserves of natural gas by about 40 parts per thousand.

The release of isotopically depleted CO_2 by burning fossil fuels is visible in various records of atmospheric CO_2 and also in tree rings. This decline validates model runs for the global carbon cycle (Goudriaan 1990, 1992).

The Biosphere as a Sink for Carbon

Increase of atmospheric CO_2 increases net CO_2 assimilation and plant growth. Since industrialization, this effect by itself has stimulated the biospheric uptake rate by about 1 Pg C/year, which is about 20% of the total global emission rate of CO_2. As another comparison, this CO_2-induced absorption is just about equal to the fluxes released by large-scale deforestation. This is a coincidental compensation, but it explains why a model with just the ocean, and without the biosphere at all, still works reasonably well. This type of model will obviously fail to explain effects of deforestation or afforestation.

This role of the biosphere as a sink cannot be directly measured, but the circumstantial evidence is very strong and indicates a strong stimulation of plant growth by increased atmospheric CO_2. According to a summary by Kimball (1983), there is a mean 40% increase of dry matter in C_3-crops with doubling of

CO_2, and a 15% increase for C_4-crops. This effect of CO_2 not only occurs in crop plants, it is a general phenomenon (Lemon 1984; Strain and Cure 1985). The primary physiological effect of increased CO_2 concentration is a stimulation of the rate of CO_2 assimilation. The respiration rate might also be reduced (Amthor 1991). The CO_2-enrichment effect is maintained when growth is limited by water (Gifford 1979), which can be explained by control of both water loss and CO_2 assimilation by stomatal resistance. Any increase in ambient CO_2 will then stimulate CO_2 uptake without raising water loss. Nutrient shortages, especially of phosphorus and potassium, tend to impose absolute limits on crop growth without leaving much room for stimulation by CO_2. Nitrogen, however, differs from other nutrients because it permits a small positive CO_2 effect, even under rather severe nitrogen shortage.

Arp (1991) showed that part of the conflicting evidence in the literature on adaptation of plants during growth can be explained by differences in the pot size use in the experiments because the disappearance of the CO_2 effect on growth is associated with small pot size. In the field, a primary CO_2 stimulus can lead to better exploration of the soil; therefore, nutrient uptake may go up even under nutrient-limited circumstances. Indeed, in an experimental study, during 4 years of continued exposure of a natural salt-marsh vegetation to high CO_2, Arp (1991) found no decline in increased photosynthesis and no decline in efficiency of water use.

Deforestation and Carbon Dioxide Fertilization

The CO_2-induced sink effect must be separated from the effect of changes in land use because land use is independent of atmospheric CO_2. Precise estimation of CO_2 emission from deforestation is full of pitfalls. Direct burning of wood and plant biomass is only one source of CO_2 resulting from deforestation. Deforestation also stimulates oxidation of soil carbon. This process takes many years to reach a new steady state, and during that period the soil releases CO_2 faster than the forest soil. This increased flux was represented in the model of the carbon cycle (Goudriann and Ketner 1984) by decreased longevity of humus after land is turned into arable land. Another factor included in the model was incomplete biomass burning, leading to partial charcoal formation (10–20% of the aboveground biomass). The net effect of these factors is expressed in the total amount of simulated biospheric carbon. Comparison of total biospheric carbon for model runs with and without changes in land use (deforestation in particular) resulted in a difference because of deforestation of 90 Pg C over the period 1780–1980 (Table 1, first column for $\beta = 0$). The correct order of magnitude of this number is confirmed by a comparison with the amount of biomass (present on the total forest area) lost during this period. For tropical forests, almost 800 Mha was lost with a mean biomass of 8 kg C/m^2 (64 Pg C), and for temperate forests, 200 Mha with a mean biomass of 11 kg C/m^2 (22 Pg C).

Table 1. Simulated total amount of carbon (Pg) in the terrestral biosphere for the year 1980. In the reference case without any changes in land use and without any CO_2 effect ($\beta = 0$), the initial level of 1894 Pg C in the year 1780 would have been maintained at the steady-state level. In reality, this steady state was disturbed by changes in land use and caused a loss of 90 Pg C. The simultaneous CO_2 effect caused a gain of 51 Pg C. Had no changes in land use occurred, the CO_2 effect would have been smaller (32 Pg). The interaction of changes in land use and CO_2 enrichment has given rise to an additional carbon fixation of 19 Pg and lead to a total net effect of –39 Pg C. The numbers in bold give the most realistic effects of land use and CO_2

	$\beta = 0$	$\beta = 0.5$	Enrichment effect
No changes in land use	1894	1926	32
With changes in land use	1804	1855	**51**
Effect of land use	**–90**	–71	–39

However, deforestation and CO_2 fertilization occurred simultaneously, and exhibited a strong interaction. The positive CO_2-enrichment effect was larger when deforestation was included, or, equally valid, the biomass removed by deforestation was smaller when CO_2 fertilization occurred.The best way to separate these effects is to determine the effects of land use from model results without CO_2 enrichment (left column, Table 1). The CO_2-enrichment effect, on the other hand, is estimated with the changes in land use (middle row, Table 1).

The interaction can be understood. On one hand, deforestation enhanced the dynamics in the biosphere; therefore, the slow biospheric compartments were stirred up and could more easily absorb the additional carbon produced by CO_2 enrichment. This increased the CO_2-enrichment effect from 32 to 51 Pg. On the other hand, CO_2 enrichment reduced the net effect of deforestation by recycling carbon. The deforestation emission was 90 Pg C, but the CO_2-enrichment effect (caused by this very emission) recycled about 20% of this quantity to the biosphere, which leaves about 71 Pg C as the net emission. This net emission, however, is not a good estimate for the size of the source itself. Most recycling occurs in parts of the biosphere that are not deforested. The combined effects of CO_2 fertilization and deforestation caused a simulated biospheric carbon loss of 39 Pg from 1780 to 1980 (diagonal transition, Table 1).

Conclusion

The total emission of 249 Pg C (159 fossil and 90 biospheric) is partitioned to the three major reservoirs as 88 Pg to the ocean, 51 Pg to the biosphere, and 110 Pg remaining in the atmosphere. For the biosphere, the net result is a loss of 39 Pg of carbon, which is added to the release of 159 Pg of fossil carbon. The time course of the total emission follows an approximately exponential increase at a relative growth rate of about 2.4% per year.

The time patterns of biospheric emission and fossil-fuel emission have been dissimilar in shape. About the turn of the centrury, the biospheric emission was larger than the fossil-fuel emission, whereas today the situation has reversed. Over the past 200 years, the biosphere has been a net source, but at some time during the past decades it must have turned into a net sink. This switch is the result of the steadily increasing CO_2-fertilization effect. The dual character of the biosphere in being both a source and a sink is undeniable. The biosphere may not be quite as strong a sink as the ocean is, but it is capable of counteracting its own role as a source (Lugo and Brown 1986).

Warming, if it occurs, will almost certainly have various effects (Goudriaan and Unsworth 1990). In some places, enhanced decomposition of organic matter in the soil will prevail (most likely in the tundra regions). Simultaneously, this enhanced decay in the soil compartments may be compensated by increased formation of above-ground dry matter (Esser 1987). In other places, extension of the duration of the climatic growing season will be more important and the net carbon storage will increase. Agricultural zones will shift (Parry 1990; Leemans and Solomon 1993; Cramer and Solomon 1993). In 1993, the rising trend of atmospheric CO_2 stalled unexpectedly (Taylor 1993). Analyses of concurrent courses of carbon isotopes will be needed to reveal which CO_2 sink has become more active, or which source has suddenly reduced its activity.

Acknowledgment. Part of this chapter was prepared while on study leave as a recipient of a CSIRO McMaster Fellowship at the Centre of Environmental Mechanics of CSIRO, Canberra, Australia.

Simulated Impacts of Global Climate Change on Crops

J.W. Jones,[1] N.B. Pickering,[2] C. Rosenzweig,[3] and K. J. Boote[4]

Scientists at several laboratories have used large complex models of atmospheric processes, called general circulation models (GCMs), to estimate the possible effects of increasing concentrations of greenhouse gases in the atmosphere on global climate. These models have been used to estimate the changes under an effective doubling of CO_2, which is defined as the combined radiative absorption effects of all greenhouse gases having the same absorption characteristics as doubled CO_2, usually taken to be 600–640 μl/l. Because different gases are increasing at different rates (IPCC 1990), the atmospheric CO_2 concentration is projected to be less than 600 μl/l when this effective doubling of radiation absorption occurs (Rosenzweig and Parry 1994). Five GCM studies estimated an average increase in global surface temperature of 3.9 °C (range 2.8–5.2 °C) and an increase in global precipitation of 10.1% (range 7.1–15.8%) (Rosenzweig 1990). These predicted changes in climate variables are still highly uncertain, although there is evidence that temperatures have increased by about 0.5% since the late 19th century (Jones 1993). Nevertheless, scientific evidence of the changes in greenhouse gases, coupled with the predictions of their effect on climate, has caused major concerns.

Agriculture is highly dependent on weather, and therefore changes in global climate could have major effects on crop yields and world food supply. Potential effects are difficult to assess not only because of the uncertainty in the magnitude of changes in climate variables, but also because of uncertainties in crop responses to combinations of CO_2, weather, soils, and management factors, the spatial variation of these factors, and the uncertainties in the evolution of global social, political, and land-use systems. Assessments are needed to provide decision-makers with the information they need to develop appropriate plans to reduce the expected climate changes or to adapt to them. Recently, several studies have been conducted to estimate the possible impacts of climate change on crop production at regional, national, and international levels. Results suggest that

[1] Professor, Agricultural Engineering Department, University of Florida, Gainesville, FL 32611-0840, USA
[2] Visiting Assistant, Agricultural Engineering Department, University of Florida, Gainesville, FL 32611-0840, USA
[3] Agronomist, Goddard Institute for Space Studies, Columbia University, NY, USA
[4] Professor, Agronomy Department, University of Florida, Gainesville, FL 32611-0840, USA

S. Peng et al. (Eds.) Climate Change and Rice

yields of many food crops would decrease in some regions and increase in others under the climate-change scenarios produced by GCMs, and that production in tropical areas would be affected more than in temperate regions (Rosenzweig and Parry 1994). Dynamic crop-simulation models have been used to simulate crop production under various scenarios of climate change and CO_2 levels produced by GCMs for doubled CO_2 conditions.

Overview of Recent Studies

Characterisitics of Crop Models

Ideally, crop models should accurately simulate the effects of specified weather, soil, and management conditions on growth and yield, as well as predict the use of water, nutrients, and other input resources, regardless of where the crops are grown. In addition, they should be able to estimate the direct physiological effects of CO_2 concentration on crop yield and resource use. These models require specification of soil, weather, management, and CO_2 concentrations for every point in time or space where an estimate of production is needed. It is not realistic, however, to expect models to be perfect, nor is it possible for inputs to be accurately specified for every point in space and time. Crop models have been developed with various assumptions, and their use is restricted to conditions where those assumptions are met. For example, most crop models do not include the effects of nutrients (except nitrogen) or pests. Even if they did, it would be difficult (if not impossible) to fully characterize each field with respect to its pest, micronutrient, and other conditions for simulations.

Some authors (e.g., Penning de Vries et al. 1989) defined various levels of crop simulation. At the first level, only weather variables (temperature,light, and perhaps CO_2) affect crop growth. This level represents potential yields for well-irrigated, well-fertilized crops. The second level includes water limitations in addition to weather effects. For this level, water-holding characteristics of the soil are needed as inputs, and the model must include a soil-water balance, including calculations of evapotranspiration, infiltration, runoff, and the effects of water stress. If the effects of CO_2 concentration are included, the interactions between CO_2, water use, and crop growth and yield must be included. Second level models are useful in many regions of the world where water is the major limiting factor, and in studies where irrigation management is important. Level three crop models include nitrogen effects. These models are needed where soil nitrogen is limiting, and when studies are needed to optimize management of N fertilizer. Interactions of nitrogen with water and weather conditions must be included. A fourth level, which is seldom addressed, includes other nutrients, pests, diseases, and other stresses.

Jones and Ritchie (1990) provided a list of crop models that have the capability of simulating water-limited production (level 2) for use in studies of irrigation management. Reynolds and Acock (1985) listed 25 crop models, half of which included simulation of soil-water status and water-limited growth and yield. Reynolds and Acock (1985) emphasized the characteristics needed by crop models to predict responses to increasing CO_2 concentrations. In their list, only 1 of the 25 crop models included response to CO_2. Since 1985, the direct effects of CO_2 on crop growth have been incorporated into several crop models. The primary effects of CO_2 on plants, reviewed by Allen (1990), are on photosynthesis (and therefore dry-matter growth rates) and on transpiration because of changes in stomatal conductance. The incorporation of CO_2 effects into the models of crop growth used in most climate-change studies have included only these two effects and made the assumption that other direct effects on plants were small. Other direct CO_2 effects have also been found but are not as well documented. For example, Baker et al. (1989) reported changes in rates of appearance of soybean leaves under elevated CO_2 concentrations as well as higher nonstructural carbohydrate content in leaves and stems.

Many attempts have been made to evaluate the ability of crop models to accurately simulate crops growing in a region under different weather conditions and different management practices. Minimum data requirements for testing crop models have been recommended (IBSNAT 1988). Although there are many difficulties with model-validation efforts, they are very important to establish a level of confidence that model predictions are reasonable and that responses to treatment factors can be predicted. In many cases, models predict yields fairly well for experimental fields, but overestimate yields for farmer fields where pest and nutrient limitations may not be as well managed. Therefore, if crop models are applied over large regions, one should expect simulated yields to be higher than actual yields, even under current climate conditions. Inaccurate estimations result from a lack of detailed information for each field and from an inability to model plant responses to every real-world factor. Previous validation efforts have been restricted to historic weather conditions. There is a critical need to test models using experimental data under conditions of elevated temperatures and CO_2. The use of crop models to study the impacts of climate change on agriculture must rely on estimates of relative changes in production, rather than absolute values of yield under future climates.

Various research groups have conducted studies to estimate the possible effects of climate change on agricultural production using different crop models (Table 1). These models use daily total solar radiation, maximum and minimum temperatures, and precipitation as inputs (some require relative humidity and wind speed), and estimate the day-to-day development and growth of the crops. Most have a soil-water balance component. They operate for uniform fields characterized by their weather and soil-input information. These models were originally developed to study the effects of weather variability on crop production and to improve crop management under uncertain environments. The models originally did not include the direct physiological effects of CO_2, and they had to be modified to estimate the combined effects of climate change and CO_2 on growth, development, and yield.

Table 1. Some crop models used to simulate the effects of climate change on agricultural production in different regions of the world

Model	Crop(s)	Reference
CERES-Wheat	Wheat	Ritchie (1985), Ritchie and Otter (1985), Rosenzweig (1990)
AFRCWHEAT2	Wheat	Porter (1993), Weir et al. (1984)
WOFOST	Wheat, maize	van Diepen et al. (1989), Wolf (1993)
CERES-Maize	Maize	Jones and Kiniry (1986), Ritchie et al. (1989), Rosenzweig (1990)
SOYGRO	Soybean	Wilkerson et al. (1983), Jones et al. (1989), Curry et al. (1990)
SIMRIW	Rice	Horie (1988)
CERES-Rice	Rice	Godwin et al. (1990), Singh and Ritchie (1993)
EPIC	Maize,wheat, sorghum, soybean	Williams et al. (1984), Easterling et al. (1992, 1993)
GAPS	Maize, wheat, sorghum, soybean	Buttler and Riha (1989)
PNUTGRO	Peanut	Boote et al. (1992), Hoogenboom et al. (1990)
Sinclair model	Soybean, maize	Sinclair and Rawlins (1993)

Wheat and Maize Yields in the Southern Great Plains

Rosenzweig (1990) used CERES-Wheat and CERES-Maize models to simulate the effects of climate change under doubled CO_2 conditions on yields of wheat and maize grown in the Southern Great Plains of the United States. These two models use daily weather data to estimate crop development, growth, and yield as affected by solar radiation, temperature maximum and minimum, and precipitation. They include a soil-water balance and the effects of water stress on growth and yield. These models were modified to account for the direct effects of CO_2 on daily growth (by modifying daily net assimilation and efficiency of light use) and on evapotranspiration (ET) (Peart et al. 1989). Efficiencies of light use were assumed to increase 25% in wheat and 10% in maize for doubled CO_2 concentrations. Potential plant transpiration was modified by using the Penman-Monteith relationship between potential ET and stomatal conductance, and by decreasing stomatal conductance under doubled CO_2 using relationships published by Rogers et al. (1983a). Climate scenarios from two GCMs were used in the study: the Goddard Institute for Space Studies or GISS model (Hansen et al. 1983), and the Geophysical Fluid Dynamics Laboratory or GFDL model (Manabe and Wetherald 1987).

Simulated yields decreased for both crops under the climate-change scenarios, primarily because of the shorter life cycle of the crops under elevated temperatures. Water stress was more severe under climate change because higher temperatures caused increased ET. Simulated yield decreases for irrigated crops were less than for dryland crops. Direct effects of CO_2 partially offset negative effects of temperature increase, but not entirely. Some yield recovery was simulated if varieties were changed.

Soybean and Maize Yields in the Southeastern United States

A soybean growth model (SOYGRO, Wilkerson et al. 1983; Jones et al. 1989) and the CERES-Maize model were used by Curry et al. (1990) to study the effects of climate change on soybean and maize yields for 19 locations in the Southeastern United States. The soybean model is similar to CERES-Maize in that it requires daily weather data and uses the same soil-water balance model. It was also modified by Peart et al. (1989) to simulate the effects of CO_2 on canopy photosynthesis and transpiration. Curry et al. (1990) simulated these crops for 30 years of historical weather and 30 years for the GISS and GFDL scenarios, with and without irrigation. They found considerable differences in results for the two GCM scenarios. Under the GISS scenario with direct effects of CO_2, soybean yields increased in 12 locations and decreased in 7 locations under rainfed conditions. Under the GFDL scenario, yields declined in 17 locations. The GFDL scenario had significantly less precipitation during the summer months than the GISS scenario. With irrigation, soybean yields increased in most locations, and averaged about 15%. Rainfed maize yields were reduced more than soybean yields. The beneficial effects of CO_2 were considerably less for maize (a C_4 crop) than for soybean (a C_3 crop). They also found that irrigation requirements increased considerably and that efficiency of water use was reduced for maize under both scenarios.

Wheat Yields in Europe

Two wheat models have been used to simulate climate-change effects on wheat production in Europe. The WOFOST model (van Diepen et al. 1989) simulates daily increments of crop growth using daily weather data. It simulates potential production based on weather variables alone, and can also simulate effects of water-limitation on production. Wolf (1993) used WOFOST with weather and soil data for 20 locations across Europe and three GCM scenarios to simulate the effects of climate change on yields of winter wheat. He modified the model to simulate CO_2 effects on photosynthesis, transpiration, and specific leaf area and found that water-limited yields of winter wheat increased under all scenarios (by 1 to 2.3 t/ha) when the direct effects of CO_2 were included.

Semenov et al. (1993) used the wheat model AFRCWHEAT2 to estimate the effects of climate change on yields of winter wheat in the United Kingdom and durum wheat in France. This model uses daily weather and a daily time step to simulate growth of the crop. The CO_2 effects are included in the model through changes in maximum photosynthesis rate, photosynthetic efficiency at low light levels, and stomatal resistance. Simulated wheat yields increased under the climate-change scenarios, but yield variability also increased.

Simulating Crop Responses to Climate Change Using Selected Historic Weather

Although most studies have developed daily weather data using information from the GCMs, the MINK study (Easterling et al. 1992, 1993) took a different approach. The MINK region includes four midwestern states in the United States (Missouri, Iowa, Nebraska, and Kansas). This region was selected because it experienced 10 years of high temperatures and drought during the 1930s and was referred to as the Dust Bowl (Easterling et al. 1992). During the 1930s, temperatures were 1.1–1.6 °C higher during the summer months than during the period 1951–1980 and precipitation was 21–59 mm lower. The MINK study used the EPIC crop model (Williams et al. 1984) to simulate wheat, maize, sorghum, and soybean using weather data from the 1930s as an analog for climate change, and compared the results with simulated results using 1951–1980 weather data for the same region. EPIC, originally developed to simulate erosion and its impacts in the United States, includes a generic crop model that simulates growth and yield for a number of crops in response to daily weather data, soil conditions, and management. Easterling et al. (1992) modified EPIC to respond to direct effects of CO_2, using an approach that was similar to that used by Peart et al. (1989). Simulated yields of dryland maize, sorghum, and soybean declined 25% during the warm period without CO_2 effects. Increased CO_2 only partially compensated for these losses. Their results provide an interesting look into a famous historic period. However, the analogue scenario that they used does not reflect the full temperature range for doubled CO_2 climate projected by the IPCC (1990, 1992) of 1.5–4.5 °C.

A Global Study of Climate Change Impacts Based on Crop Model Results

A global study was recently completed by Rosenzweig and Parry (1994) in which they evaluated the potential impact of climate change on global food supply. This study was the first attempt to include international trade and population changes as well as the effects of climate change on agricultural production. The major finding of that study was that there appears to be a large disparity in vulnerability to climate change between developing and developed countries. Their analysis, which included direct effects of CO_2 on crops, showed that simulated yields were mostly reduced for the tropical latitudes, although temperature increases were projected to be less than in temperate regions in most GCM scenarios. Under the GCM scenarios global grain production declined somewhat and caused prices to increase. In some mid- and high-latitude countries, grain production was projected to increase. The implications of this study are that people in the already impoverished tropical and subtropical countries face increased risks of malnutrition and starvation under these climate-change scenarios.

The economic analysis in this study relied on a world trade model of linked agricultural sectors. However, the major biophysical inputs to this global model were from a global crop-simulation study (Rosenzweig and Iglesias 1994). Agricultural scientists from 18 countries used a set of compatible crop models to estimate changes in crop yields at 112 different sites. Each of the scientists compared model simulations with data collected in their own countries to test the credibility of results in their regions under current weather conditions. Simulations were then carried out in regions representing about 70% of the current world production of wheat, maize, and soybean, and about 48% of current rice production. The models were those incorporated in the Decision Support System for Agrotechnology Transfer or DSSAT (IBSNAT 1989), and include CERES-Maize, CERES-Wheat, CERES-Rice, SOYGRO, and CERES-Barley. Yield results from each country simulation were presented in a uniform way for integration into the global analysis conducted by Rosenzweig and Parry (1994).

Evaluating the Need for More Mechanistic Model Components

The models used in previous climate-change studies have relied on simplified descriptions of various processes, and as such there is some question as to whether they correctly represent real-world responses. We studied the need for more mechanistic detail in crop models, assuming the goal is to estimate yield responses to climate change. A simulation experiment was performed to test the hypothesis that simulated response to climate change is different when a mechanistic hourly light interception – photosynthesis model component is used instead of a daily canopy photosynthesis model. A second set of computer simulations was made to evaluate the effects of simulating hourly canopy temperatures and using them to drive crop processes. In these simulations, hourly calculations of canopy energy balance and ET were included instead of daily calculations of ET as in the original model.

Components of Crop Model

Photosynthesis

The soybean growth model, CROPGRO-Soybean, was used in this study. This model is basically the same as SOYGRO V5.42 (Jones et al. 1989), but it has additional options for simulating hourly photosynthesis based on the photosynthesis submodel developed by Boote and Pickering (1994) and implemented by Pickering et al. (1994c). This submodel computes light interception for incomplete canopies and calculates photosynthesis based on light capture by sunlit and shaded leaves. The leaf calculations are based on the leaf photosynthesis model of Farquhar et al. (1980) and Farquhar and von Caemmerer (1982). This leaf model defines the maximum photosynthesis rate as the maximum of two equations, one where internal CO_2 levels are low and Rubisco kinetics regulate gas exchange, and the second where internal CO_2 levels are high and RuBP regeneration and electron transport control assimilation rates. Pickering et al. (1994c) ignored the first equation. Temperature and CO_2 effects on maximum leaf photosynthesis and quantum efficiency were modeled using the equation of Farquhar and von Caemmerer (1982) for limiting RuBP. In addition, temperature affects the specificity of Rubisco for CO_2 versus O_2 in the model (based on the work of Jordan and Ogren 1984). Temperature also affects maximum leaf photosynthesis rate under saturated CO_2 levels in the model (based on the data of Harley et al. 1985). A constant ratio of internal to external leaf CO_2 concentrations was assumed. Therefore, this leaf model includes basic reponses to light, temperature, and CO_2. It is used to compute hourly canopy photosynthesis rates, which are summed to compute daily total canopy photosynthesis. This version of the soybean model will be referred to as the LEAF model.

In contrast to this new mechanistic photosynthesis model, the original SOYGRO model computed daily canopy photosynthesis from daily light and average daytime temperature values (Wilkerson et al. 1983). The temperature response curve is shown in Fig. 1 for this simple daily model compared with the integrated daily canopy photosynthesis response to temperature based on the mechanistic LEAF model (Pickering et al. 1994c). The original method was retained in the CROPGRO-Soybean model. An exponential equation was added to increase or decrease daily canopy photosynthesis when CO_2 concentrations vary from a standard value of 350 μl/l. This exponential equation was based on simulations from the mechanistic LEAF model. This version of the soybean model will be referred to as the CANOPY model.

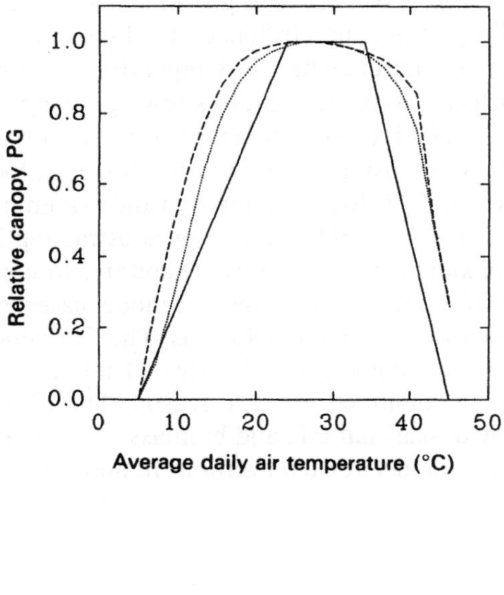

Fig. 1. Relative effect of temperature on daily canopy gross photosynthesis in the CANOPY (— OLD SOYGRO), LEAF model without the energy balance (··· NEW-NO EB), and the LEAF model with the energy balance (- - - NEW WITH EB) at 350 μl/l

Energy Balance

Crop models usually assume that plants are at the same temperature as the air, and that all processes are driven by air temperature. There are times when this assumption may result in considerable error (e.g., when plants are small and their temperatures are influenced by soil temperature, which may be several degrees warmer or cooler than air temperature). Also, as CO_2 concentrations increase, the relationships between air and plant temperature may change because of the change in transpiration and its effect on the energy balance of the crop.

Pickering et al. (1994c) incorporated a detailed canopy energy balance into the LEAF version of CROPGRO-Soybean (based on the model of Jagtap and Jones 1989). This energy-balance model computes ET and canopy temperature by partitioning the canopy into a soil surface and sunlit and shaded leaf surfaces. A series of simulations were run to estimate the differences caused by using computed canopy temperature instead of using air temperature. These runs are referred to as EB, which refers to the use of the hourly energy balance.

Comparing LEAF with CANOPY Model Versions

For the first set of simulation experiments,only the method of calculating daily canopy photosynthesis was varied. All other relationships remained the same; therefore, differences in simulated results are caused entirely by differences in the LEAF and CANOPY methods of computing daily photosynthesis. Comparisons were made for six locations (Gainesville, FL and Des Moines, IA, USA; Pergamino, Argentina; Anguedon, Ivory Coast; Griffen, Australia; and Hyderabad, India) using 10 years of weather data for each location. These locations represent a wide range of environments. Latitudes range from 34 °S to 46 °N, with one location near the equator, and temperature, solar radiation, and precipitation differ greatly among locations. Cultivars were selected for each location to achieve a growing season of about 100 days. For Des Moines, Pergamino, and Griffen, earlier than normal cultivars were selected so that they would mature when the temperature was decreased by 4 °C. Planting dates were selected to correspond to known growing seasons at each location.

Yield response to temperature was obtained by simulating each location for 10 years using actual weather for each location, then simulating 10 years by adding 2 °C to daily minimum and maximum temperatures. This was repeated by adding 4, 6, and 8 °C and by substracting 2 and 4 °C from daily minimum and maximum temperatures. Therefore, to compare temperature responses of the two models, 6 locations × 2 model cases × 7 temperature cases × 10 years per location resulted in 840 runs. The CO_2 concentration was set at 350 μl/l. Crops were assumed to be well-irrigated, and changes in ET and irrigation requirements were computed for each season. The 10-year averages of season length, grain yield, seasonal ET, and biomass versus temperature change were calculated for each model at each location. In addition, changes in these crop variables when temperatures were increased by 4 °C were compared for the two models across locations.

Yield response to CO_2 change was simulated by setting CO_2 concentrations to 200, 350, 500, 650, 800, and 950 μl/l for each of the 6 locations and 10 years at each location. Temperature was not changed for these simulations. The 10-year averages of season length, grain yield, seasonal ET, and biomass versus CO_2 for each model at each location were calculated. The change in these crop variables when CO_2 was increased from 350 to 500 μl/l were compared for the two models across locations.

Comparing Energy Balance Results Under Varying Vapor Pressure Conditions

The LEAF version of CROPGRO-Soybean was used for additional simulations in which the energy balance was used to compute canopy ET. The unique feature of this model version is that it computes canopy temperature hourly, and crop processes respond to this temperature rather than to air temperature. Preliminary results showed that there was little difference when the energy balance was used, assuming that daily dew-point temperature was equal to minimum temperature. Therefore, as temperature increased, so did dew-point temperature and absolute humidity of the air. As long as this was the case, there were only small differences between the LEAF model with and without the energy balance. Therefore, we made a series of runs to compare results under different air humidities by setting daily dew-point temperature to minimum daily temperature minus 2, 4, 6, 10, 12, and 14 °C and by increasing humidity by setting daily dew-point temperature to minimum temperature plus 2 and 4 °C. These runs were made for only one location (Hyderabad, India) for 10 years of actual weather data assuming a CO_2 concentration of 350 μl/l. The original soybean model did not include vapor pressure effects on ET; therefore, comparison of results with the original model were not possible.

Results

Temperature Studies

Simulated responses by both models to temperature change were remarkably similar. There was, however, a difference of about 300–400 kg/ha between the LEAF and CANOPY models across most temperature changes for Iowa. The high values for the CANOPY model are caused because this model predicts slightly higher photosynthesis values for high daily light lelvels at this location. At Hyderabad, this bias was smaller because of the lower light levels. The other location that showed this large bias was Griffen, which also had high light levels

(Table 2). Over all locations, simulated yields decreased by 9.5% for the CANOPY and 7.3% for the LEAF model when temperature was increased by 4 °C. Table 2 summarizes the simulated effects of a 4 °C increase in temperature on season length, yield, ET, and biomass for each location. Yield decrease was about the same as biomass decrease when temperature increased by 4 °C. Season length decreased in locations with cooler climates and increased in warmer climates. There were no differences between models on seasonlength because the same temperature function for development was used in both models. Differences in seasonal ET between the two models under increased temperatures were caused by differences in leaf-area development between the two models, which affected canopy-light interception and therefore ET. In general, seasonal ET increased by about 11% for the 4 °C temperature increase.

Carbon Dioxide Studies

Simulated yield responses to CO_2 for Iowa again show the high light bias, but the bias was consistent for CO_2 concentrations above 350 μl/l. Atl Hyderabad, the differences between the two models were about 2% greater at high CO_2 concentrations than at 350 μl/l. When averaged across locations and years, yield response to an increase in CO_2 concentration from 350 to 500 μl/l was 22.8% for the CANOPY and 20.2% for the LEAF. Seasonal ET decreased about 2% for increased CO_2 in both models (Table 3).

Dew-Point Temperature Changes Using the Energy-Balance Model

Assuming that dew-point temperature equaled the daily minimum temperature, the use of simulated canopy temperatures resulted in about a 1–2-day delay in maturity because canopy temperatures averaged about 1 °C cooler than air temperature over the season. This difference had very little effect on yield. When dew-point temperature was set to minimum temperature minus 14 °C, there was a delay in maturity of about 5 days because canopy temperature averaged 3–4 °C lower than air temperature (Table 4). This would be a fairly dry climate. The major difference in these simulations was seasonal ET. There was more than a 100% increase in ET as the dew-point deviation from minimum temperature changed from 0 to –14 °C. Yield changed only slightly (about 130 kg/ha at the lowest dew point). It was assumed that simulated crops were well irrigated and that altered vapor pressure deficit would have no direct effect on leaf photosynthesis, and therefore that an increase in ET did not affect yields. However, if rainfed crops are simulated for the same conditions, there would most likely be large yield decreases if dew-point temperatures were lower relative to daily minimum temperatures. Water-use efficiency (yield or biomass divided by ET) decreased considerably as dew-point temperatures were decreased.

Table 2. Summary of simulated season lengths, yields, seasonal evapotranspiration, and biomass using the mechanistic (LEAF) and daily (CANOPY) photosynthesis models in the CROPGRO-soybean model as affected by temperature change. Each value is the average over 10 years of simulated crop seasons, using historic (base) weather as well as historic weather with 4 °C added to both minimum and maximum temperatures each day

	Base weather		Plus 4 °C		Change (%)	
	Canopy	Leaf	Canopy	Leaf	Canopy	Leaf
Season length (days)						
Gainesville, FL (UFGA)	117.9	117.9	120.6	120.6	2.3	2.3
Pergamino, Argentina (UBPE)	97.2	97.2	85.8	85.8	−12.5	−12.5
Ivory Coast (ANGU)	117.2	117.2	121.4	121.4	3.5	3.5
Griffen, Australia (GRIF)	104.2	104.2	95.8	95.8	−8.4	−8.4
Hyderabad, India (ITRC)	106.2	106.2	107.3	107.3	1.0	1.0
Des Moines, Iowa (DEIA)	93.1	93.1	84.3	84.3	−9.9	−9.9
Average	106.0	106.0	102.5	102.5	−3.99	−3.99
Standard deviation	9.26	9.26	15.07	15.07	6.42	6.42
Yield (kg/ha)						
Gainesville, FL (UFGA)	3519	3331	3027	2859	−15.0	−15.2
Pergamino, Argentina (UBPE)	3086	2748	2822	2549	−8.9	−7.5
Ivory Coast (ANGU)	3046	2832	2709	2522	−11.7	−11.6
Griffen, Australia (GRIF)	3719	3206	3578	3226	−3.9	0.6
Hyderabad, India (ITRC)	3800	3602	3302	3204	−14.0	−11.7
Des Moines, Iowa (DEIA)	2998	2574	2901	2620	−3.3	1.7
Average	3361	3049	3056	2830	−9.47	−7.27
Standard deviation	329.80	359.00	298.10	293.10	4.59	6.40
Cumulative ET (mm)						
Gainesville, FL (UFGA)	517	457	574	549	10.4	18.5
Pergamino, Argentina (UBPE)	354	304	367	317	3.5	4.1
Ivory Coast (ANGU)	373	346	430	401	14.1	14.9
Griffen, Australia (GRIF)	566	485	626	549	10.0	12.4
Hyderabad, India (ITRC)	397	376	441	419	10.6	10.9
Des Moines, Iowa (DEIA)	382	316	407	342	6.3	7.8
Average	432	381	474	430	9.14	11.44
Standard deviation	80.1	68.3	93.2	91.3	3.39	4.66
Biomass (kg/ha)						
Gainesville, FL (UFGA)	7659	7251	6929	6631	−10.0	−8.9
Pergamino, Argentina (UBPE)	5308	4671	4681	4145	−12.6	−11.9
Ivory Coast (ANGU)	6019	5395	5675	5173	−5.9	−4.2
Griffen, Australia (GRIF)	7307	6307	6492	5906	−11.8	−6.6
Hyderabad, India (ITRC)	6197	5783	5697	5415	−8.4	−6.6
Des Moines, Iowa (DEIA)	5059	4297	4617	4076	−9.1	−5.3
Average	6258	5617	5682	5224	−9.63	−7.25
Standard deviation	954	988	851	910	2.21	2.55

Table 3. Summary of simulated season lengths, yields, seasonal evapotranspiration, and biomass using the mechanistic (LEAF) and daily (CANOPY) photosynthesis models in the CROPGRO-soybean model as affected by increased CO_2 concentrations. Each value is the average over 10 years of simulated crop seasons, using historic (base) weather and 350 μl/l CO_2 level as well as historic weather with 500 μl/l CO_2 concentration

	Base (350 μl/l)		500 μl/l		Change (%)	
	Canopy	Leaf	Canopy	Leaf	Canopy	Leaf
Yield (Kg/ha)						
Gainesville, FL (UFGA)	3519	3331	4344	3942	21.0	16.8
Pergamino, Argentina (UBPE)	3086	2748	3880	3332	22.8	19.2
Ivory Coast (ANGU)	3046	2832	3853	3554	23.4	22.6
Griffen, Australia (GRIF)	3719	3206	4624	3931	21.7	20.3
Hyderabad, India (ITRC)	3800	3602	4840	4477	24.1	21.7
Des Moines, Iowa (DEIA)	2998	2574	3800	3171	23.6	20.8
Average	3361	3049	4224	3734	22.76	20.23
Standard deviation	330	359	406	437	1.09	1.86
Cumulative ET (mm)						
Gainesville, FL (UFGA)	477	456	465	445	−2.6	−2.4
Pergamino, Argentina (UBPE)	354	304	342	296	−3.4	−2.7
Ivory Coast (ANGU)	373	345	363	339	−2.7	−1.8
Griffen, Australia (GRIF)	566	485	549	474	−3.0	−2.3
Hyderabad, India (ITRC)	356	305	353	304	−0.8	−0.3
Des Moines, Iowa (DEIA)	382	316	369	308	−3.5	−2.6
Average	418	368	407	361	−2.67	−2.01
Standard deviation	78.2	73.9	75.3	71.4	0.89	0.81
Biomass (kg/ha)						
Gainesville, FL (UFGA)	7659	7251	9593	8691	22.4	18.1
Pergamino, Argentina (UBPE)	5308	4671	6709	5698	23.3	19.8
Ivory Coast (ANGU)	6019	5395	7662	6848	24.0	23.7
Griffen, Australia (GRIF)	7307	6307	9087	7721	21.7	20.2
Hyderabad, India (ITRC)	6197	5783	7909	7295	24.3	23.1
Des Moines, Iowa (DEIA)	5059	4297	6460	5314	24.3	21.2
Average	6258	5617	7903	6927	23.34	21.00
Standard deviation	954	988	1141	1154	0.98	1.94

Discussion

Crop models have been the major tool to estimate climate-change effects on crop yields. Results from crop-model studies are being aggregated for regional, national, and global studies. Different models have been used to simulate crop responses to climate change. These models have mostly operated at a daily time-step and have used rather simple calculations of photosynthesis and other processes of crop growth. Results from the simulation study reported in this chapter suggest that there would be only small differences in simulated results if a more sophisticated photosynthesis model was used instead of the simple relationships used by most models. There are interactive effects of temperature and CO_2 on photosynthesis, such that the optimal temperature for photosynthe-

Table 4. Summary of simulated season length, biomass, yield, seasonal ET, and irrigation requirements using the mechanistic LEAF photosynthesis and energy-balance models in CROPGRO-soybean as affected by daily dew-point deviation from minimum temperature. Positive values represent higher dew-point temperatures than minimum temperatures and therefore more humid conditions, whereas negative values represent more arid conditions

Dew-point deviation (°C)	Season length (days)	Biomass (kg/ha)	Grain yield (kg/ha)	Cumulative ET (mm)	Irrigation required (mm)
4	108.4	5936.4	3628.1	328.8	120.5
2	108.1	5934.1	3637.9	371.3	148.1
0	108.4	5958.5	3668.7	421.4	182.6
–2	108.5	5981.5	3682.0	473.4	217.2
–4	109.2	109.2	3718.5	521.8	261.0
–6	109.7	6114.6	3731.9	564.3	286.0
–8	110.3	6173.6	3748.9	597.6	316.1
–10	111.1	6237.7	3747.7	627.8	347.9
–12	112.1	6324.8	3763.7	656.2	374.0
–14	113.0	6395.6	3758.1	679.7	396.8

sis tends to increase as CO_2 increases. However, the effects of such interactions appear to be rather small when simulated along with other important growth processes that are affected by temperature (e.g., reproductive development).

One might have expected the use of canopy temperature rather than air temperature to result in greater differences in season length and yield. There are probably conditions where this is true. However, canopy temperatures could be warmer than the air because of soil-surface influence, particularly in humid regions, and result in more rapid development than would be predicted by air temperatures. The most striking result from these studies is the change in water use, which would have direct impact on rainfed crop yields.

The new findings reported in this chapter should not be interpreted as a validation of the models or the results from previous climate-change studies in which they were used. The findings do suggest that previous studies of climate change on crop production using crop models would not have been very different had they used mechanistic photosynthesis models. Climate variables affect other processes besides photosynthesis in crops, and these effects are equally important in determining crop development, growth, and yield. Temperature relationships used in most crop-growth model are based on limited data at high temperatures. Research is now producing new information on crop responses to high temperatures and CO_2 concentrations in the field. There is a need to integrate crop modeling and experimental efforts so that relationships in the models can be improved, particularly for high temperature and high CO_2, and model results can be validated under projected climate-change conditions.

Open-Top Chambers for Field Studies of Rice Response to Carbon Dioxide and Temperature: System Design

L.B. Collins,[1] P.H. Jones,[2] K.T. Ingram,[3] and R.C. Pamplona[4]

There is a large body of literature on the individual effects of CO_2 and temperature on rice physiology. In summary, CO_2 concentration is directly related to biomass production (Baker et al. 1988), crop development rate (Baker et al. 1988), leaf-level water-use efficiency (Allen et al. 1988), and grain yield (Yoshida 1976; Cure 1985). For tropical areas, increased temperature leads to faster crop development (Nishiyama 1976), higher respiration rates (Munakata 1976), spikelet sterility (Yoshida et al. 1981; Mackill et al. 1982), and reduced grain yield (Imai et al. 1983).

Most experiments on the effects of increases in temperature or CO_2 on rice have been conducted either in closed chambers (Acock and Allen 1985; Baker et al. 1988) or by short-term exposures in the field (Imai and Murata 1978a,b; Akita 1980; Imai et al. 1983). Furthermore, most of these experiments were conducted in temperate or subtropical environments. Although closed-chamber and short-term exposure experiments have contributed significantly to our understanding of how physiological processes respond to environmental changes (Yoshida 1981), they may not accurately reflect the performance of a field-grown rice crop in the tropics.

More important, because temperature and CO_2 are expected to increase together, there are likely to be significant interactions among these climate factors (Imai and Murata 1979b). Though there are few data on the interactive effects of elevated CO_2 and temperature, speculated interactions based on their individual physiological effects are mostly negative. For example, increased temperature hastens crop development, thereby shortening the time from planting until maturity and reducing the total time for photoassimilation and yield development. Elevated CO_2 on the other hand increases the rate of CO_2 uptake, thereby offsetting, at least in part, the negative effect of increased temperature. Unfortunately, elevated CO_2 also hastens crop development (Baker et al. 1988)

[1] Former Graduate Assistant, Agricultural and Biological Engineering Department, University of Florida, Gainesville, FL 32611, USA
[2] Associate Professor, Agricultural and Biological Engineering Department, University of Florida, Gainesville, FL 32611, USA
[3] Former Scientist, International Rice Research Institute, P.O. Box 933, 1099 Manila, Philippines
[4] Engineer, International Rice Research Institute, P.O. Box 933, 1099 Manila, Philippines

S. Peng et al. (Eds.) Climate Change and Rice

and seems likely to exacerbate the negative effects of increased temperature on total duration of crop growth.

Preliminary experimental data support this view and indicate that increased temperatures cause large reductions in rice yield, which are not compensated for by increases in CO_2 (US DOE 1989). Extrapolation of these data to real-world field conditions suggest the possibility that *tropical rice grain production of current cultivars could decline about 7 or 8% for each 1 °C rise in temperature, seriously affecting world food supply* (US DOE 1989). Although these results suggest dramatic effects of climate change on rice production, they are based on only one experiment under highly controlled conditions for one rice cultivar. Additional cultivars must be studied under actual tropical conditions with variable temperatures to verify the above data and to verify whether they can be extrapolated to real-world rice production. Because of these uncertainties, it is important to know the direction and magnitude of interactions between temperature and CO_2 as they affect rice growth, development, and yield.

To address the need for new, experimentally based knowledge on the responses of rice to climate, the United States Environmental Protection Agency provided funding beginning in 1990 to support a research program on the effects of elevated CO_2 and temperature conditions on rice. The program specified season-long field experiments that were to be conducted at IRRI under tropical conditions with the overall purpose of gaining a fundamental understanding of the possible effects of global climate change on wetland rice grown under natural conditions. The experimental program was to study several rice cultivars at ambient and doubled CO_2 concentrations and at temperatures from ambient to ambient plus 6 °C. This chapter describes and discusses the prototypic environmental control system developed at the University of Florida (Gainesville, FL), to conduct the elevated CO_2 and temperature studies at IRRI (Los Baños, Philippines).

System Design

Many types of systems have been used to study the impact of superambient CO_2 (Allen et al. 1992) and other gases on plant communities. At one extreme are intensive, expensive, precisely controlled closed systems that continuously recondition and recirculate the air in controlled environment growth cabinets, phytotrons, or naturally sunlit, outdoor growth chambers (Jones et al. 1984; Pickering et al. 1994a). At the other extreme are open-field dosing or free-air CO_2 enrichment (FACE) systems (Allen 1992) that are extensive, moderately expensive, and less precise (Hileman et al. 1992a) than closed chamber systems.

Closed chamber systems tightly regulate temperatures and CO_2 concentrations with minimal dependence on ambient conditions and can continuously monitor gas exchange (Jones et al. 1985c). Unfortunately, closed chamber systems are generally small and can significantly alter canopy microclimate compared with plants grown in field conditions. In contrast, FACE systems have essentially no effect on microclimate and can be applied to fairly large cropping areas. However, temperature control in a FACE system is not feasible. Between these extremes are open-top chambers (OTCs) that can provide moderately precise environmental controls in relatively natural, field-like conditions.

Open-top chambers that flush composition conditioned air through plant communities have been used successfully to study plant response to pollutants (Heagle et al. 1973; Olszyk et al. 1980) and CO_2 enrichment (Rogers et al. 1983a). In addition to their closer approximation to field conditions compared with closed systems, OTCs generally have lower operating and construction costs per unit chamber area. This allows researchers simultaneously to operate more individual chambers and permits statistical replication of treatments. Given the EPA-IRRI rice program budget, timelines, and the research goals, OTC technology was considered to be more appropriate than closed chamber systems.

From an engineering perspective the choice of OTC technology required that certain design decisions be made to meet the research goals. The foremost design issue was how to implement elevated temperature controls. Although the successful use of OTCs to control CO_2 over seasonal time frames is well documented, OTCs have not been used for regulated elevated temperature studies. Most researchers report that unregulated air temperatures inside OTCs range from 1 to 2 °C above ambient (Allen et al. 1992). Heagle et al. (1989) found a positive correlation between insolation and temperature rise in OTCs. Drake et al. (1989) found that in their salt marsh OTCs both nighttime and daytime temperatures fluctuated about 2 °C above ambient. The IRRI research project called for controlled diurnal temperatures up to 6 °C above ambient.

A corollary to the temperature-control problem is the problem of chamber microclimate. The literature on OTCs frequently postulates the importance of horizontally well-mixed air to provide homogeneous temperature and CO_2 concentration in cross section (Hileman et al. 1992b), but vertical microclimate is seldom discussed. Ordway (1969) noted the problem and raised concerns that airflow through the vegetation in an OTC was unnatural. Most OTC designs force air into the system around the perimeter of the chamber near the bottom and out through the top (Rogers et al. 1983b). This arrangement creates unnatural vertical temperature, CO_2, and humidity profiles within OTCs compared with field conditions (Lemon et al. 1971). This problem could be amplified in the IRRI OTCs, if heated air were introduced at the bottom just above the water. The dynamics between latent and sensible energy would be difficult to regulate in terms of both profiles and set-point temperature control. The profile issue also raises questions about where measurements of treatment variables should be made.

Based on these considerations, the IRRI OTCs were designed with an exhaust fan to pull air vertically down through the canopy and exhaust it out the bottom of the chambers. A horizontal stream of temperature and CO_2-conditioned air was designed to be introduced and mixed into the downward-moving vertical air stream above the canopy. The design called for sensors to be placed just above canopy height in the downward-moving air stream. This design was expected to provide two primary advantages. First, the uppermost, sunlit, portion of the canopy would be exposed to the most closely treatment-regulated and monitored air. Second, profiles of canopy microclimate (especially temperature and humidity profiles) would be more natural than in an OTC with a conventional air-handling arrangement.

System Description

Physical Chamber Units

Each open-top chamber is shaped like an octagonal cylinder with a small frustrum on top (Fig. 1). Chambers are 2.0 m in height from the ground to the top of the frustrum, and about 3.5 m^2 in cross section at the bottom. The total volume of the chambers including the frustrum is approximately 7.2 m^3.

The framework for each chamber is constructed of PVC members joined with directionally adjustable, aluminum fittings (Nu-Rail Corporation). Eight pieces of PVC, 31.7 mm in diameter and 0.74 m in length, are connected with eight of the adjustable fittings (each set to 135 °) to form an octagonal ring. Eight vertical PVC members, 31.7 mm in diameter and 1.9 m in length, are used to support three octagonal rings horizontally at 0.1, 1.0, and 1.9 m above ground.

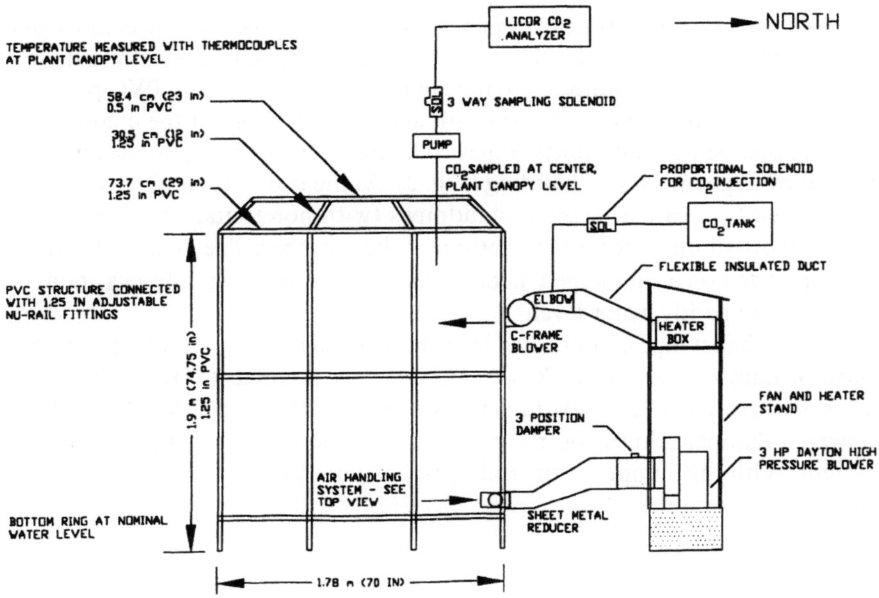

Fig. 1. A schematic side view of an IRRI OTC and its associated control equipment

The frustrum (or octagonal truncated cone) is formed by using 45° elbows that slant inward toward the center of the cylinder. Eight PVC members, 31.7 mm in diameter and 0.30 m in length, project inward and are connected with fittings to the top octagonal ring constructed of PVC members, 12.7 mm in diameter and 0.58 m in length.

Vertical PVC supports extend 0.10 m below the bottom octagonal ring. These extensions support the chamber on the bottom of the rice paddy and nominally position the bottom octagonal ring at the water level in each paddy. To secure the chambers, wooden stakes are driven through the hollow PVC vertical supports and into the soil beneath the paddy to a depth of about 0.30 m.

Each chamber's frustrum and six of its octagonal walls are covered with 5-mil Mylar film (Dupont). Octagonal walls facing north and south are covered with clear acrylic panels, 6.4 mm thick, 0.86 m wide, and 1.45 m in height. To allow access to the plants inside the chamber, the panel on the south side is attached to the PVC frame in a sliding frame. The solid acrylic panel on the north side of the chamber is installed to provide structural support for injection air handling and other devices attached to the chamber.

Primary Air-Handling System

OTCs require that fresh air be flushed through the system to maintain near-ambient concentrations of atmospheric gases and temperatures. In this new system, air is pulled through the chambers from top to bottom. An exhaust pipe formed with a PVC cross (0.1 m diameter) at water level is positioned to pass through the bottom of the north panel. The exhaust pipe is connected to an exhaust manifold that follows the inside contour of the chamber along its north side. The manifold is connected to nine parallel, variable length, PVC pipes (0.05 m diameter) that extend north to south across the bottom of the chamber. The spacing between the horizontal pipes is 0.2 m and was chosen to accommodate typical row spacing in rice planting (Fig. 2). At the end of each pipe, a 90° elbow is attached to create a series of standpipes (with horizontal openings oriented upward) distributed around the bottom of the chamber. The manifold and pipes are supported on a submerged metal frame to position the standpipes about 0.10 m above the paddy water level.

The exhaust pipe is connected by a sheet-metal reducer to a motorized, three position damper (Durozone, Model 38041) which is connected to the inlet side of a 220 V, three-phase, 2237 W, high pressure blower (Dayton, Model 7C561). The powered damper allows control of the volumetric flow rate of air through the system. With the damper in the full open position, the blower provides three air exchanges per minute.

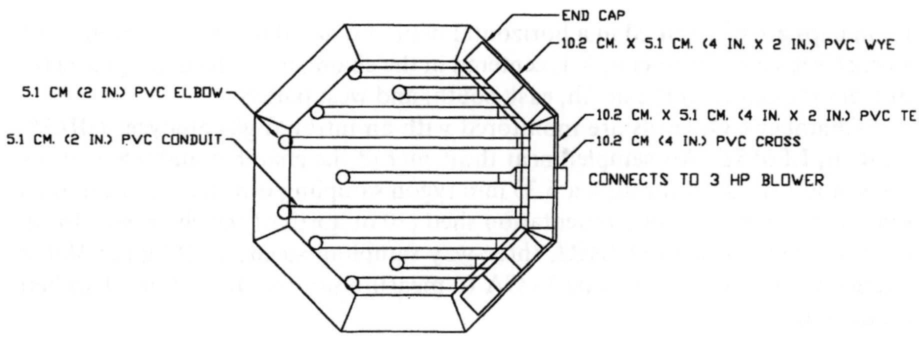

Fig. 2. A top view of the exhaust system of the chamber

Secondary Air-Handling and Conditioning System

Three ports through the north acrylic panel at heights of 0.80, 1.20, and 1.60 m are provided. At any given time, a horizontally oriented, 186 W, C-frame blower with a volumetric flow rate of 2.55 m³/min (Grainger, Model 4C763) is attached to the port immediately above the top of the rice canopy. The blower operates continuously to introduce a secondary air stream into the chamber and has three purposes. First, it provides horizontal mixing across the vertically downward moving primary air stream. Second, the secondary air stream passes over regulated heating coils prior to entering the chamber to raise overall canopy air temperature as required. Third, the secondary air stream is used to inject CO_2 into the primary air stream to create desired super-ambient CO_2 treatment conditions.

The secondary air stream is heated with a 54000-W heater coil (Whirlpool, Model 279698) mounted in a neoprene enclosure (Stahlin) (0.46 m long × 0.30 m wide × 0.22 m deep) lined with high-temperature insulating material. The C-frame blower continuously pulls ambient air into the enclosure, across the heating coil, and through a flexible, insulated hose (0.15 m diameter × 1.00 m long) into the open-top chamber.

To maintain super-ambient CO_2 treatments in the open-top chamber, an inpolene tube connects a CO_2 tank to a proportional control valve at the chamber. Tubing from the exhaust side of the proportional control valve injects CO_2 into the (continuously flowing) secondary air stream near the inlet side of the blower. At the IRRI research site, an 18-Mg, central, refrigerated CO_2 container system is installed to supply 20 OTCs through individual inpolene tubes (6.7 mm diameter) connected to a manifold at the CO_2 tank.

Sensors

To evaluate temperature distribution in the chamber, a set of five nonaspirated copper-constantin thermocouples shielded with aluminum-foil-covered styrofoam cups are distributed in a horizontal plane 1.6 and 0.6 m above the ground. In each set, one thermocouple is centered in the chamber and four are placed 0.3 m from the center of the north, east, south, and west panels.

Chamber CO_2 levels are monitored with an infrared gas analyzer (IRGA) (Li-Cor, LI 6552). Air sampled near the center of the chamber and 0.6 m above the ground is pulled through a 6.35 mm tygon sampling tube by a continuously operated pump to an instrumentation shed (10 m north of the chamber). In the multi-chamber system at IRRI, three-way sampling solenoids (Skinner Valve, Model V53LB2100) allow one IRGA to measure samples from four chambers sequentially.

Data Acquisition and Control System

Data from the sensors are recorded using a programmable CR10T microprocessor and two AM416 multiplexers for analog input. Data from the CR10T is periodically uploaded through an addressable serial network (Campbell Scientific multidrop interface MD-9 and SC532A) to a host computer (Tatung i486) and stored on the hard drive of the computer. In additon to handling data acquisition, the CR10T handles real-time control of chamber temperature and CO_2 level through a SDM-CD16 (Campbell Scientific) digital output device and through two AO4 analog output devices (Campbell Scientific). The data acquisition and environmental control system is shown schematically in Fig. 3. At IRRI, the data acquisition equipment is housed in an air conditioned instrumentation trailer approximately 30 m from the OTCs.

Thermocouple measurements taken every 10 s provide feedback for temperature control in the chambers. Temperature control requires that two devices be regulated: the motorized damper and the 5400-W heating element. Although the blower for the primary air stream operates continuously, the volumetric airflow rate depends on the position of the damper.

The damper is controlled to be in one of three positions: closed, one-third open, or fully open. If the rate of airflow through the chamber is reduced, heating and CO_2 injection requirements are likewise reduced. This design is intended to allow the control program to choose the lowest flow rate that allows chamber temperatures to be kept below the desired experimental treatment level. After the airflow rate is determined, the program uses an AO4 and a proportional controller (Douglas Randal, Model R40B) connected to an input module (Douglas Randal, RDB) to regulate the heating element to maintain the desired temperature.

CO_2 injection rate is regulated with a proportional solenoid valve (Skinner Valve, Model BT2EV0012). At startup, the control program initializes the gross valve position to correspond to damper position. When the damper position changes, the base position of the valve is immediately reinitialized. From its base position, direct CO_2 measurements taken every 5 min with the IRGA provide feedback control to fine tune the valve position. The control program used a

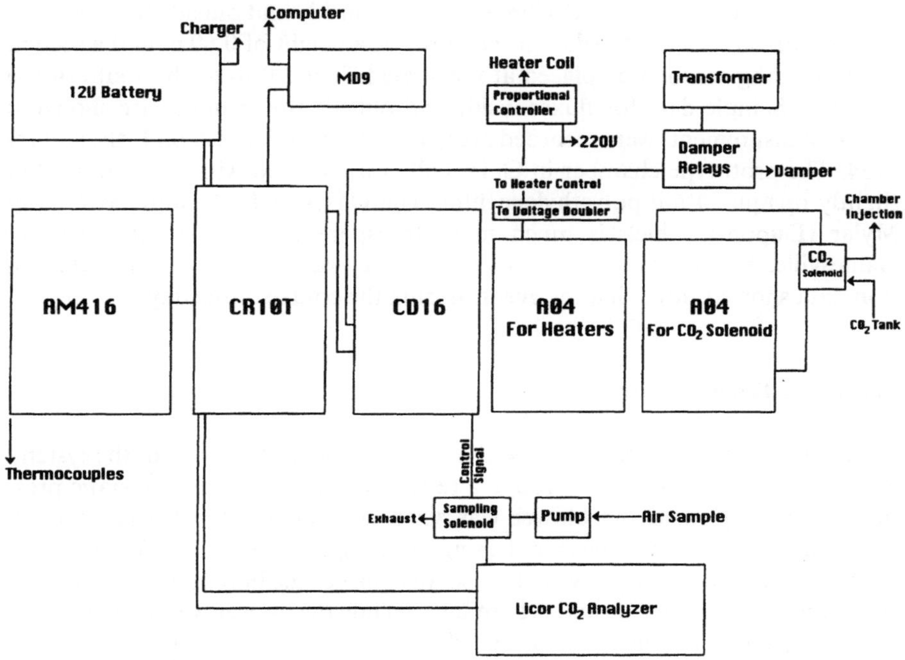

Fig. 3. A schematic representation of the data acquisition and environmental control system

simple proportional control algorithm to maintain the desired CO_2 treatment level at some increment above ambient.

Performance Testing

Initial tests of chamber performance were conducted on the prototype chamber at the University of Florida Energy Research Park (Gainesville, FL) to characterize environmental conditions in the chamber: specifically, to evaluate differences between the chamber and open-field environments, to determine spatial homogeneity of the chamber environment, and to evaluate the ability of the control system to maintain treatment conditions. All tests were performed with St. Augustine grass turf growing in the chambers.

Light Transmission

All chambers reduce light flux to the canopy because of the translucent material covering the chamber and the structural members. Light energy sensors (Li-Cor

190s) were used to measure incident energy flux inside and outside the chamber. One sensor was centered inside the chamber at a height of 0.60 m and a second corresponding sensor was placed at the same height 2.0 m to the south of the chamber. Sample data for fluxes inside and outside the chamber are shown in Fig. 4. Measurements were recorded every 10 min during the day on 19th January 1994. The prototype chamber in Gainesville was covered with translucent, but slightly opaque, 4-mil polyethylene film. The OTCs at IRRI were covered in Mylar (Dupont), which is much more transparent than the film used in Gainesville. Average transmission during the day was about 74%. The transmission data showed three shadow events during the course of the day.

Characteristics of Airflow

Volumetric airflow rate through the chamber is basic to the design of the system. Because the system pulls air through a restrictive exhaust manifold, static pressure is very high and the blower was sized accordingly. On 8 March 1993, volumetric flow rate was measured using an Omega Rotary Vane Anemometer on the exhaust side of the blower. The exhaust port of the blower was rectangular and measured 0.10 m wide and 0.15 m high. While the blower was in continuous operation, the anemometer was moved through five grid positions over the exhaust area (one at each corner and one at the center). The series of measurements was repeated with the damper in each of its three positions: full open, 1/3 open and "closed". Measurements were recorded and averaged to estimate total volumetric flow rates of 0.38 m^3/s, or 3.17 air exchanges, with the damper fully open, 0.27 m^3/s in the one-third open position, and 0.16 m^3/s in the closed position.

Temperature distribution was measured in the chamber to evaluate whether the air being pulled vertically through the chamber and mixed with the horizontal blower provided a sufficiently homogeneous chamber environment. The two sets of five horizontally distributed thermocouples were used to measure temperature distribution in the chamber. Temperatures were measured through several

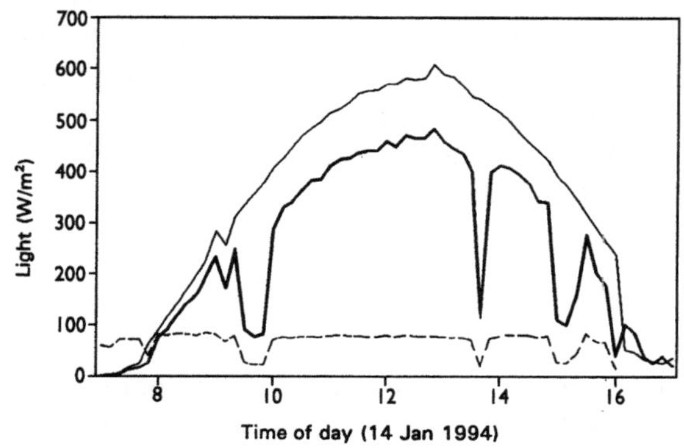

Fig. 4. Characteristics of sunlight transmission in the chamber: ambient is *dark line*, chamber is *light line*, and percentage transmission is *broken line*

diurnal cycles with the dampers kept in the full open position and the temperature setpoint equal to ambient (no heating). The purpose was to determine the degree of heat-up that might occur with the greatest possible volumetric flow rate of ambient air passing through the chamber. A second purpose was to determine how variable the temperature was horizontally and vertically. Measurements were taken on 29 and 31 July 1993, the averages of the five thermocouples at 1.6 m were 1.12 and 1.08 °C above ambient, respectively. For daytime hours only, the averages were 2.30 and 2.23 °C above ambient, respectively. Among the five thermocouples at 1.6 m (during the daytime on 29 July 1993), the average difference between individual thermocouple measurements and the average for all five ranged from –0.73 to +0.47 °C. More work is being done to characterize horizontal and vertical variations.

Temperature Controls

Figure 5 shows temperature controlled to 4 °C above ambient. The data are for 19 January 1994 when temperature was measured every minute and averaged over 10-min intervals throughout a 24-h period. Chamber temperature was based on the average of the five thermocouples. Data were averaged because the ambient temperature was measured just above the frustrum and was affected by chamber temperature. As a result, ambient temperature values fluctuated rapidly over a range of several degrees. Because the control setpoint was based on the fluctuating values, it was not possible to separate deviations from the setpoint caused by fluctuations in ambient temperature from those caused by the control system. Keeping these circumstances in mind, controls were reasonable. Average chamber temperature over the entire cycle was 3.91 °C above ambient.

For a second series of measurements the shielded thermocouple used to measure ambient temperature was moved 3 m away from the chamber and kept at a height of 2 m. Chamber temperature was controlled to ambient +6 °C. In additon to testing the stability of the ambient temperature measurement, the general purpose was to test the capacity of the elevated-temperature control subsystem. Measurements were taken over a complete diurnal period on 21

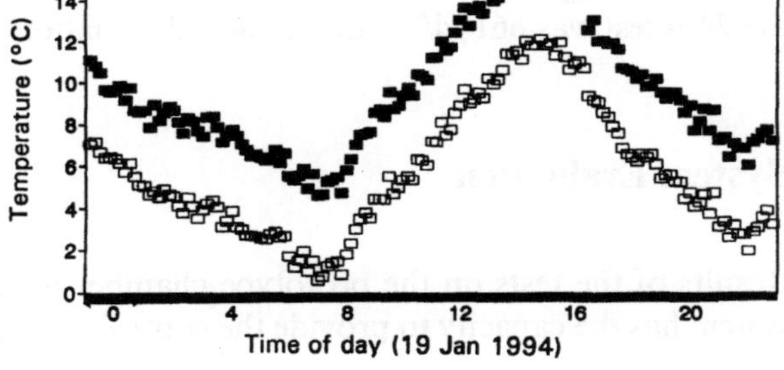

Fig. 5. Chamber controls for ambient +4 °C: ■ chamber and □ ambient

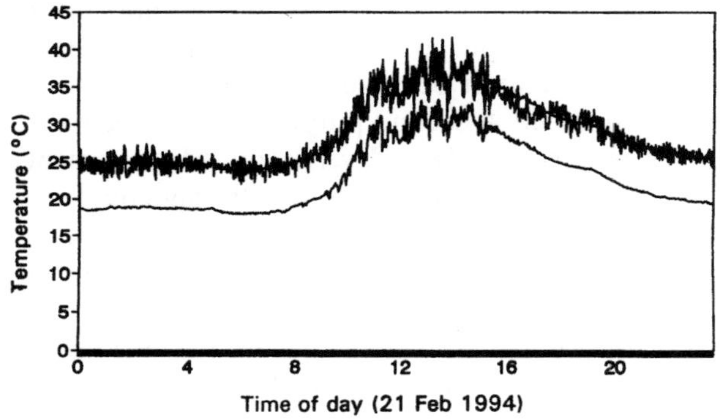

Fig. 6. Chamber controls for ambient +6 °C: chamber is *upper jagged line*, setpoint is *line through chamber readings*, and ambient is *lower jagged line*. Average temperature difference 5.92 °C; SD 1.27 °C

February 1994. Figure 6 shows measured temperatures taken at 1-min intervals inside and outside the chamber. Ambient values were more stable, especially under low light conditions. Average chamber temperature over the 24-h period (1400 readings) was 5.92 °C above ambient with a standard deviation of 1.27 °C. During the daylight portion of the test period, skies in Gainesville were partly cloudy and it was windy with gusts to about 20 km/h. Based on observations of the shielded thermocouples being buffeted inside the chamber during the test period, it is clear that the windy conditions were affecting the chamber environment. Under these relatively extreme conditions the system worked well.

Carbon Dioxide Injection System

Elevated CO_2 controls were tested over a 24-h period on 24 February 1994. During the test, the CO_2 concentration was controlled to 660 μl/l.. The CO_2 was injected into the horizontal air stream entering the chamber at a height of 0.8 m, and the sample pickup was located in the center of the chamber at a height of 0.6 m. The sample was pulled through a 1.5-l baffling chamber at the rate of 6 l/min and through the Li-cor IRGA at 2 l/min. Results (Fig. 7) indicate that the capacity of the system to deliver CO_2 was adequate and its controls were able to maintain a target CO_2 level within 10% of the setpoint. Average CO_2 level during the 24-h test was 661 μl/l with a standard deviation of 28 μl/l for 288 readings.

System Evaluation

Results of the tests on the prototype chamber in Gainesville indicate that the system has the capacity to provide the controls originally specified. It controlled

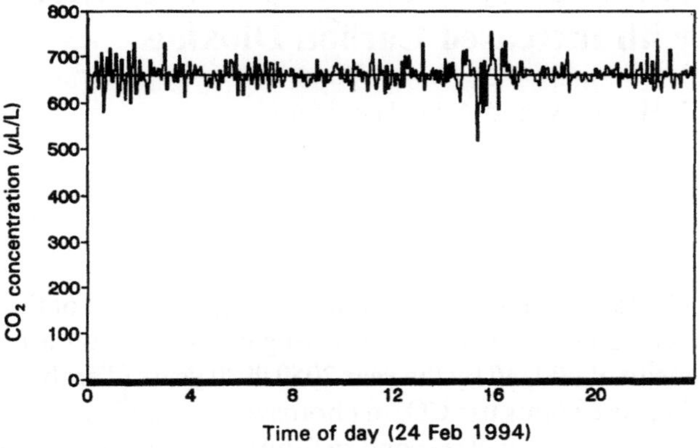

Fig. 7. Chamber CO_2 controls at 660 μl/l: chamber CO_2 level is *jagged line* and CO_2 setpoint is *straight line*. Average CO_2 level 661 μl/l; SD 28 μl/l

chamber temperature to 6 °C above ambient, even under windy conditions, and it provided double ambient CO_2 levels. Stability was reasonable for CO_2, but it could be improved for temperature. Temperature control can probably be improved to some extent by fine tuning the control algorithms. In terms of its basic design, the vertical volumetric flow rate of air through the system needs to be increased by increasing either the number or the cross sectional area of standpipe openings in the bottom of the chamber, or by increasing the size of the exhaust fan. Likewise, the horizontal airflow into the chamber needs to be baffled for better spatial distribution. In addition to these design issues, initial experience with the fielded chambers at IRRI has led to other redesign suggestions involving the frustrum.

There are two primary advantages of the new OTC design. First, this design provides simple and reasonably effective controls of elevated temperature unlike other OTCs that have no temperature controls. Second, this design provides more field-like canopy microclimate profiles than OTCs that push air upward through the canopy. The current design does have one significant disadvantage compared with other OTCs. The new exhaust system creates high static pressures that require much more powerful fans to pull air downward through the canopy than in OTC that push air upward through the canopy. As a practical result, the volumetric flow rate tends to be lower in the newly designed OTCs. Redesign should alleviate the problem. With these caveats in mind, the system has the potential to be a useful research tool for studying $CO_2 \times$ temperature interactions. Finally, it should not be assumed that this is a simple system because it is based on OTC technology. The system requires considerable infrastructure investment and technical support to remain fine-tuned and to be operated to its fullest potential.

Intraspecific Variation in Growth and Photosynthetic Responses of Rice with Increased Carbon Dioxide

L.H. ZISKA[1] and A.H. TERAMURA[2]

Global atmospheric CO_2 is increasing as a result of the burning of fossil fuels and deforestation. There is a growing consensus that atmospheric CO_2 will reach a mean of 600 μl/l by the year 2050 ± 20 years (Tarabalka et al. 1985; Strain 1987). Because plants use CO_2 in photosynthesis and growth, there is continued interest in how plants will respond to future increases in CO_2. It is clear from a number of experiments that increasing CO_2 will have significant effects on photosynthesis and growth (Kimball 1983). However, there is a wide range of interspecific variation for temperate and tropical, cultivated and native plant species (Kimball 1983; Cure and Acock 1986; Ziska et al. 1991).

Little information is available to determine if this interspecific variation also applies to intraspecific variation. If genotypes within a population differ in their response to CO_2, such information could be used to select specific strains or cultivars with optimal production in a future, enhanced CO_2 environment. Presumably, as with interspecific variation, optimal performance under elevated CO_2 would be linked to three key parameters: the degree of photosynthetic enhancement and whether initial stimulation of photosynthesis is maintained over time, changes in biomass partitioning, especially changes in root-shoot ratio or increased allocation of biomass into reproductive structures, and total crop biomass through greater leaf production or ability to compete with weeds.

To date, the only study to examine intraspecific variability has been with natural populations of *Plantago lanceolata* grown at ambient and twice ambient CO_2 concentrations (Wulff and Alexander 1985). Data from this experiment indicated the existence of substantial genetic variability in this species in response to increased CO_2. However, to our knowledge, no data are as yet available on genotypic variation in an agronomically important species. As CO_2 continues to rise, future rice breeders may want to select lines based on optimal physiological and economic performance with increased CO_2.

For the present study, we examined those characteristics that would optimize performance in an elevated-CO_2 environment using two contrasting rice cultivars: a short indica rice, IR36 from the Philippines, and a tall, japonica rice, Fujisaka-5 from Japan. Our purpose in exposing these cultivars to a future CO_2

[1] International Rice Research Institute, P.O. Box 933, 1099 Manila, Philippines
[2] College of Natural Sciences, University of Hawaii, Honolulu, HI 96822, USA

S. Peng et al. (Eds.) Climate Change and Rice

environment was twofold: to characterize variation in the stimulation of photosynthesis at high CO_2 concentrations, and to quantify differences in reproductive characteristics, biomass enhancement, and biomass partitioning.

Materials and Methods

Fujisaka-5 produces few tillers and obtains heights exceeding 100 cm. In contrast, IR36 is short (<60 cm) and produces many tillers. Seeds of each cultivar were sown in 20-l pots in unshaded greenhouses on 17 July 1990 at the Duke University Phytotron, NC. Large pots were used to prevent potential feedback limitations to photosynthesis resulting from restricted root growth (Robbins and Pharr 1988; Thomas and Strain 1991). Plants were grown in a medium of pea gravel, sand, and peat (v:v:v 2:2:1). All pots were placed on carts, and carts were rotated weekly to reduce bias from greenhouse spatial heterogeneity. Pots were watered four times a day to saturation with half-strength Hoagland solution (Downs and Hellmers 1978), but pots were not flooded.

Twenty plants of each cultivar were exposed to either ambient CO_2 (360 ± 40 μl/l) or elevated CO_2 (660 ± 53 μl/l) from seed germination through harvest. Five replicates of four plants each were randomly distributed within a treatment. The CO_2 concentrations were maintained using a CO_2 injection system. Growing conditions in each of the two phytotron greenhouses were: 31/23 °C day-night temperature, 65% relative humidity, and total daily photosynthetic photon flux (PPF) at 80–85% ambient. Afternoon ambient light on cloudless days at Durham, NC, typically approaches 1800–2000 μmol m^{-2} s^{-1}.

Initial gas-exchange measurements were obtained for the flag leaf of four replicate plants from each CO_2 treatment at 64 days after sowing (DAS). Panicle initiation and development occurred 56–70 DAS. Leaf CO_2 flux was measured in an open gas-exchange system (Model LCA 3, ADC Corporation, Hoddesson, UK) that allowed for independent control of CO_2 and humidity reaching the leaf. Measurements of light-saturated CO_2 assimilation were made for each cultivar and CO_2 treatment at 64 and 79 DAS to determine if the enhancement of photosynthesis at high CO_2 was changing with time. All gas exchange parameters were calculated according to the parameters of von Caemmerer and Farquhar (1981).

The response of CO_2 assimilation (A) to internal CO_2 (C_i) was determined by initially measuring A at the external growth CO_2 concentration (C_a) and PPF of 1800 μmol m^{-2} s^{-1}. No increases in net CO_2 assimilation were observed at PPF greater than 1800 μmol m^{-2} s^{-1}; therefore, this PPF was assumed to be saturating. External CO_2 concentration was then reduced to 130 μl/l and A was measured at increasing C_a values of 130, 260, 390, 650, 1040, and 1430 μl/l. Average leaf temperature during measurements was 33.7 °C. Natural sunlight was supplemented with a Westinghouse 300-W cool-beam floodlight on cloudy days. The airstream was humidified to a given dew point to maintain a vapor pressure deficit less than 2.0 kPa within the chamber. Dark respiration rates were measured at 21:00–24:00 on six to eight replicate flag leaves from each cultivar and CO_2 treatment using the same CO_2 analyzer.

To determine if differences in CO_2 assimilation occurred as a result of stomatal limitation, rates of CO_2 assimilation obtained with the gas analyzer were compared with O_2 evolution rates obtained at saturating PPF in an oxygen electrode at a super-saturating concentration of 50 000 μl/l CO_2. Because little change was observed in O_2 evolution between 50 000 and 100 000 μl/l, 50 000 μl/l CO_2 was assumed to be saturating. Light at 1800 μmol PPF m^{-2} s^{-1} was supplied by a Bjorkman lamp connected to a Hansatech LS-2 light source (Hansatech Ltd, Kingslyn, UK) during measurement. Temperature of the electrode was maintained at 34 °C by circulating water through a water bath. Chlorophyll was determined for the leaf discs used in O_2 evolution measurements using the procedure of Knudson et al. (1977).

To determine if elevated CO_2 changed photochemical efficiency, the ratio of variable to maximal leaf fluorescence (F_v/F_m) was measured using a portable fluorescence meter (Plant Stress Meter, Biomonitor, Charleston, SC) on ten leaves per cultivar per treatment from 13:00 to 16:00 over a 3-day period 1 week after photosynthetic measurements. All plants were harvested 113 DAS (approximately 50% grain fill). Plants were separated into panicles, vegetative shoots, and roots, dried at 65 °C for 7–10 days, and then weighed. Subsamples of 20 leaves per cultivar per treatment chosen at random were placed in moistened plastic bags to prevent leaf rolling and leaf area was determined with an area meter (Model 3100, Li-Cor Corporation, Lincoln, NE). These leaves were then dried for 48 h at 65 °C and total leaf area per plant and specific leaf weight were estimated by dividing leaf dry weight by leaf area (r^2=0.85). Statistical differences in growth or photosynthetic parameters were determined by analysis of variance and different means were separated by Student-Newman-Keuls multiple range test at the p=0.05 level.

Results and Discussion

Photosynthetic Parameters

Measurements for rice grown and measured at 360 or 660 ml/l at 64 DAS indicate a significant enhancement of photosynthesis in both cultivars by elevated CO_2 (Table 1). No decline in the stimulatory effect of elevated CO_2 was observed for A measurements at 79 DAS (Table 1). The enhancement effect of elevated CO_2 persisted and was approximately 50% for both cultivars even at 79 DAS. This suggests that in these two cultivars, elevated CO_2 has a long-term stimulatory effect on photosynthetic rate that does not appear to rapidly change with time.

The response of A over a range of C_i values varied depending on the growth C_a (Fig. 1). Leaves grown and developed at elevated CO_2 had a lower initial slope

Table 1. Changes in light-saturated A (μmol CO_2 m^{-2} s^{-1}) ± standard error on two dates for two rice cultivars grown at 360 and 660 μl/l CO_2 during the growing season (n = 4–6)

Cultivar	Growth CO_2 (μl/l)	Photosynthesis	
		64 DAS	79 DAS
Fujisaka-5	360	28.9 ± 0.3	29.1 ± 1.5
	660	42.4 ± 0.5	43.0 ± 2.7
IR36	360	23.2 ± 0.9	22.6 ± 2.1
	660	36.2 ± 1.3	33.2 ± 2.3

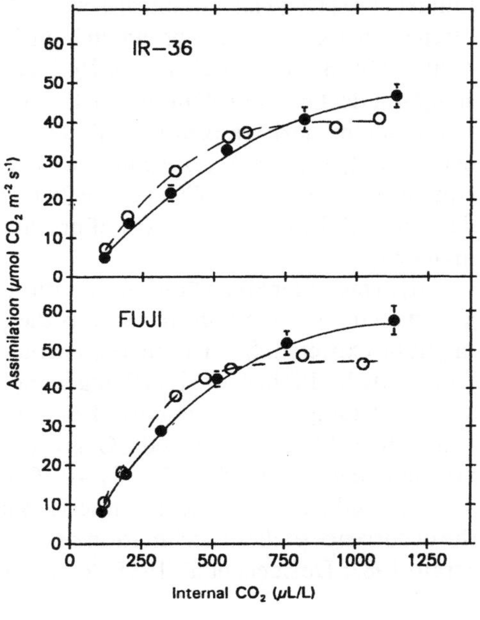

Fig. 1. The response of CO_2 assimilation to intercellular CO_2 (C_i) at saturating light intensity in two rice cultivars grown at 360 μl/l (O) and 660 μl/l (●) (n = 3–4)

of the A versus C_i curve, but saturated at both higher A and Ci when compared with leaves grown and developed at ambient CO_2 (Fig. 1). Assuming saturating PPF, A at low C_i values in C_3 plants is determined by the activity and concentration of Ribulose 1,5-bisphosphate carboxylase/oxygenase (Rubisco) (Farquhar et al. 1980; Motto 1990). As C_i increases and Rubisco becomes saturated, A is limited by either regeneration of the substrate, ribulose-bis-phosphate (RuBP), or inorganic phosphate (P_i) (Woodrow and Berry 1988; von Caemmerer and Farquhar 1981). The difference in A response to C_i at elevated and ambient CO_2 suggests a possible reallocation of resources (probably N) from the limitation imposed by Rubisco to that imposed by RuBP/P_i (Sage et al. 1989). The A versus C_i response shown here for leaves grown in elevated CO_2 would be consistent with optimal N utilization. The net result is an enhanced ability to use CO_2 for those plants grown at a higher CO_2 concentration. A similar result has been observed for certain tropical plants exposed to elevated CO_2 (Ziska et al. 1991).

To further test this idea in rice, leaf discs from each CO_2 concentration were examined for O_2 evolution at super-saturating concentrations of CO_2 (50 000 μl/l). Differences in O_2 evolution at saturated CO_2 will not be affected by stomatal limitation to CO_2 uptake or by differences in photorespiration, but instead, by internal changes in the mechanism of CO_2 fixation. Significant increases in O_2 evolution were observed for both IR36 and Fujisaka-5 at high relative to ambient CO_2, suggesting an increased sensitivity or acclimation with an enhanced CO_2 environment (Table 2).

Part of the change in the mechanism of photosynthesis at elevated CO_2 could involve a reduction in the amount of leaf N or protein (primarily Rubisco) per unit leaf area (Rowland-Bamford et al. 1991; Baker et al. 1992d). Foliar nitrogen often decreases for plants grown in high CO_2 concentrations (Wong 1979; Curtis et al. 1989; Besford et al. 1990). Reductions in leaf protein at high CO_2 could suggest a decline in protein production or faster degradation, and a subsequent reduction in dark respiration (Ryan 1991; Ziska and Bunce 1993). In this experiment, leaf dark respiration was reduced in both cultivars for elevated compared with ambient CO_2 (Table 2). Growth data are inadequate, however, to determine if the respiratory cost of producing or maintaining tissue is reduced at high CO_2.

If, in fact, there is an "up" acclimation of photosynthesis within leaves grown at elevated CO_2, then an increase in the photochemical efficiency at high CO_2 might be expected. Such an increase would be consistent with reduced ability to regenerate RuBP because RuBP regeneration depends principally on the capture and processing of light energy. For both cultivars, photochemical efficiency tended to be higher in elevated CO_2 grown leaves, although these differences were not significant (p=0.15 for IR36 p=0.10 for Fujisaka-5) (Table 2).

Up acclimation of photosynthesis to high CO_2 is somewhat surprising given the occurrence of *down* acclimation or desensitization in the literature (Wulff and Strain 1983; DeLucia et al. 1985; Reekie and Bazzaz 1989; Yelle et al. 1989a,b).

Table 2. Changes in photosynthetic characteristics for two rice cultivars grown at 360 and 660 μl/l CO_2. The O_2 evolution rates were obtained at 50 000 μl/l CO_2 for discs obtained from flag leaves (n = 8–10); Respiration rates are based on nighttime CO_2 efflux (20:00–24:00) from leaves of two rice cultivars (n = 8–10). Photochemical efficiency is based on the ratio of F_v/F_m (variable to maximal fluorescence), obtained from afternoon measurements of dark adapted leaves (n = 8–10). Additional details are given in the text[a]

Cultivar	Growth CO_2 (μl/l)	O_2 evolution (μmol O_2 m^{-2} s^{-1})	Dark respiration (μmol CO_2 m^{-2} s^{-1})	Photochemical efficiency (F_v/F_m)
Fujisaka-5	360	56.8 ± 4.5	0.32 ± 0.02	0.68 ± 0.03
	660	63.4 ± 1.3*	0.22 ± 0.03*	0.72 ± 0.01NS
IR36	360	43.2 ± 1.4	0.47 ± 0.06*	0.66 ± 0.02
	660	48.6 ± 1.3*	0.25 ± 0.03	0.69 ± 0.02NS

[a] The * indicates significance at the p=0.05 level (S–N–K test) for a given measurement, NS not significant, all values ± 1 standard error.

The appearance of desensitization may be dependent on the balance between photosynthetic production of carbohydrates (source) and their consumption by growth processes (sink) (Neales and Incoll 1968; Herold 1980). Many of the experiments that demonstrate photosynthetic desensitization to high CO_2 may have imposed limitations to growth because of low light (Yelle et al. 1989 a,b) or by growing plants in small pots that may have limited root development (Reekie and Bazzaz 1989). In a setting where growth is not artificially limited, photosynthetic acclimation to high CO_2 does not necessarily occur (Ziska et al. 1990). If the photosynthetic response to increased CO_2 is dependent on the long-term physiological demand for carbon, limitations to growth may have limited photosynthetic capacity, with a subsequent reduction in the ability to physiologically exploit additional CO_2.

Biomass and Partitioning

Elevated CO_2 significantly increased total plant biomass in both rice cultivars (Tables 3 and 4). The increase was similar in both cultivars, 27% for IR36 and 32% for Fujisaka-5. Increased biomass in both cultivars was principally caused by increases in stem and root weight (Table 3). In contrast, leaf area tended to decrease with increased CO_2, but because specific leaf weight increased, no differences in leaf weight were observed with enhanced CO_2 in either cultivar. Roots responded more to CO_2 than did shoots; consequently, root-shoot ratio increased at high CO_2. However, this result was significant only for Fujisaka-5 (Table 3). Observed differences in biomass allocation to roots with high CO_2 suggest an enhanced ability to extract minerals under nutrient-limiting conditions for Fujisaka-5 relative to IR36.

Given the nature of this experiment, it is not possible to extrapolate with confidence how elevated CO_2 will change economic yield. However, changes in tiller number, average panicle weight, and panicle number suggest some potential changes in rice with long-term exposure to elevated CO_2 (Table 4). These indicators of reproductive potential varied intraspecifically with enhanced CO_2. For example, tiller number significantly increased in Fujisaka-5, although panicles per tiller and panicle weight remained unchanged (Table 4). In contrast IR36 had greater panicle weight, but no change in tiller number or panicles per tiller. The harvest index (HI), calculated as panicle weight divided by total plant weight, was not significantly affected by elevated CO_2 for Fujisaka-5, but HI increased 40% in IR36 (Table 4). Differences in reproductive characteristics between these two cultivars suggest that rice could be screened with respect to optimal yield in a future, elevated-CO_2 environment.

Table 3. Changes in biomass allocation and vegetative characteristics of Fujisaka-5 and IR36 at 360 and 600 μl/l CO_2. Data were obtained at final harvest (113 DAS) [a]

	Growth CO_2 (μl/l)	Root (g/plant)	Leaf (g/plant)	Stems (g/plant)	Total vegetative (g/plant)	R–S	Leaf area (m^2/plant)	SLW (g/m^2)
Fujisaka-5	360	30.2 ± 2.0	82.4 ± 0.8	40.3 ± 1.3	153.2 ± 2.6	0.24 ± 0.01	1.70 ± 0.02	36.5 ± 2.1
	660	63.9 ± 5.4	84.6 ± 1.2	55.0 ± 1.4	203.9 ± 4.4	0.46 ± 0.04	1.60 ± 0.03	43.5 ± 1.5
Change (%)		111.6% *	2.7% NS	36.5% *	33.1% *	92.0% *	–5.9% NS	19.2% NS
IR36	330	61.1 ± 4.9	100.8 ± 2.5	63.0 ± 1.7	225.2 ± 3.2	0.37 ± 0.02	1.73 ± 0.04	39.9 ± 1.3
	660	84.1 ± 6.7	109.0 ± 2.7	74.4 ± 2.0	268.2 ± 4.2	0.46 ± 0.04	1.60 ± 0.04	48.2 ± 1.9
Change (%)		37.6 % *	8.1 % NS	18.1% *	19.1% *	24.3% NS	–7.5%	20.8% *

[a] R–S root-shoot ratio, SLW specific leaf weight, all values are ± standard error, the * indicates significance at the p=0.05 level (S–N–K).

Table 4. Changes in reproductive characteristics (± standard error) in Fujisaka-5 and IR-36 with enhanced CO_2 [a]

Cultivar	Growth CO_2 (μl/l)	Tillers/plant	Panicles/plant	Average panicle weight (g)	Total reproductive weight (g)	Total Plant weight (g)	HI
Fujisaka-5	360	32.3 ± 1.7	1.04 ± 0.05	2.29 ± 0.13	75.3 ± 3.4	228.6 ± 13.2	0.33 ± 0.03
	660	39.5 ± 1.2	1.04 ± 0.05	2.42 ± 0.08	97.1 ± 4.2	301.0 ± 11.2	0.32 ± 0.05
Change (%)		22.3% *	2.0% NS	5.7% NS	28.9% *	31.7% *	–3% NS
IR36	360	95.0 ± 4.2	0.67 ± 0.03	0.63 ± 0.03	39.5 ± 3.7	264.7 ± 17.4	0.15 ±0.03
	660	91.3 ± 3.4	0.73 ± 0.03	1.04 ± 0.06	69.3 ± 4.8	337.5 ± 26.1	0.21 ±0.02
Change (%)		–3.9% NS	8.9% NS	65.1% *	75.4% *	27.5% *	40.2% *

[a] HI harvest index (calculated as total panicle weight divided by total plant weight), the * indicates significance at the p=0.05 level (S–N–K test)

Conclusion

Because elevated CO_2 stimulated photosynthesis until panicle initiation and there was a possible reallocation of resources to "up" acclimate photosynthetic response, because elevated CO_2 substantially enhanced biomass, especially for roots, and because reproductive characteristics varied substantially between cultivars for their response to elevated CO_2, we conclude: (1) that leaf photosynthesis and plant growth will respond favorably to elevated CO_2 for certain cultivars, and (2) that sufficient intraspecific variability exists to select cultivars that will maximize biomass or reproductive yield with future increases in atmospheric CO_2.

Physiological Response of Rice to Carbon Dioxide, Temperature, and Nutrients

K. Imai

Carbon dioxide, together with increases in other greenhouse gases, may induce climate change in the future. Therefore, CO_2 affects plant life directly through photosynthesis and indirectly through climate change. Elevated atmospheric CO_2 concentrations promote growth and yield of rice by way of the enhancement of net photosynthetic rate, but the degree of promotion is affected by concomitant factors such as light intensity, temperature, and mineral nutrition (Imai 1988, 1993; Baker and Allen 1993b).

Rice is one of the staple crops in the world. In many areas, however, rice is cultivated without or with small amounts of fertilizer, which limits productivity. If rice yield interacts with fertilizer at elevated CO_2 conditions, the prediction and measurement of the production process is becoming more important because of concurrent population growth in such areas. Under the predicted global greenhouse warming conditions, the day-night temperature difference may decrease (D. Bachelet, pers. comm.), that is, night temperature will increase more than day temperature, especially in rice-growing areas, because of the large specific heat of water.

This chapter reviews the responses of rice to elevated CO_2, temperature, and fertilizer as the physiological basis of yield formation, although experimental observations are still unsatisfactory. This discussion assumes: an adequate water supply by rain or irrigation, that solar radiation will not be changed significantly, and that there will not be unusual damage of rice by pests.

Gas Exchange

Photosynthesis

Net leaf photosynthetic rate (Pn) of rice is substantially increased by temporary elevated CO_2 concentration (30–50% at 700–1000 μl/l) when compared with the normal CO_2 level (about 350 μl/l, Imai and Murata 1978c; Imai and Okamoto-Sato 1991). This is caused by the suppression of photorespiration by high CO_2

Institute of Agriculture and Forestry, University of Tsukuba, Tsukuba, Ibaraki 305, Japan

S. Peng et al. (Eds.) Climate Change and Rice

relative to O_2 in the atmosphere and is progressed, within limit, under higher light intensity and temperature conditions (Imai and Murata 1978c, 1979a,b), rather than the suppression of dark respiration rate (Amthor 1991; Imai and Nomura 1992).

When rice plants are subjected to higher atmospheric CO_2 concentrations, leaf Pn often declines from the expected rates within a few days. This is called photosynthetic acclimation to high CO_2 (Imai and Murata 1978b,c). Even if plants are returned to normal CO_2 conditions, their photosynthetic capacity may not recover until new leaves develop (Imai and Murata 1978b,c). In cotton leaves, the suppression recovers within 4–5 days (Sasek et al. 1985).

Acclimation may be induced by: the suppression of Rubisco activity, the excessive accumulation of starch in chloroplalsts, the limited RuBP regeneration because of insufficient inorganic phosphate influx, or the limited sink size of plants relative to enhanced photosynthesis (Wong 1979; Rowland-Bamford et al. 1991; Conroy and Hocking 1993; Imai 1993). In these cases, factors related to carbohydrate metabolism are mutually related, and N and P are more important at high CO_2. Leaves tend to accumulate carbohydrate more than mineral nutrients at high CO_2, thereby elevating the C-N ratio and diluting mineral contents. The latter may be coupled with lowered mass flow through decreased transpiration (Conroy and Hocking 1993). If nutrients are supplied adequately (e.g., by solution culture), acclimation may be minimized (Imai and Okamoto-Sato 1991; Imai and Nomura 1992; Hotta and Imai 1994). In fact, acclimation occurs rapidly and mashes the CO_2 effect when rooting space of cotton is limited, such as by small pots (Thomas and Strain 1991).

In rice, substantial responses may occur to P nutrition because there are large areas of ricefields that are deficient in available soil P (Tanaka 1984), and this element is the most immobile of the major plant nutrients (Mengel and Kirby 1987). At the elevated CO_2 concentrations predicted for the future, the promotion of Pn by CO_2 may be either increased or decreased by the P requirement of rice plants (Adachi and Imai 1991; Conroy and Hocking 1993). We observed, within limits, positive CO_2 × temperature interaction on leaf photosynthesis of rice plants (Imai and Okamoto-Sato 1991), but Baker and Allen (1993b) observed relatively insensitive interactive responses under dense populations.

Respiration

Baker et al. (1992d) reported that the specific dark respiration (Rd) of a rice population was suppressed with increasing CO_2 concentration across the range 160–900 μl/l. However, most of the effect was with exposures of 160 and 250 μl/l. If this occurs generally, suppressed Rd may increase net production of rice, unless low Rd adversely affects energetic processes. When young rice plants (Haun index: 6–8) were grown under 350 or 700 μl/l CO_2, Rd was higher in 700 than in 350 μl/l, and Rd was closely correlated with previous daytime Pn. Specific leaf weight of rice was higher in 700 μl/l CO_2 plants, but effects of CO_2 on Rd did not change whether expressed on leaf area or dry weight basis (Hotta and Imai 1994).

Mineral nutrition may affect response of rice to elevated CO_2. At elevated CO_2, relative increase of the C-N ratio may decrease the specific respiration rate, especially at later growth stages when maintenance respiration is the largest fraction of respiration (Amthor 1991). Respiration effects are probably explained by increased levels of inert carbon in high-CO_2 plants. Imai and Nomura (1992) observed a slight suppression of leaf Rd at high CO_2 and high P conditions, but the reason was not known. At higher termperatures, rice Rd may acclimate to its environment and become lower than the rate expected from instantaneous measurement (Amthor 1991).

Transpiration and Efficiency of Water Use

Transpiration is controlled by evaporative demand around plant surfaces (e.g., saturation deficit, temperature, wind) (Kanda et al. 1984; Imai and Okamoto-Sato 1991) and stomatal aperture. As high CO_2 decreases stomatal aperture (or stomatal conductance), transpiration may decrease if CO_2 increases and if leaf temperature is constant (Imai and Murata 1978c; Imai and Okamoto-Sato 1991). Transpiration may also decrease somewhat at high CO_2 because of decreased stomatal density (Imai 1977), although the opposite observation on stomatal density has also been reported (Rowland-Bamford et al. 1990). However, leaf temperature increases at high CO_2 under sunlight (Imai and Murata 1976). High leaf temperatures may partly counteract the expected suppression of transpiration (Morrison and Gifford 1984), especially at higher ambient temperatures (Baker and Allen 1993b). In spite of higher leaf temperatures, water-use efficiency (the ratio of photosynthesis to transpiration) increases at high CO_2 mainly because of increased photosynthesis (Imai 1988; Baker et al. 1990b; Imai and Okamoto-Sato 1991).

Vegetative Growth

Leaf Development

In general, leaf-area development is not accelerated as much as dry matter is increased by high CO_2 (Imai and Murata 1976; Imai et al. 1985; Baker et al. 1990a). Instead, specific leaf weight often increases at high CO_2 as a result of thicker leaves or increased starch or nonstructural carbohydrate content of leaves (Imai 1988; Nomura et al. 1993). Recently, we found that development of bundle-sheath extensions in lamina and the number of xylem-vessel in midrib were increased and the distance between bundle sheaths was shortened by high CO_2 (Nomura et al. 1993). This is apparently the formation of a robust leaf, but substantial starch accumulation in both mesophyll and bundle-sheath chloroplasts and partial destruction of the thylakoid lammellae were observed in plants grown under 700 μl/l CO_2. These symptoms of damage were not improved by sufficient P supply (Nomura et al., unpubl. observ.). Development of the bundle-sheath extensions may not necessarily be a desirable response if it lengthens the effective path of CO_2 diffusion inside a leaf.

At moderately high temperature (e.g., 33/26 °C), rice ontogenesis is shortened. At high CO_2, the flag leaf appears at one position lower on the stem when compared with that of ambient CO_2-grown plants. This is induced earliness. When combined with high temperature, earliness may be pronounced and rice yields may be reduced because of smaller plants, unless an increase in effective tiller can compensate for a shorter vegetative growth period (Imai et al. 1985, 1994).

Tillering

Rice plants grown at high CO_2 have more tillers than plants grown at ambient CO_2, especially when high CO_2 is combined with high N because the supply of carbohydrate is increased (Imai and Murata 1978a; Imai et al. 1985). Because individual leaf size is not much enlarged and the leaf number on a stem tended to decrease with elevated CO_2, tiller development is important for leaf-area development. If high nighttime temperatures accompany high CO_2, late tillering may occur (Imai et al. 1985). Late tillering can alter dry-matter partitioning at maturity and lower harvest index (Imai et al. 1994).

Rooting

Number and length of crown roots increase with elevated CO_2 because of increased photosynthate partitioning to roots (Imai 1977). These seedlings are suitable for earlier transplanting and establishment in a ricefield (Tajima 1966). A large root system may protect plants from lodging at high CO_2 and may improve water and nutrient absorption.

Production of Dry Matter

During the vegetative stage, production of dry matter of rice is closely related to leaf-area development, which is promoted by tillering. Response of dry-matter production to various CO_2 concentrations is similar among young rice cultivars (Imai and Murata 1977). Elevated CO_2 promotes dry-matter production more than grain yield through the enhancement of Pn (or net assimilation rate) and leaf area, even if temperatures remain substantially high (Imai et al. 1985, 1994). At elevated CO_2, light intensity positively affects Pn and increased temperature promotes both Pn and leaf area (Imai and Murata 1979b).

Among nutrients, nitrogen is the most important (Imai and Murata 1978a). Insufficient phosphorus also limits production of dry matter at elevated CO_2 (Adachi and Imai 1991; Imai and Nomura 1992; Imai et al. 1994). Mineral-use efficiency of various rice cultivars under elevated CO_2 should be compared.

Reproductive Growth and Yield

Flowering

In rice, onset of reproductive growth is promoted by high CO_2 (several days to 1 week earlier flowering). Earlier flowering in high CO_2 is accompanied by the flag leaf from one lower nodal position compared with the position in ambient CO_2. This may be accelerated if higher temperature and P conditions are combined (Imai et al. 1985, 1994) but the effect of N is obscure.

Grain Yield

Grain yield of rice is generally promoted by high CO_2. Yields increase 30% or more with doubled CO_2 in the absence of other limiting factors (Yoshida 1976; Akita 1980; Imai et al. 1985, 1994; Baker and Allen 1993b). Under high CO_2, yields may be great if rice has a large sink capacity like root and tuber crops (Imai and Coleman 1983; Imai et al. 1984).

At moderately high temperature (33/26°C), yield increased 70% compared with yields at 28/21 °C under sufficient solar input (Imai et al. 1985). If temperatures, probably nighttime temperature, are low enough to permit reproductive process during grain filling, prolonged maturation may lead to higher grain yield; but if temperature is higher under high CO_2, which is predicted for the future, and if substantial N is applied, many small panicles may be produced at upper nodes (Imai et al. 1985), or dry-matter partitioning may shift toward vegetative parts (Imai et al. 1994). This may, as a whole, diminish the beneficial effect of high CO_2 on grain yield. When rice plants were grown in small pots that limited soil volume, the promotion of grain yield by doubled CO_2 was limited to 10–15% (Imai, unpubl.).

Yield Components

Yield components for rice are panicle number (effective tiller), filled grains per panicle, and grain weight. All components increase at high CO_2, but in general, the contribution of single grain weight is less than the other two because the husk physically limits the size of grains and husk characters are strongly under genetic control (Imai et al. 1985). Grain number per panicle may increase because high

CO_2 overcomes factors that often cause flower degeneration during ear development. Maturity percentage may increase because of sufficient photoassimilate supply (Yoshida 1976; Akita 1980; Imai et al. 1985). But, when combined with high nighttime temperature, maturity percentage may decrease because of greater dry-matter partitioning to vegetative parts (Imai et al. 1994). Baker et al. (1992b,c) reported a decline of yield and maturity percentages at higher temperatures when rice was grown at elevated CO_2; but it is not clear whether spikelet fertilization, or grain filling, or both were blocked by high temperatures. The threshold temperature that induces infertile spikelets is about 35°C, but variation among cultivars is not well studied.

Grain Quality

It is important to know whether grain quality will be imporved or will deteriorate in response to elevated CO_2. Conroy and Hocking (1993) pointed out that grain N and protien contents in Australian wheat have successively decreased over the last 23 years and this decline is partly mediated by increasing CO_2 concentration. In our ongoing work on rice under high CO_2 conditions, grain Mg content and Mg-K ratio decreased, and grain N content increased as nighttime temperature increased, and P nutrition did not alter that situation. These facts indicate reduced grain quality of rice (Horino 1990). Such changes should be corrected by cultural practices (e.g., imporved fertilization) or by breeding cultivars with grain quality characters that are insensitive to high temperatures.

Preparing for Global Change

The increase in atmospheric CO_2 concentration during the past 240 years (about 80 μl/l) should have increased rice production. Yield and quality of rice will be affected by temperature and nutrients at higher CO_2 levels. At elevated CO_2, high temperatures with low day-night differences may alter dry-matter partitioning and lead to less promotion of yield. In such a case, cultivars with fewer tillers and larger panicles may be useful. Also, water temperature during rice growth may have substantial effects on leaf, tiller, and panicle development, because growing points stay close to, or under, the water until internode elongation at later reproductive stages. Temperature effects on rice may be larger than we have expected from increases in air temperature.

The physiological responses of rice that should be studied are: source-sink relations and dry-matter partitioning, plant morphology that will receive maximum solar energy, critical conditions that may induce deteriorations of vegetative and reproductive growth, fertilizer and water-use efficiency, and grain quality under different CO_2, temperature, and fertilizer conditions.

Carbon Dioxide and Temperature Effects on Rice

L.H. ALLEN JR,[1] J.T. BAKER,[2] S.L. ALBRECHT,[3] K.J. BOOTE,[2]
D. PAN,[2] and J.C.V. VU[1]

The rise in atmospheric CO_2 has been documented continuously since 1958 by Keeling et al. (1989) and currently the concentration of CO_2 in air is about 360 μl/l. The concentration could increase to about 670–760 μl/l by 2075 mainly because of the burning of fossil fuels (Rotty and Marland 1986; Trabalka et al. 1986). General circulation models predict that global warming will result from rising CO_2 and other greenhouse gases (Washington and Meehl 1984; Manabe and Wetherald 1987; Wilson and Mitchell 1987; Hansen et al. 1988).

The effects of rising CO_2 and elevated temperatures on tropical plants have received less attention than the effects on temperate species (Hogan et al. 1991). Because both CO_2 and temperature have large effects on plants, especially those with the C_3 photosynthetic pathway, it is important to quantify effects of these climatic variables on C_3 food crops such as rice.

Grain yields of rice are increased by CO_2 enrichment (Imai and Murata 1976, 1979a, b; Imai et al. 1985; Baker et al. 1990a, 1992a,b). Increased grain yield is often (but not always) associated with increased tillering and number of panicles (Imai et al. 1985; Baker et al. 1990a). Yields of IR30 declined by 10% for each 1°C rise in day-night temperature above 28/21 °C, and elevated CO_2 had little effect in ameliorating this temperature response (Baker and Allen 1993a). Sharp decreases in the number of filled grains per panicle accompanied these yield decreases.

Following five studies on IR30, a study was undertaken (Baker et al. 1994a) on IR72, one of the cultivars used in CO_2 x temperature experiments at IRRI. The objective of this review is to summarize our experiments on the effects of temperature and CO_2 on photosynthesis, growth, and grain yield of the rice cultivars IR30 and IR72.

[1] United States Department of Agriculture, Agricultural Research Service, University of Florida, Gainesville, FL 32611-0840, USA
[2] Agronomy Department, University of Florida, Gainesville, FL 32611-0840,USA
[3] United States Department of Agriculture, Agricultural Research Service, Columbia Plateau Conservation Research Center, Pendleton, OR 97801,USA

S. Peng et al. (Eds.) Climate Change and Rice

Carbon Dioxide and Temperature Studies on Rice Cultivar IR72

Controlled Environment Chambers

Rice (IR72) was grown from germination to harvest in eight outdoor, sunlit, controlled-environment chambers, also known as Soil-Plant-Atmosphere-Research (SPAR) chambers (Baker et al. 1994a). These chambers have been described by Baker et al. (1990a,b,c, 1992a,b,d) and by Baker and Allen (1993a,b). Environmental setpoints were maintained by single-chamber controller/data-loggers (Campbell Model CR-10T) interfaced to a PC host processor (Pickering et al. 1994a,b).

Plant Culture

Prior to planting, the soil in each chamber was fertilized with P and K, both at 8.97 g/m^2. On 20 July 1992, rice was seeded into hills spaced 20 cm × 20 cm in nonflooded soil. At the second leaf stage, the plants were thinned to four plants per hill or 100 plants/m^2. Fertilizer N (as urea) was applied at a rate of 12.6 g N/m^2 at 7 days after sowing (DAS). Soil vats were flooded to a 50-mm depth above the soil surface beginning at 8 DAS. Shade cloths were placed at plant height outside each chamber to provide a canopy light environment similar to field conditions. Additional 6.3 g N/m^2 as urea was applied at 31 and 63 DAS for a total 25.2 g N/m^2. To measure root growth, three cylindrical, 1.6-l pots with a depth of 0.5 m and a diameter of 0.2 m were filled with soil and placed in the chambers. One hill of four plants was grown in each pot.

Treatments

In four chambers each, CO_2 concentration was maintained at 330 or 660 μl/l during the daytime by controlled injection of CO_2 gas. Chambers were vented briefly at night whenever CO_2 rose to 150 μl/l above setpoint. Chamber dry-bulb air temperatures were controlled to follow a near-sinusoidal, diurnally varying setpoint. The baseline setpoints were: 24.5, 24.0, 23.5, 23.5, 23.0, 23.0, 23.0, 23.5, 25.0, 27.0, 29.0, 30.5, 31.5, 32.0, 32.0, 31.5, 31.0, 30.0, 29.0, 28.0, 27.0, 26.0, 25.5, and 25.0 °C for each hour from 00:00 through 23:00. An algorithm was developed to fit the hourly setpoints to a smooth curve. The two elevated temperature treatments were 3° and 6 °C higher than baseline. Day-night maximum/minimum air temperatures were 32/23 °C, 35/26 °C, and 38/29 °C in one chamber of each CO_2 concentration, with time-weighted averages of 27°, 30 °C, and 33 °C, respectively. A replicated high-temperature treatment (38/29 °C) was maintained at each CO_2 level. Dew-point air temperatures were maintained at 18°, 21°, and 24 °C in the 32/23 °C, 35/26 °C, and 38/29 °C dry-bulb air temperatures, respectively. Measured values of dry bulb and dew-point temperatures remained within 0.3 °C of setpoint (Baker et al. 1994a). Additional details for controlling environmental setpoints are given by Pickering et al. (1994a,b).

Measurements of Leaf Photosynthesis

Beginning 90 DAS, photosynthetic rates of leaves were determined using a LiCor model LI-6000-13 0.25-l cuvette interfaced to a LiCor model LI-6250 Portable Photosynthesis System. A constant photosynthetic photon flux density (PPFD) of 1400 μmol m^{-2} s^{1} was provided with a light source from an Hansatech oxygen electrode instrument placed at 35 cm above the leaf cuvette. Photosynthetic assimilation rate (A) versus intercellular CO_2 (C_i) response curves were obtained for leaves of each CO_2 and temperature treatment, using seven compressed air cylinders containing calibrated CO_2 concentrations that ranged from 100 to 1040 μl/l. Moisture was added to the airstream by bubbling the compressed air through a flask of water at ambient temperature. A fan was used to ventilate the outside of the cuvette. Each leaf was equilibrated for 5 min in the curvette prior to making A versus C_i measurements, starting with the highest CO_2 concentration (1040 μl/l).

Developmental Measurements

Main-stem Haun leaf stage was determined on five tagged plants of five separate hills from measurements made two to three times each week (Haun 1973). Reciprocals of linear regressions of leaf stage versus DAS were used to determine rates of leaf appearance or phyllochron intervals (Baker et al. 1990c).

Growth and Yield Measurements

Rice plants were sampled at 29, 46, 70, 86 and 120–126 DAS. One plant from 12 separate hills was sampled at 46 and 86 DAS. At 120–126 DAS (final harvest), all plants from 11 separate hills (44 total plants) were sampled from each chamber. At 29 and 70 DAS, and at final harvest, the plants (including roots) from one pot of each chamber were sampled. Live leaves, tillers, and panicles were counted, and laminar area per plant was measured. After oven drying at 70 °C for 48 h, dry weights were determined for leaf laminae, stems including leaf sheaths, and panicles. Grain yield, grains per panicle, and individual grain mass were determined from the final sampling after over drying and threshing each panicle. For plants grown in pots, shoots were measured as described. Root-length density was determined from nine 5-cm depth layers of each pot. Root biomass was determined after oven drying at 70 °C for 48 h.

Statistical Analyses

Because this study was partially replicated (two replications of the 38/29 °C treatment at both CO_2 levels), the measured error of these replications was used to test effects of CO_2, temperature, and CO_2 × temperature using F-values from the Type III mean square for developmental rates, yield, and yield components. Temporal trends in above-ground biomass, root biomass, tillers per plant, and leaf-area index were compared with the standard deviation of replicated treatments.

Results and Discussion

Rates of Leaf Appearance

For all treatments, Haun scale growth units versus DAS showed two different rates of leaf appearance: a faster rate before panicle initiation, and a slower rate from panicle initiation to panicle emergence. Five leaves, including the flag leaf, appeared on all main stems during the slower rate of leaf appearance. A decrease in the rate of the leaf appearance of rice near panicle initiation has been reported previously (Yoshida 1977; Vergara 1980).

Phyllochron intervals during the vegetative and reproductive phases spanned 4.1–4.7 and 8.9–11.1 days per leaf, respectively (Table 1). All linear regressions had r^2 values greater than 0.97. Temperature increases from 32/23 °C

Table 1. Leaf appearance rates (phyllochron interval) during vegetative and reproductive growth and final main-stem leaf number for flooded rice (IR72) plants grown in controlled environment chambers at Gainesville, Florida, in two CO_2 concentrations and three temperature treatments. (After Baker et al. 1994a).

CO_2 (μl/l)	Temperature[a] (°C)	Phyllochron interval		Total no. of main-stem leaves
		Vegetative (days/leaf)	Reproductive (days/leaf)	
330	32/23(27):18	4.6	11.1	17.6
	35/26(30):21	4.7	9.4	18.0
	38/29(33):24	4.1	8.9	18.1
660	32/23(27): 18	4.6	10.7	18.6
	35/26(30):21	4.1	9.0	18.5
	38/29(33):24	4.1	9.0	18.3
F-values[b]				
CO_2		0.6 NS	1.3 NS	26.9*
Temperature		2.6 NS	40.2*	0.6 NS
CO_2 × temperature	0.7 NS	1.0 NS	5.5 NS	

[a] Temperature treatments are expressed as maximum daytime/minimum nighttime (daily average) air temperature: dew-point temperature.
[b] The * represents significance at the 0.10 probability level, NS not significant.

to 38/29 °C caused phyllochron intervals to decrease about 2 days per leaf during the reproductive phase. Enrichment with CO_2 resulted in a small increase in the final number of mainstem leaves but did not significantly affect phyllochron interval (Table 1). Imai and Murata (1979b) found small increases in leaf appearance rates of rice plants exposed to 1000 versus 300 μl/l CO_2. There were no CO_2 × temperature effects on leaf development (Table 1).

Growth and Yield

Changes in above-ground biomass during the season for the CO_2 and temperature treatments are shown in Fig. 1. Final aboveground biomass (Table 2) increased with CO_2 enrichment by 0, 13, and 47% with increasing temperatures of 32/23 °C, 35/26 °C, and 38/29 °C, respectively. The enhancement of CO_2-stimulated biomass growth by increasing temperatures agrees with reports of Idso et al. (1987) for other vegetation, and with the Biomass Growth Modification Ratio analysis of Allen (1994). Overall, effects of CO_2 enrichment on final aboveground biomass were not statistically significant (Table 2).

Trends in response to CO_2 and temperature of both root and shoot biomass of rice grown in pots were similar to trends in shoot biomass of plants grown outside the pots. The root-shoot ratio decreased with time during the season. Root-length density was closely related to root biomass throughout the 5-cm layers of the soil profile in pots (data not shown).

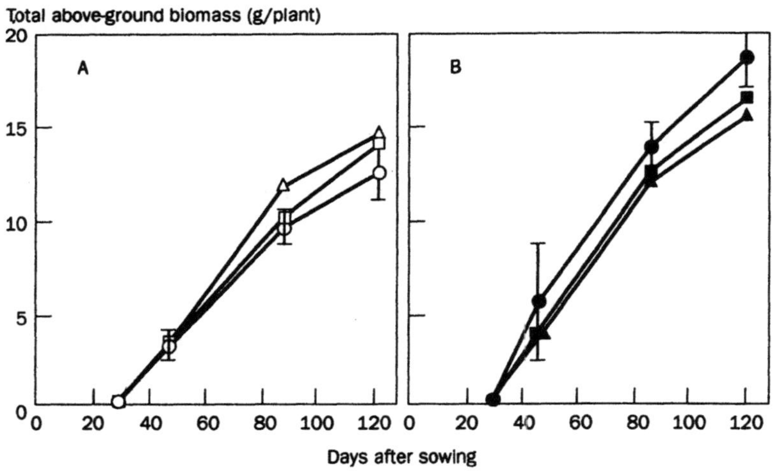

Fig. 1A, B. Total aboveground biomass throughout the season for rice plants grown in three temperature regimes and two CO_2 concentrations. *Vertical bars* indicate the standard deviation of replicated treatments for the 38/29 °C temperature treatment. (After Baker et al. 1994b). **A** 330 μl/l △ 32/23 °C; □ 35/26 °C; ○ 38/29 °C. **B** 660 μl/l ▲ 32/23 °C; ■ 35/26 °C; ● 38/29 °C

Table 2. Grain yield, components of yield, total above ground biomass, and harvest index for flooded rice (cv. IR72) plants grown in controlled environment chambers at Gainesville, Florida, in two CO_2 concentrations and three temperature treatments. (After Baker et al. 1994a)

CO_2 (μl/l)	Temperature[a] (°C)	Grain yield (Mg/ha)	Grain yield (g/plant)	Panicles (no./plant)	Filled grain (no./panicle)	Grain mass (mg/seed)	Total above-ground biomass (g/plant)	Harvest index[b]
330	32/23(27):18	7.6	7.6	6.7	53.2	21.6	15.8	0.48
	35/26(30):21	7.3	7.3	6.3	55.5	21.1	15.1	0.48
	38/29(33):24	1.4	1.4	5.6	14.6	18.4	13.1	0.11
660	32/23(27):18	7.3	7.3	6.9	48.9	21.4	15.8	0.45
	35/26(30):21	7.1	7.1	6.2	57.3	20.8	17.1	0.43
	38/29(33):24	2.6	2.6	5.8	22.5	19.2	19.3	0.13
F-values[c]								
CO_2		0.1 NS	–	0.3 NS	0.1 NS	0.1 NS	6.7 NS	0.1 NS
Temperature		11.2*	–	7.0 NS	6.8 NS	11.8*	0.3 NS	16.1*
CO_2 × temperature		0.2 NS	–	0.3 NS	0.1 NS	0.5 NS	2.6 NS	0.1 NS

[a] Temperature treatments are expressed as maximum daytime/minimum nighttime (daily average) air temperature: dew-point temperature.
[b] Based on 11 hills per chamber. The values are slightly different if the average grain yield is divided by the average total aboveground biomass.
[c] The * indicates significance at the 0.10 probability level, NS not significant.

Tillers per plant and leaf-area index showed no consistent trends throughout the season among the CO_2 or temperature treatments (Baker et al. 1994a). The CO_2 enrichment resulted in no significant differences in either grain yield or yield components (Table 2). The lack of CO_2 effects on tillering (Baker et al. 1994a) may partly explain the lack of CO_2 effect on yield. In previous studies, increased tillering (more panicles) contributed profoundly to yield increases under CO_2 enrichment (Imai et al. 1985; Baker et al. 1990a). However, the number of panicles has been less important in other reports of yield responses to CO_2 enrichment (Baker et al. 1992a,b).

Declining solar radiation during late October and early November may have contributed to the lack of CO_2 enrichment effects on yield. Baker et al. (1992a) found little CO_2 stimulation of IR30 rice yield when the crop was planted on 6 October 1988, but reported a 60% increase in yield with a doubling of CO_2 with rice planted on 14 July 1989 (Baker et al. 1992b).

Grain yield and harvest index decreased with increasing temperature, especially from 35/26 °C to 38/29 °C (Table 2). All three components of yield tended to decrease with increases in air temperature, although only individual grain mass was affected significantly. Tashiro and Wardlaw (1989) found that grain mass of the rice cultivar Calrose remained nearly constant across a range of 21.7–26.7 °C, followed by a decline in grain mass of 4,4% per 1 °C rise in air temperature from 26.7–35.7 °C.

Leaf Photosynthesis Responses

Leaf photosynthetic CO_2 assimilation rates of rice, determined for plants grown at the three temperature regimes (but measured at ambient temperature), averaged 18.8 and 30.4 μmol CO_2 uptake/m^2 of leaf area per second for the 330 and 660 μl/l treatments, respectively, an increase of about 60% caused by elevated CO_2. A pair of leaf A versus C curves obtained at 101 DAS from each CO_2 treatment at the growth temperature of 35/26 °C are shown in Fig. 2. Similar pairs of curves (not shown) were obtained for treatments of 38/29 °C on 107 DAS and 32/23 °C on 109 DAS. Each pair of leaf A versus C_i curves were similar within each temperature regime and date of measurement. There were essentially no differences in A between the two CO_2 treatments at each specific C_i level (Fig. 2). Extrapolations of the response curves to zero values of A gave a leaf CO_2 compensation point of about 60 μl/l CO_2.

Rowland-Bamford et al. (1991) reported decreases in both amount and activity of Rubisco with increasing CO_2 in IR30 leaves growth across the range 160–900 μl/l. However, across our smaller range of 330–660 μl/l CO_2, the leaf A versus C_i response curves gave no indication of a downward acclimation of photosynthetic capacity when measurements were made. One possibility is that the amount and activity of Rubisco do not limit photosynthesis under CO_2 enrichment as much as under ambient CO_2. More work is needed, under a range of CO_2 treatments, to explore the interaction effects of sink capacity, nitrogen

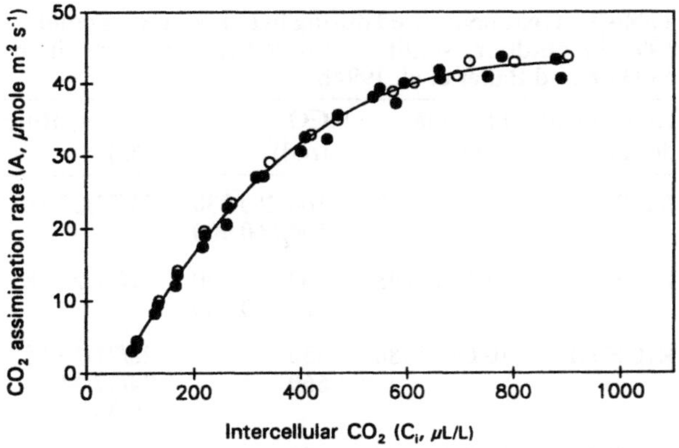

Fig. 2. Photosynthesis CO_2 assimilation rate (A) as a function of intercellular CO_2 (C_i) for single, attached, fully expanded leaves of rice plants grown at CO_2 concentrations of 330 µl/l (○) or 660 µl/l (●) and dry bulb air temperatures (day-night) of 35/26 °C. Measurements were made on 29 October 1992 at an irradiance of 1200–1300 µmol (photons) m^{-2} s^{-1}.

nutrition of leaves, and other internal CO_2 fixation processes on photosynthetic behavior and crop yield.

Carbon dioxide enrichment slightly increased the number of main-stem leaves, but did not significantly affect phyllochron interval, grain yield, or components of grain yield. Increasing temperatures shortened the phyllochron interval by about 2 days per leaf during resproductive development. Increasing temperatures reduced grain yield, individual seed mass, and harvest index. Based on leaf A versus C_i response curves, the rice plants in this experiment did not acclimate either up or down across the long-term CO_2 enrichment range of 330–600 µl/l, at least when measurements were made.

Carbon Dioxide by Temperature Studies on Rice Cultivar IR30

Controlled Environment Chambers and Experimental Treatments

In five experiments conducted from 1987 to 1990, rice (IR30) was grown in six outdoor, sunlit, controlled-environment chambers (Table 3). The design and performance of these chambers were described by Jones et al. (1984). The 1987–1990 chambers were similar to the chambers used in the 1992 experiments; only major differences that affected experimental treatments will be mentioned. In RICE I and II experiments, flood-water temperature was maintained using electrical resistance heat (Table 3). In Rice III, IV, and V, flood-water temperature was controlled using a heat exchanger consisting of submerged copper-pipe manifolds. During RICE I and II experiments, subambient CO_2 levels (160 and 250 µl/l, Table 3) were obtained by injecting CO_2-free air during periods of low

Table 3. Treatments used during five experiments conducted on rice (IR30) from 1987 to 1990 in outdoor, sunlit, controlled-environment chambers. (After Baker and Allen 1993a,b and Baker et al. 1994b)

Experiment designation	Planting data	CO_2 (μl/l)	Temperature[a] (°C)	Comment
RICE I	22 Jan 1987	160 250 330 500 660 900	31/31/27:18	Photoperiod extended to >18 h until 2 March
RICE II	23 June 1987	160 250 330 500 660 900	31/31/27:18	Natural photoperiod
RICE III	10 Oct 1988	330 660	28/21/25:15 34/27/31:18 40/33/37:21	Full factorial: CO_2× temperature
RICEIV	14 July 1989	660	25/18/21:10.5 28/21/25:12.0 31/24/28:13.5 34/27/31:15.0 37/30/34:16.5	One additional chamber at 330 μl/l and 28/21/25:12.0 °C
RICE V	8 June 1990	330 660	28/21/25:17	Full factorial: 2 CO_2 × 3 N fertilizer rates

[a] Temperature treatments are expressed as daytime dry-bulb air temperature/nighttime dry bulb air temperature/water temperature: dew-point temperature.

daytime canopy photosynthetic rates. Plant growth chambers were decribed in detail by Baker et al. (1990a,b,c, 1992a,b,d, 1994b,c) and Baker and Allen (1993a,b, 1994).

Summaries of the CO_2 and temperature setpoints for the five experiments are shown in Table 3. RICE I and RICE II were designed to study the effects of CO_2 across subambient to superambient ranges at a fixed temperature for two reasons: to develop response curves rather than merely to contrast responses to ambient and doubled-ambient CO_2, and to demonstrate responses to CO_2 that could have existed during the last glacial maximum when the CO_2 levels were as low as 180–200 μl/l, as well as to demonstrate responses to rising atmospheric CO_2. RICE III and RICE IV were designed to investigate the effects of a wide range of elevated temperatures at current and doubled CO_2 levels (Table 3). RICE V was designed to study the effect of nitrogen fertilizer on CO_2 response at a fixed temperature. Because numbers of available growth chambers restricted studies to no more than two factors at a time, with limited replication, we examined overall responses of rice to CO_2 and temperature based on all five experiments.

Plant Culture

Rice was seeded by hand in 11 rows 0.18–m apart and thinned to 235 plants/m^2 after emergence. Floodwater was applied at the second leaf stage and then specific control setpoints were initiated for each chamber (Table 3). Soil was fertilized with P and K prior to planting, and with N as urea prior of flooding. Additional urea-N was applied later in 2 or 3 split portions.

Stages of leaf development of five replicate mainstems were determined after Haun (1973). Growth and yield attributes (including panicle numbers, number of grains, total above-ground biomass, average mass per grain, and grain yield) of each of 30 plants (RICE 1) or 45 plants (RICE II to RICE V) were measured from each chamber.

Results and Discussion

Development

In all five experiments, plots of Haun-scale growth units showed the two characteristic rates of leaf appearance for vegetative and reproductive phases of growth (e.g., Fig. 3A). All linear regressions of leaf stage versus DAS used for defining the phyllochron intervals had r^2 values greater than 0.97 (Table 4).

In RICE II, the number of main-stem leaves at panicle initiation and final number of mainstem leaves decreased across the CO_2 treatment range 160–500 μl/l and remained similar at 500–900 μl/l (Table 4). Panicle initiation and boot stage occurred 10–12 days earlier in the superambient CO_2 treatments compared with the 160 μl/l treatment (Baker et al. 1990; Baker and Allen 1993b). Plant

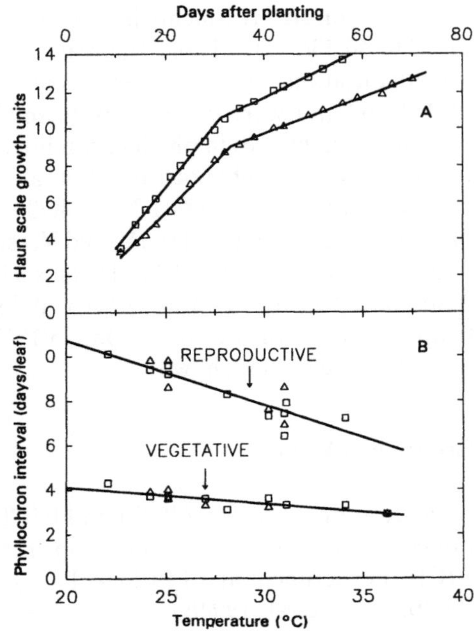

Fig. 3. A Main-stem Haun scale growth units versus days after planting for rice plants in two different CO_2 and temperature regimes from the RICE III experiment. □ 660 μl/l 40/33/37 °C; Δ 330 μl/l 28/21/25 °C. **B** Phyllochron interval versus temperature treatment for all five rice experiments. □ 660 μl/l, Δ 330 ml/l. Vegetative data are plotted against flood-water temperature and reproductive data are plotted against average day-night temperature adjusted for thermo-period. (After Baker et al. 1994b)

Table 4. Effects of CO_2 concentration on the phyllochron interval and main-stem leaf stage under two different photoperiods (see Table 3). After Baker et al. 1990c and Baker and Allen 1993b)

CO_2 treatment (μl/l)	Vegetative phase phyllochron interval (days/leaf)[a]		Main-stem leaf stage at panicle initiation (no. of main-stem leaves)[a]		Final main-stem leaf number (no. of main-stem leaves)[a]	
160	4.1	3.6	12.1	12.9 NS	15.4	15.3 NS
250	4.1	3.7	12.0	12.0 NS	14.6	14.2 NS
330	3.6	3.3	11.8	11.6 NS	14.8	13.6**
500	3.4	3.8	12.1	10.7**	14.8	13.0**
660	3.3	3.6	12.2	10.5**	14.6	12.8*
900	3.6	3.6	12.0	10.5**	14.6	12.8*
$LSD_{0.05}$	0.6	NS	NS	0.8	NS	0.8

[a] Data in left column are from experiment I, Data in right column are from experiment II. The * and ** indicate that the data in paired columns are significantly different at the 0.10 and 0.05 probability levels, respectively, NS not significant.

developmental rate was accelerated by increasing CO_2 up to 500 μl/l because of a reduction in the number of main-stem leaves formed prior to panicle initiation. On the other hand, in the RICE I experiment, effects of CO_2 on number of main-stem leaves at panicle initiation (Table 4) and on timing of the boot stage (Baker et al. 1990c) were small. The lack of phenological response to CO_2 may be because daily photoperiod was extended until midnight with lights from 22 January to 2 March 1987 (39 DAS) forcing the plants to remain vegetative (Baker et al. 1990c; Baker and Allen 1993b). Low light levels early in the year could have contributed to the lack of a CO_2 effect.

Temperature treatments in the five rice experiments, ranging from 25/18/21 °C to 40/33/37 °C (Table 3), greatly influenced developmental rate and total growth duration; whereas, the effects of CO_2 enrichment across the limited range of 330–660 μl/l were comparatively small (Baker et al. 1992a,b, 1993b). No consistent differences in phyllochron interval between 330 μl/l and 660 μl/l were observed, whereas phyllochron interval increased with increasing temperature across the range of 25/18/21 °C to 40/33/37 °C, especially for the reproductive phase (Fig. 3B). In RICE III, IV, and V, estimated dates for anthesis ranged from 0 to 6 days earlier in the 660 μl/l versus 330 μl/l treatments (Baker et al. 1994b). Also, time to anthesis was shortened by approximately 10 days across the temperature treatment range from 25/18/21 °C to 34/27/31 °C for the RICE IV experiment (data not shown).

Growth and Yield

Average grain yields for RICE I and RICE II versus CO_2 are plotted in Fig. 4A. For comparison, Fig. 4B shows canopy net photosynthesis (Pn) on 61 DAS in the RICE II experiment at a PPFD of 1500 μmol photons m^{-2} s^{-1} (Baker and Allen 1993b; Baker et al. 1994b,c). Data of both figures were normalized to the 330

Fig. 4A, B. Rice-grain yields and net photosynthesis of canopy in response to CO_2 concentration normalized to the values obtained from the 330 µl/l CO_2 treatment. The percentage variation explained by the rectangular hyperbola model was 90 and 96% for grain yield and photosynthesis, respectively. (After Baker and Allen 1993b and Baker et al. 1994b)

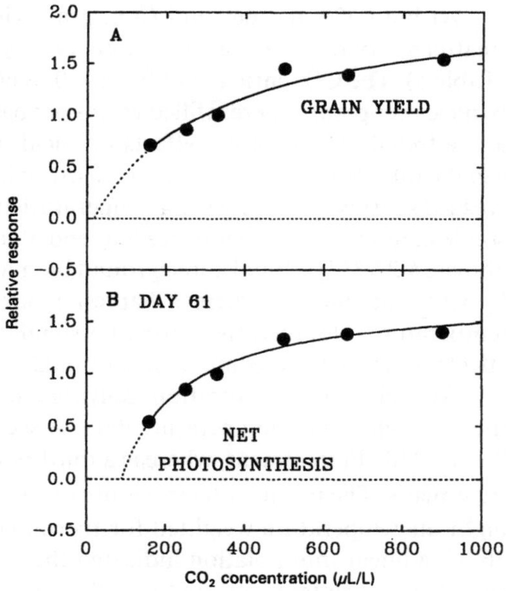

µl/l CO_2 treatment responses. A nonlinear rectangular hyperbola (Allen et al. 1987) was fitted to these canopy photosynthesis and seed yield data:

$$Y = (Y_{max} \times [CO_2]) / ([CO_2] + Km) + Yi, \quad (1)$$

where Y is grain yield or Pn relative to values at 330 µl/l CO_2, $[CO_2]$ is CO_2 concentration, Y_{max} is the asymptotic response limit of (Y – Yi) at very high $[CO_2]$, Yi is the intercept on the y axis, and Km is the value of $[CO_2]$ where $(Y - Yi) = Y_{max}/2$. Parameters of $Y_{max} = 2.24$, Km = 284 µl/l, and Yi = – 0.13 were obtained for relative grain yield (Fig. 4A) with an asymptotic relative yield response ceiling $(Y_{max} + Yi) = 2.11$ at infinite $[CO_2]$. For Pn, fitted parameters were $Y_{max} = 3.96$, Km = 70.8 µl/l, and Yi = –2.21, with a relative Pn response ceiling of 1.75 at infinite $[CO_2]$. The calculated increases in response at 660 versus 330 µl/l were 44% for grain yield and 36% for Pn (Baker and Allen 1993b).

Across a relatively wide temperature range from 25/18/21 °C to 37/30/34 °C, Baker et al. (1992b) found a broad temperature optimum for biomass production in the midtemperature ranges. Plants grown at 40/33/37 °C were found to be near the upper temperature limit for survival. High-temperature spikelet sterility of rice is induced almost exclusively on the day of anthesis (Satake and Yoshida 1978) when temperatures greater than 35 °C for more than 1 h induce a high percentage of sterility (Yoshida 1981). In the 40/33/37 °C treatments, plants in the 330 µl/l CO_2 chamber died during internode elongation, whereas plants in the 660 µl/l chamber produced small, abnormally shaped panicles that were sterile (Baker et al. 1992a). Therefore, elevated CO_2 may increase slightly the maximum temperature at which rice plants can survive.

At both 330 and 660 μl/l CO_2, grain yield was highest in the 28/21/25 °C treatment followed by a decline to zero yield in the 40/33/37 °C treatment (Table 5). The CO_2 enrichment from 330 to 660 μl/l increased yield by increasing panicles per plant, whereas filled grains per panicle and individual seed mass were less affected. Temperature effects on yield and yield components were highly significant. Panicles per plant increased, whereas filled grains per panicle decreased sharply with increasing temperature treatment. Individual seed mass was stable at moderate temperatures but tended to decline at temperature treatments above 34/27/31 °C. Final aboveground biomass and harvest index were increased by CO_2 enrichment, whereas harvest index declined sharply with increasing temperature. Notably, there were no significant $CO_2 \times$ temperature interaction effects on yield, yield components, or final aboveground biomass.

At each CO_2 concentration, polynomial regression equations were fitted to the rice yield versus temperature data. A second-degree polynomial fit the data for the 330 μl/l treatment, whereas a third-degree polynomial fit the 660 μl/l CO_2 treatments. The mean air temperature of each study was the average of the day–night air temperature weighted for thermoperiod. Solving for yields at 26 and 36 °C, a linear interpolation indicated that yields would decline about 10% for each 1 °C rise in temperature for both CO_2 treatments (Baker and Allen 1993a). Although future increases in atmospheric CO_2 should benefit rice yields, large negative effects are likely if temperatures also rise.

Canopy Photosynthetic Rates

At 60 DAS, which was after canopy closure, the Pn versus PPFD responses were linear and did not approach light saturation in any of the five experiments, probably because of the erect leaf orientation of cultivar IR30 and high plant populations used in these studies. As in Fig. 4B, at high light, Pn increased with increasing CO_2 up to 500 μl/l and was similar from 500 to 900 μl/l.

Canopy Pn responses to PPFD at 60 DAS in the RICE IV experiment were similar for all temperature treatments (25/18 to 37 °C/30 °C) exposed to 660 μl/l CO_2 (Fig. 5). Canopy Pn of the 330μl/l treatments were about 25% less than those `of the 660 μl/l CO_2 treatments. Studies of canopy Pn in cotton, and soybean (Jones et al. 1985a), have shown only small differences across wide ranges of temperature. Evaporative cooling may lower foliage temperature below air temperature (Allen 1990; Pickering et al. 1994c), or the integrating light-capture system of the whole canopy may broaden the tempeature response for the whole-canopy photosynthetic rates to reduce the effects of air temperature on canopy Pn.

Table 5. Grain yield, components of yield, total aboveground biomass, and harvest index for two CO_2 concentrations extracted from five separate rice experiments. (After Baker and Allen 1993a,b and Baker et al. 1994b)

CO_2 (μl/l)	Temperature[a] (°C)	Mean air temperature[b] (°C)	Grain yield (t/ha)	Panicle/plant (no/plant)	Filled grain (no/.panicle)	Grain mass (mg/seed)	Biomass (g/plant)	Harvest index	Experiment no.
330	28/21/25	24.2	7.9	5.1	34.5	17.4	7.3	0.47	RICE III
	28/21/25	25.1	6.6	3.9	39.6	17.5	6.5	0.44	RICE IV
	28/21/25	25.1	8.0	4.0	47.5	18.5	8.1	0.43	RICE V
	31/31/27	31.0	5.2	5.9	23.0	17.1	5.5	0.42	RICE I
	31/31/27	31.0	4.3	5.4	19.0	17.9	7.2	0.26	RICE II
	34/27/31	30.2	4.2	7.7	15.2	16.2	5.6	0.43	RICE III
	40/33/37	36.2	0.0	–	–	–	–	–	RICE III
660	25/18/21	22.1	8.4	4.4	46.0	18.2	7.8	0.47	RICE IV
	28/21/25	24.2	8.4	5.0	37.7	18.0	7.9	0.46	RICE III
	28/21/25	25.1	10.4	4.2	58.3	18.3	8.9	0.50	RICE IV
	28/21/25	25.1	10.1	4.4	54.2	19.0	9.3	0.47	RICE V
	31/31/27	31.0	6.8	6.9	25.1	17.2	7.5	0.40	RICE I
	31/31/27	31.0	6.4	6.0	24.8	18.4	9.3	0.29	RICE II
	34/27/31	30.2	4.8	6.5	18.5	16.7	6.3	0.32	RICE III
	34/27/31	31.1	3.4	7.5	12.9	16.3	8.1	0.18	RICE IV
	37/30/34	34.1	1.0	8.0	3.0	14.2	7.1	0.06	RICE IV
	40/33/37	36.2	0.0	–	–	–	–	–	RICE III
F-values[c]									
CO_2			4.2*	4.3*	0.4 NS	0.1 NS	13.1**	4.6*	
Temperature			51.8**	22.6**	51.1**	27.0**	3.2*	32.8**	
CO_2 × temperature			1.9 NS	1.7 NS	1.3 NS	0.3 NS	0.2 NS	1.9 NS	

[a] Temperature treatments are expressed as daytime dry-bulb air temperature/nighttime dry-bulb air temperature/rice water temperature.
[b] Mean air temperature is the average of day and night temperature adjusted for thermoperiod.
[c] The ** and * indicate significance at the 0.01 and 0.05 probability levels, respectively, NS not significant.

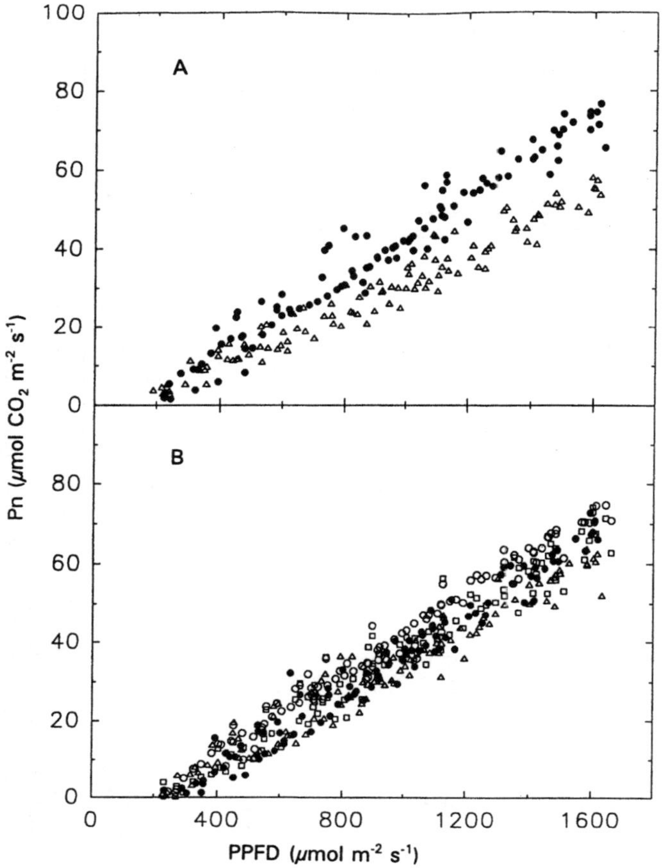

Fig. 5A,B. Net photosynthetic rate of canopy (*Pn*) versus incident photosynthetic photon flux density (*PPFD*) for two daytime CO_2 treatments and five daytime temperature treatments on day 60 of the RICE IV experiment. (After Baker and Allen 1993b). **A** Δ 330 μl/l 28/21 °C; ● 660 μl/l 28/21 °C. **B** All 660 μl/l; Δ 25/18 °C; ● 31/24 °C; □ 34/27 °C; O 37/30 °C

Photosynthetic Acclimation to Carbon Dioxide at the Canopy Level

Rice canopy Pn increased with increasing CO_2 concentration from 160 to 500 μl/l, followed by only small differences among treatments from 500 to 900 μl/l. To test for acclimation of canopy photosynthetic capacity, Baker et al. (1990c) used rice that had been grown at 160, 250, 330, 500, and 900 μl/l CO_2 during the RICE II experiment. For half-day periods (mornings) on 62, 63, and 64 DAS, common CO_2 setpoints of 160, 330, or 660 μl/l were imposed on all chambers on those respective days. Within each of these short-term CO_2 exposure comparisons, Baker et al. (1990b, 1994c) showed that the short-term canopy Pn decreased with increasing long-term CO_2 treatment (Fig. 6). Among short-term CO_2 exposures, relative effects were greatest at 160 μl/l CO_2 (Fig. 6). For example, when compared at a common short-term CO_2 exposure of 160 μl/l, canopy Pn of the chamber containing the 900 μl/l long-term CO_2 treatment was only about one-third of that of the 160 μl/l long-term CO_2 treatment (Fig. 6).

Evidence for leaf acclimation to CO_2 of rice and soybean has been compared within the protein and Rubisco domains (Baker and Allen 1994). Leaf soluble protein, leaf nitrogen concentration, and the percentage of total leaf soluble protein represented by Rubisco protein are shown in Table 6 for 34 DAS soybean (Campbell et al. 1988) and 75 DAS rice (Rowland-Bamford et al. 1991), both grown at a wide range of CO_2 concentrations. For rice, Rubisco activity expressed on a leaf-area basis decreased by 66% across the 160–900 μl/l long-term CO_2 treatments (Rowland-Bamford et al. 1991). A major cause of this decline in Rubisco activity was a 32% decrease in the amount of Rubisco relative to other soluble proteins (Table 6).

Dark Respiration Rates in the Canopy

Plant respiration may be decreased by both short- and long-term exposure to high CO_2 concentrations (Bunce 1990; Amthor 1991). Baker et al. (1992d) monitored nighttime canopy dark respiration rates [R_d, μmol (CO_2) m^{-2} (ground

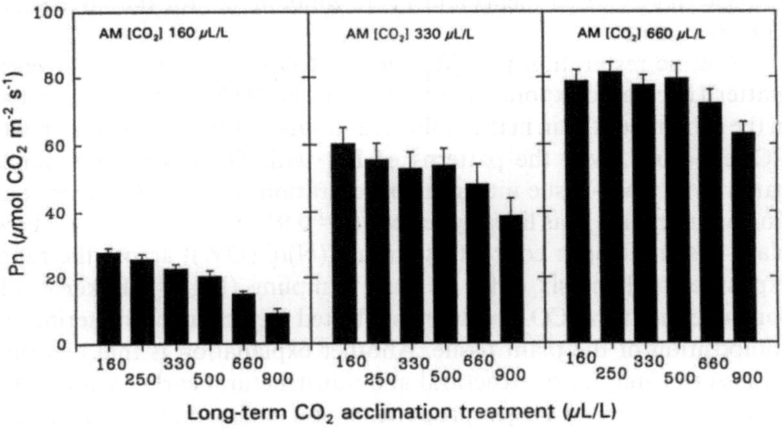

Fig. 6. Comparison of net photosynthetic rate of canopy (*Pn*) versus long-term [CO_2] acclimation treatment for rice canopies grown at subambient (160 and 250 μl/l), ambient (330 μl/l), and superambient (500, 660, and 900 μl/l) [CO_2] treatments in 1987. The Pn estimates were obtained during a short-term [CO_2] change-over study in which the [CO_2] was maintained during the morning hours in all six long-term [CO_2] treatments at 160, 330, and 660 μl/l on days 62, 63, and 64 days after planting, respectively. The Pn was estimated from linear regression equations of Pn versus photosynthetic photon flux density (PPFD) with PPFD set to 1500 μmol photons m^{-2} s^{-1}. The *vertical bars* represent 95% confidence intervals. (After Baker et al. 1990b, 1994b)

Table 6. For soybean, leaf-blade soluble protein expressed on a leaf-blade area basis and percentage Rubisco protein expressed on a leaf-blade soluble protein basis for 34-day-old soybean plants grown under a wide range of CO_2 concentrations. (After Campbell et al. 1988 and Baker and Allen 1994a). For rice, nitrogen content of leaves expressed on a leaf-area basis and percentage Rubisco protein expressed on a leaf soluble-protein basis for 75-day-old rice plants grown under a wide range of CO_2 concentrations. (After Rowland-Bamford et al. 1991 and Baker and Allen 1994)

Soybean			Rice		
CO_2-growth Concentration (μl/l)	Leaf-blade Soluble (g/m²)	Rubisco protein (%)	CO_2-growth concentration (μl/l)	Leaf nitrogen (mmol/m²)	Rubisco protein (%)
160	2.5	56	160	95	62
220	3.2	54	250	90	59
280	2.6	–	330	81	54
330	2.3	57	500	62	49
660	2.3	54	660	78	43
990	2.3	55	900	64	42

area) s^{-1}] of rice exposed to daytime CO_2 ranging from 160 to 900 μl/l. Similar to photosynthetic rates, R_d increased with CO_2 exposure from 160 to 500 μl/l but leveled off across the 500–900 μl/l range. The R_d of the ambient and superambient CO_2 treatments reached a broad maximum about 30–50 DAS; whereas, the broad maximum of the subambient treatments occurred later at about 50–70 DAS. Maximum values of R_d were about 6, 8, and 9 μmol m^{-2} s^{-1} for the 160, 250, and 330 μl/l CO_2 treatments and 11–12 μmol m^{-2} s^{-1} for the three superambient CO_2 treatments.

Specific respiration rate [R_{dw}, μmol (CO_2) s^{-1} kg^{-1} (total aboveground dry matter)] decreased exponentially with DAS at all CO_2 exposures, and was higher in the subambient than in the ambient and superambient CO_2 treatments. At each CO_2-exposure level, the patterns of R_{dw} with DAS were very similar to the patterns of plant-tissue nitrogen concentration with DAS (Baker et al. 1992d). Furthermore, R_{dw} was linearly related ($r = 0.91, p = 0.01$) to total above-ground plant-tissue nitrogen concentration [mg (N)/g (DW)] across the range of CO_2 exposures and the six dates of plant sampling (Fig. 7). Baker et al. (1992d) concluded that the CO_2 treatments affected R_{dw} mainly by altering the protein composition of the plant tissue. Another explanation is that elevated CO_2 increased the amount of structural and nonstructural carbohydrates in the plant tissues; therefore, a larger proportion of the dry matter was sequestered in nonprotein materials. This study showed no respiration acclimation to long-term CO_2 enrichment that could not be explained by nitrogen concentration of the plant tissues.

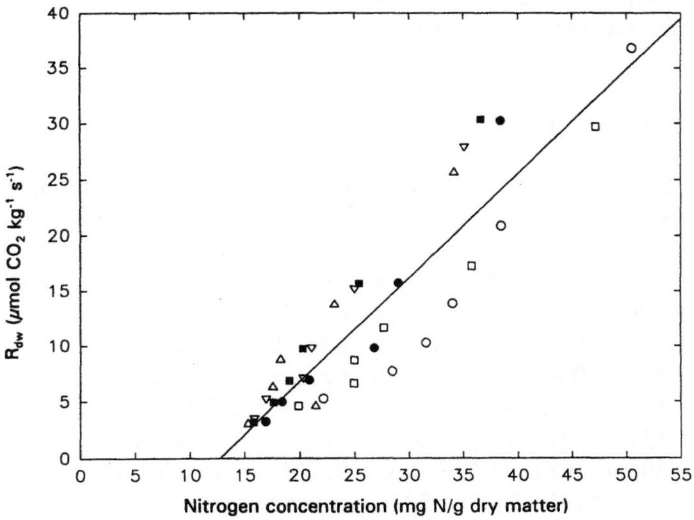

Fig. 7. Specific respiration rate versus nitrogen concentration of above-ground biomass for rice exposed to subambient [160 μl/l (○) and 250 μl/l (□)], ambient [330 μl/l (●)], and superambient [500 μl/l (■), 660 μl/l (▽), and 900 μl/l (△)] CO_2 concentrations. (After Baker et al. 1992d)

Daily Canopy Evapotranspiration and Photosynthesis

Both plant transpiration and direct evaporation from the floodwater surface contribute to evapotranspiration (ET) of rice growing in the SPAR chambers. After canopy closure, much of the daytime ET can be attributed to transpiration. The diurnal trend of ET followed the diurnal trend in solar irradiance, especially during the two studies (RICE I and RICE II, Table 3) in which air temperature and water temperature were held to constant setpoint values. After canopy closure, maximum ET rates were about 35% greater from rice grown at 160 μl/l than at 900 μl/l CO_2 (Baker et al. 1990c). However, ET rates were similar when all chambers were exposed for one-half day to the same CO_2 concentration (data not shown, Baker et al. 1990b). These similar ET rates when rice was exposed to the same CO_2 levels demonstrate the effect of the instantaneous exposure level of CO_2 on stomatal control of transpiration.

Temperature also has a very large effect on the magnitude of diurnal ET rates, mediated primarily through vapor pressure deficit of the air. Solar irradiance also has a large effect on ET through both the energy inputs to the canopy and through the stomatal opening response to light. Midday maximum ET rates at 29 DAS were about 75 and 35% higher for rice grown at 40/33/37 °C and 34/27/31 °C, respectively, compared with rice grown at 28/21/25 °C (Baker et al. 1993a). At 29 DAS, direct evaporation from the flood water is likely still an important component of ET. The interacting effects of CO_2 concentration on stomata under a range of temperature and vapor-pressure deficit conditions was modeled and discussed by Allen (1990).

Water-Use Efficiency

Stomatal conductance decreases with increasing CO_2 concentration, which can reduce both leaf and whole canopy transpiration. However, CO_2 enrichment may also increase the leaf-surface area of the canopy for transpiration, thereby offsetting some of the water savings (Allen et al. 1985; Jones et al. 1985b). At 62 DAS in the RICE II experiment, daytime water-use efficiency (WUE) ranged from 1.22 to 3.90 μmol (CO_2) mol^{-1} (H_2O) across the CO_2 treatments from 160 to 990 μl/l (Baker et al. 1990c and Baker and Allen 1993b, 1994). Calculated WUE increased with increasing CO_2 because of the decline in ET [from 608 to 470 mol (H_2O) m^{-2} per day] across this CO_2 range and the increase in Pn [from 0.74 to 1.80 mol (CO_2) m^{-2} per day with CO_2 up to 500 μl/l.

At 58 DAS in the 28 °C air temperature treatment of the RICE IV experiment, CO_2 enrichment from 330 to 660 μl/l increased daytime total CO_2 uptake by 30% and decreased ET by 22% with a concomitant 67% increase of WUE from 1.42 to 2.37 μmol (CO_2) $mmol^{-1}$ (H_2O). Daytime ET was more than doubled from the 25 °C to the 37 °C daytime temperature treatment [from 360 to 910 mol (H_2O) m^{-2} per day]. Calculated WUE declined from the 28 °C to the 37 °C daytime treatment [from 2.37 to 1.06 μmol (CO_2) mol^{-1} (H_2O)] because of the sharp increase in ET and the relatively stable Pn across this temperature range (Baker an Allen 1993a,b, 1994).

Conclusion

Total growth duration of rice was shortened by 10–12 days across a CO_2 concentration treatment range of 160 to 500 μl/l because of a shortened vegetative phase of development and a reduction in number of main-stem leaves formed during this period. Photosynthesis, growth, and final grain yield increased with CO_2 from 160 to 500 μl/l, but these responses tended to flatten out from 500 to 900 μl/l. Carbon-dioxide enrichment from 330 to 660 μl/l increased grain yield mainly by increasing panicles per plant; whereas, increasing temperature above 28/21/25 °C resulted in decreased grain yield largely because of a decline in filled grains per panicle. Grain yields were highest at a weighted mean temperature of 26 °C, and declined by about 10% per each 1 °C rise in temperature above 26 °C.

Long-term exposure to subambient through superambient CO_2 concentrations changed the canopy photosynthetic capacity of cultivar IR30 rice. Canopy photosynthetic rates measured at a common CO_2 concentration decreased with increasing long-term CO_2-growth treatment. This response appears to differ from conclusions reached in the study with IR72. Concomitantly, both Rubisco protein content and activity of leaves declined with increasing long-term CO_2-growth treatment. Downward acclimation of the photosynthetic machinery in rice may have decreased the productivity potential, but it did not lead to any actual yield losses under CO_2 enrichment. Specific dark respiration rates of rice grown at subambient through superambient levels of CO_2 were linearly related to plant-tissue nitrogen.

Increases in WUE with increasing CO_2 were realized by both substantial increases in CO_2 uptake and somewhat smaller decreases in water loss. Increasing temperatures greatly increased water use and decreased photosynthetic WUE. Based on grain yield, WUE decreased drastically with increasing temperature not only because of increasing ET, but also because of sharply decreasing seed production.

Rising atmospheric CO_2 is likely to benefit rice production by increasing photosynthesis, growth, and grain yield while reducing water use and increasing WUE. In warm areas of the world, however, possible future global warming may result in both substantial yield decreases and increased water requirements.

Acknowledgments. Supported in part by US EPA Interagency Agreement No. DW 12934099–0 and US DOE Interagency Agreements No. DE-AIO1-81ER60001 and DE-AI05–88ER69014 with USDAARS.

Mention of proprietary products is for the convenience of the reader only, and does not constitute endorsement or preferential treatment by the United States Department of Agriculture or the University of Florida.

Interactive Effects of Elevated Carbon Dioxide and Temperature on Rice Growth and Development

K.T. Ingram,[1,2] P.A. Manalo,[3] O.S. Namuco,[3] R.R. Pamplona,[1,4] and W.M. Weerakoon[1,5]

Interpretations of available data on global change vary dramatically. Some researchers argue convincingly that we have no evidence for global warming (Christy and McNider 1994). Others argue that increasing atmospheric CO_2 will benefit agriculture (Idso 1990). On the other hand, global circulation models support the hypothesis that increased concentrations of atmospheric greenhouse gases will cause global warming of about 0.3 °C per decade over the next century (IPCC 1990). Opposing interpretations of the same data arise because short-term weather fluctuations are larger than predicted long-term changes, and because natural cataclysmic events, such as the 1991 eruption of Mount Pinatubo, may counteract or mask long-term climate trends.

It is indisputable, however, that human activity has dramatically increased concentrations of atmospheric greenhouse gases during the past century (Boden et al. 1992). Furthermore, global warming resulting from increases in greenhouse gases would provide the first incidence in recorded history in which human activity may affect global climates. Just as human activity may unwittingly cause global change, informed decisions and activities may either prevent, reverse, or accelerate global change. Although global warming is still a hypothesis, we must understand the consequences of global warming, so that we can act appropriately if the future proves it true.

There has been substantial research on rice responses to high temperatures (Yoshida et al. 1981; Mackill et al. 1982; Zheng and Mackill 1982) and to elevated CO_2 (Imai et al. 1985; Baker et al. 1990a; Rowland-Bamford et al. 1991). However, there has been little research on either the interactive effects of elevated temperature and CO_2 or on the variation among rice genotypes in their responses to these climate change variables.

[1] International Rice Research Institute, P.O. Box 933, 1099 Manila, Philippines
[2] Current address: Peanut CRSP, Georgia Station, Griffin, GA 30223-1797, USA
[3] Agronomy, Plant Physiology, and Agroecology Division, International Rice Research Institute, P.O. Box 933, 1099 Manila, Philippines
[4] Current address: PhilRice, Munoz, Nueva Ecija, Philippines
[5] Current address: Crop Science Department, Oregon State University, Corvallis, OR 97333, USA

S. Peng et al. (Eds.) Climate Change and Rice

Responses to High Temperatures

High temperatures can reduce crop productivity in two ways, through more frequent acute heat stress events or through chronic effects of higher average temperatures. One or both types of stress can occur in any season or location. Acute heat-stress injury generally has a threshold response with damage increasing rapidly over a small range of temperature change. Most chronic effects have temperature-response curves that may be modeled as a product of an exponentially increasing rate function and an exponential decay function (Conn and Stumpf 1972). Because acute and chronic heat stresses affect crop productivity through different processes that must be considered separately.

Acute Heat-Stress Injury

Rice is most susceptible to damage by high-temperature injury during flowering. Daytime temperatures of 38 °C or greater during flowering impair pollen shed and fertilization (Mackill et al. 1982; Zheng and Mackill 1982). Mackill et al. (1982) found that the tolerance to heat stress during flowering was related to the ability of anthers to shed pollen. Heat-tolerant cultivars either escape stress by shedding pollen during cooler morning hours, or they have true tolerance and shed pollen despite high temperatures.

Temperatures up to 41 °C have little direct impact on leaf CO_2 assimilation. Instead, high temperatures may indirectly benefit leaf CO_2 assimilation (Egeh et al. 1994). If high temperatures reduce the filled spikelet fraction, then plants are sink-limited and panicles remobilize and translocate less N and C from leaves to growing grains. Therefore, sink limitations caused by high temperatures at flowering may allow leaves to retain N and C, with a resultant delay in canopy senescence and prolonged CO_2 assimilation.

Fortunately, there is good genotypic variation among rice cultivars for tolerance to high temperature stress during flowering. Under phytotron greenhouse conditions, we grew 50 rice genotypes in pots (20 cm diameter, 55 cm tall) containing 8 kg sieved Maahas clay soil at 37/29 °C under ambient and elevated (600 μl/l) CO_2. Under ambient CO_2, the filled spikelet fraction ranged from 0 to 0.67 with an average of 0.216 (Table 1). Under elevated CO_2, the filled spikelet fraction ranged from 0 to 0.66 with an average of 0.354 (Table 1). These values compare with a filled spikelet fraction of 0.8 to 0.85 under optimal conditions. Elevated CO_2 significantly increased the filled spikelet fraction for 30 genotypes and significantly reduced the filled spikelet fraction for only three genotypes (Table 1). The ability of elevated CO_2 to mitigate high temperature effects on the filled spikelet fraction was inversely related to apparent heat tolerance under ambient CO_2. At ambient CO_2, the greatest filled spikelet fraction was 0.39 among genotypes for which elevated CO_2 increased the filled spikelet fraction. However, the smallest filled spikelet fraction was 0.43 among genotypes for which elevated CO_2 decreased the filled spikelet fraction.

Table 1. Average filled spikelet fraction of rice genotypes grown at 37/29°C day-night temperature and ambient and elevated atmospheric CO_2 concentrations. Plants were grown in a phytotron greenhouse in pots (50 cm tall × 20 cm diameter) containing 8 kg of dry soil

Effect of elevated carbon dioxide	Number of genotypes	Ambient CO_2 (filled/total spikelets)	600 μl/l CO_2 (filled/total spikelets)
Positive	30	0.084	0.322
Not significant	17	0.434	0.414
Negative	3	0.484	0.333
Total	50	0.216	0.354

The mechanism by which elevated CO_2 mitigates heat stress is not clear. Heat stress is frequently associated with water-deficit stress, because warm air has greater desiccating power than cool air. Further, elevated CO_2 increases leaf water-use efficiency. Therefore, it is possible that elevated CO_2 improves panicle water status enough to allow pollen to be shed under high temperature conditions.

In these phytotron greenhouse experiments, plants were exposed to the daytime temperature for about 6 h. In nature, daily maximum temperature generally lasts only 1–2 h. Therefore, for equal stated maximum temperatures, stress may be more severe in these experiments than under natural conditions.

Chronic High-Temperature Effects on Growth and Development

Five cultivars (IR20, IR30, IR46, IR64, and IR72) were grown in pots containing 8 kg of dry soil in closed, naturally lighted chambers with control of temperature, relative humidity, and CO_2. Chambers had 1.25 × 1.25 m floor space, enough for five pots with one plant per pot of each cultivar at each of four day-night temperature regimes 29/21 °C, 33/25 °C, 37/29 °C , and 41.33 °C. The CO_2 concentration was held constant at 500 μl/l and vapor pressure deficit at 1.2 kPa.

For the average of the five genotypes, rate of tiller appearance and number of tillers at 72 days after sowing (DAS) did not significantly differ over from 29/21 °C to 37/29 °C, but further temperature increase to 41/33 °C significantly reduced tiller development. The temperature response of leaf appearance was similar to that of tiller appearance. Phyllochron interval increased significantly as temperature increased from 37/29 °C to 41/33 °C (Table 2). For all five cultivars, time from sowing to flowering decreased from 29/21 °C to 37/29 °C and then increased as temperature increased to 41/33 °C (Fig. 1). Therefore, organogenesis showed plateau in temperature response from 29/21 °C to 37/29 °C, whereas phenological development was fastest above 37/29 °C.

For all cultivars except IR72, shoot dry weight at flowering decreased as temperature increased from 29/21 °C to 41/33 °C, with the optimum near 29/21 °C (Table 3). Shoot dry weight of IR72 had a higher optimum temperature,

Table 2. Effect of different temperature levels on phyllochron interval of different rice cultivars grown at 500 μl/l CO_2 [a]

Temperature (°C)	IR20 (days/leaf)	IR30 (days/leaf)	IR46 (days/leaf)	IR64 (days/leaf)	IR72 (days/leaf)	Mean
29/21	4.64	4.50	4.17	4.42	4.22	4.39b
33/25	3.97	4.59	3.87	4.48	3.89	4.16b
37/29	3.91	4.31	4.52	4.31	3.99	4.20b
41/33	5.46	5.36	4.42	4.21	4.10	4.71a
Mean	4.49	4.69	4.24	4.35	4.05	–

Source	*df*	*MS*
Rep	3	0.15
Temp	3	1.23**
Error (a)	9	0.16
Cultivar	4	0.94*
Temp × Cultivar	12	0.54 NS
Error (b)	48	0.36

[a] In a column, means followed by the same letter are not significantly different at the 5% level by least significant difference (LSD). To compare two cultivar means $LSD_{(0.05)} = 0.42$ days/leaf. NS not significant, * significant at 5%, ** significant at 1%.

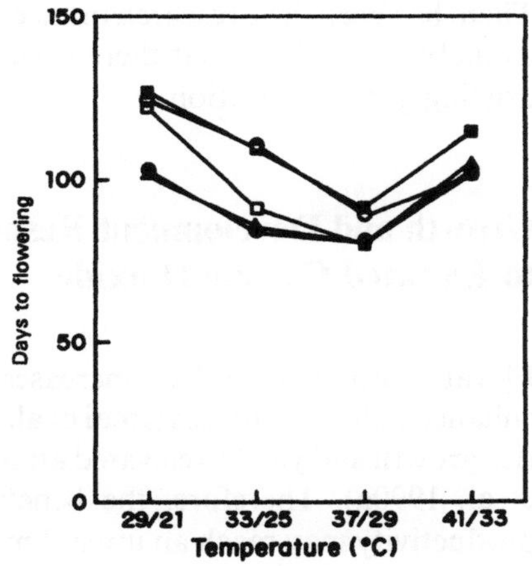

Fig. 1. Effect of temperature on days to flowering of different rice cultivars. IRRI Phytotron, WS 1992: ■ IR20; □ IR64; ● IR30; ▲ IR72; OIR46

near 33/25 °C. For IR20, IR46, and IR64, root dry weight declined about 50% as temperature increased from 29/21 °C to 33/25 °C, whereas root dry weight of IR30 and IR72 did not change significantly with temperature over the range studied.

The chronic effects of higher average temperatures on grain growth may have greater practical importance than the acute effects of high temperatures during flowering or the chronic effects on vegetative growth and development. Simulation models predict that rice will yield 9% less grain for each 1 °C increase in average temperature (Kropff et al. 1993a). This chronic effect of higher temperature on grain yield results mainly from decreased grain-filling duration.

Table 3. Effect of different temperature levels on total production of shoot biomass by different rice cultivars grown at 500 μl/l CO_2

Temperature (°C)	IR20 (g/plant)	IR30 (g/plant)	IR46 (g/plant)	IR64 (g/plant)	IR72 (g/plant)
29/21	85.0 a	62.4 a	134.5 a	92.2 a	61.3 bc
33/25	79.3 a	60.1 a	122.0 a	80.8 ab	89.8 a
37/29	48.4 b	46.9 ab	61.1 b	62.0 b	72.0 ab
41/33	35.9 b	26.5 b	62.7 b	33.9 c	46.1 c

Source	*df*	*MS*
Rep	3	573.47 NS
Temp	3	10203.83**
Error (a)	9	262.16
Cultivar	4	4521.70**
Temp × Cultivar	12	768.94**
Error (b)	48	280.93

[a] In a column under each cultivar, means followed by the same letter are not significantly different at the 5% level by Duncan's Multiple Range Test (DMRT). To compare two cultivar means at the same temperature, $LSD_{(0.05)}$ = 23.8 g. NS not significant, ** significant at 1%.

There has been little research on rice-grain growth responses to temperature, but available data show that there is small scope to manipulate this effect through existing genetic variation.

Growth and Development Responses to Elevated Carbon Dioxide

Elevated atmospheric CO_2 increases the growth rate and yield of rice through enhanced photosynthesis (Imai et al. 1985). Unfortunately, positive responses of rice growth and yield to elevated atmospheric CO_2 diminish above 500 μl/l (Baker et al. 1990a). Therefore, the benefits of increased atmospheric CO_2 on rice productivity may reach an upper limit at the atmospheric CO_2 levels of less than the 660 μl/l CO_2 anticipated for the next century.

The same five cultivars (IR20, IR30, IR46, IR64, and IR72) grown under different temperature regimes were grown at five CO_2 levels (350, 425, 500, 625, and 750 μl/l) all at 33/25 °C and 1.2 kPa vapor pressure deficit. Averaged across cultivars, tiller development increased as CO_2 increased from 350 to 425 μl/l, but there was no further effect of further increase in CO_2 on tiller development up to 750 μl/l CO_2. Phyllochron interval had a curvilinear response to CO_2. Leaves appeared at 12% shorter intervals at 500–600 μl/l than at 425 or 750 μl/l CO_2 (Table 4). In IR30 and IR72, CO_2 concentration did not significantly affect time from sowing to flowering. For the other three cultivars, time from sowing to flowering was greatest at 425 μl/l CO_2 and declined at higher and lower CO_2 levels (Fig. 2). Effects of CO_2 on development were similar in magnitude to temperature

Table 4. Effect of different CO_2 levels on phyllochron interval of different rice cultivars grown at 33/25 °C [a]

CO_2 (μl/l)	IR20 (days/leaf)	IR30 (days/leaf)	IR46 (days/leaf)	IR64 (days/leaf)	IR72 (days/leaf)	Mean
350	4.50	4.57	4.22	4.48	4.90	4.54 ab
425	4.69	4.76	4.32	5.23	5.18	4.84 a
500	3.97	4.58	3.87	4.48	3.89	4.16 c
600	4.42	3.97	4.16	4.28	4.23	4.20 bc
750	4.92	4.74	4.69	4.69	4.53	4.71 a

Source	*df*	*MS*
Rep	3	0.08 NS
CO_2	4	1.83**
Error (a)	12	0.28
Cultivar	4	0.39 NS
CO_2 × Cultivar	16	0.27 NS
Error (b)	60	0.21

[a] In a column, means followed by the same letter are not significantly different at the 5% level by LSD. NS not significant, ** significant at 1%.

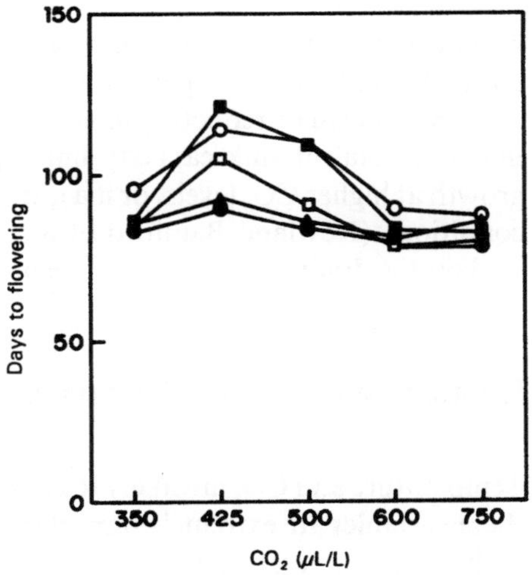

Fig. 2. Effect of CO_2 (μl/l) on days to flowering of different rice cultivars grown at 33/25 °C. IRRI Phytotron, WS 1992: ■ IR20; □ IR64; ● IR30; ▲ IR72; O IR46

effects. Also similar to temperature effects, IR30 and IR72 responded differently to CO_2 than did IR20, IR46, and IR64.

Shoot dry weight of IR30 at flowering reached a plateau above 500 μl/l CO_2 (Table 5). For the other cultivars, shoot dry weight at flowering was greatest at 425–500 μl/l CO_2 and decreased at higher CO_2 levels. By correlation analysis including all five cultivars, effects of CO_2 on time to flowering explain 51% of the variation in shoot dry weight at flowering (r=0.714**). Therefore, the direct beneficial effect of CO_2 on growth is partly negated by a shortening of time to flowering. Root dry weight data were more variable than those for shoot dry

Table 5. Effect of different CO_2 levels on total aboveground biomass of different rice cultivars grown at 33/25 °C [a]

CO_2 (μl/l)	IR20 (g/plant)	IR30 (g/plant)	IR46 (g/plant)	IR64 (g/plant)	IR72 (g/plant)
350	32.3 c	30.1 b	51.8 c	41.9 b	40.0 c
425	83.8 a	50.2 a	125.1 a	87.9 a	90.9 a
500	79.3 a	60.1 a	122.0 a	80.8 a	89.8 a
600	54.3 b	50.6 a	62.3 b	59.0 b	69.8 b
750	46.5 bc	46.2 ab	73.5 b	56.0 b	61.0 b

Source	*df*	*MS*
Rep	3	373.95 NS
CO_2	4	8657.14 **
Error (a)	12	189.55
Cultivar	4	4236.55 **
CO_2 × Cultivar	16	419.39 *
Error (b)	60	184.2

[a] In a column under each cultivar, means followed by the same letter are not significantly different at the 5% level by DMRT. To compare two cultivar means at the same CO_2 level, $LSD_{(0.05)}$ = 19.2 g. NS not significant, * significant at 5%, ** significant at 1%.

weight, but root dry weight also tended to double as CO_2 increased from 350 μl/l to 425 μl/l, and then plateau or decline above 425 μl/l CO_2.

The plateau or reduction in growth observed with increasing CO_2 above 500 μl/l may result if sink capacity limits growth, if factors other than CO_2 limit growth at higher CO_2 levels, or if a growth inhibitor accumulates under high CO_2 conditions (Rowland-Bamford et al. 1991). Additional research is needed to explain the diminishing return in response to CO_2.

Interactive Effects of Elevated Carbon Dioxide and Temperature

Temperature and CO_2 are likely to change together; therefore, the interactions of these variables are extremely important. A phytotron experiment was conducted with three temperature regimes (29/21 °C, 33/25 °C, and 37/29 °C) in factorial combinations with two CO_2 levels (330 and 660 μl/l). Six cultivars were used, three lowland indica types (IR20, IR46 and IR 64) and three traditional upland types (ITA186, Salumpikit, and Moroberekan) (Manalo et al. 1994).

Indica rices typically produce more tillers than do tropical japonica types. This was generally true for all temperature and CO_2 combinations (Fig. 3). Rate of tiller appearance was greatest in the highest temperature and CO_2 treatment combination for all genotypes, but this treatment shortened tillering duration. Therefore, in IR46 and IR64, the high CO_2-intermediate temperature treatment combination produced tillers longer and ultimately produced more tillers than the high CO_2-high temperature combination. For indica rices, increased CO_2 shortened time to panicle emergence at 37/29 °C but delayed panicle emergence

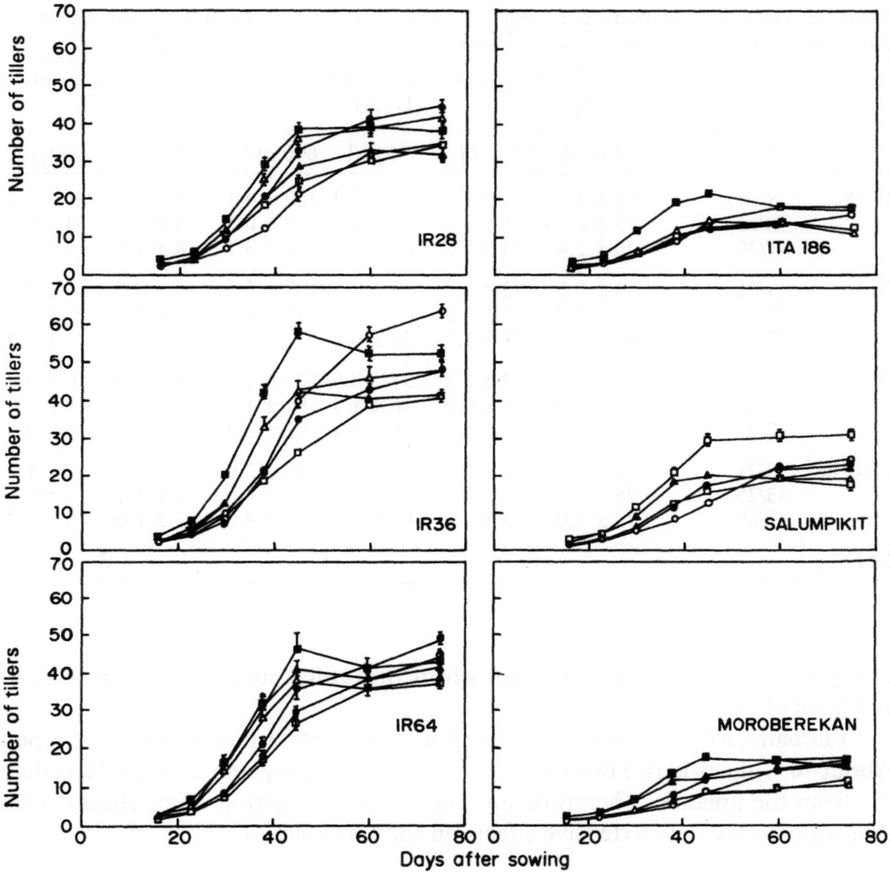

Fig. 3. Tiller development (number per plant) of six rice cultivars in response to CO_2 and temperature. IRRI Phytotron, DS 1992: O 330 μl/l 29/21 °C; Δ 330 μl/l 33/25 °C; □ 330 μl/l 37/29 °C; ● 660 μl/l 29/21 °C; ▲ 660 μl/l 33/25 °C; ■ 660 μl/l 37/29 °C

at 29/21 °C (Table 6). In the upland cultivar, however, there was no significant interaction between CO_2 and temperature on time to panicle emergence. For the indica rices, panicle-emergence duration generally increased with temperature at ambient CO_2, but decreased with temperature at elevated CO_2 (Table 6).

Indica rices are generally considered to be photoperiod-insensitive. They may, in fact, have short-day photoperiod sensitivity with a very long critical day length (Horie and Nakagawa 1990). The interaction between CO_2 and temperature on time to flowering was unexpected. These results must be confirmed, but they suggest that the photothermal theory of plant phenology is incomplete. Although there is a growing body of data showing that CO_2 affects phenology (Allen et al. 1994; Horie et al. 1994), simulations predicting rice productivity in

Table 6. Time of first panicle emergence, 50% flowering of all tillers, and flowering duration (± standard error) or rice cultivars as affected by CO_2 and temperature

	Temperature (°C)	Panicle emergence (DAS)		50% flowering (DAS)		Flowering duration (days)	
		330 μl/l	660 μl/l	330 μl/l	600 μl/l	330 μl/l	660 μl/l
IR28	29/21	90 ± 1.6	89 ± 3.0	95 ± 1.8	100 ± 4.1	5 ± 0.5	11 ± 2.2
	33/25	73 ± 0.5	71 ± 0.8	81 ± 1.8	74 ± 0.5	8 ± 1.4	4 ± 0.7
	37/29	76 ± 2.4	73 ± 2.8	84 ± 2.5	78 ± 3.1	8 ± 1.1	5 ± 0.8
IR36	29/21	84 ± 0.5	93 ± 1.0	89 ± 0.5	107 ± 2.3	5 ± 0.5	14 ± 3.2
	33/25	77 ± 2.2	73 ± 0.6	85 ± 1.6	81 ± 1.0	8 ± 0.8	7 ± 0.8
	37/29	80 ± 1.0	71 ± 1.2	87 ± 0.8	78 ± 2.1	7 ± 1.4	7 ± 1.0
IR64	29/21	88 ± 1.6	100 ± 2.2	94 ± 0.5	106 ± 1.7	5 ± 1.2	6 ± 0.8
	33/25	81 ± 1.4	84 ± 0.9	86 ± 1.4	91 ± 1.8	5 ± 0.5	8 ± 1.0
	37/29	91 ± 5.1	87 ± 2.8	99 ± 5.4	92 ± 3.2	8 ± 0.5	5 ± 0.6
ITA186	29/21	96 ± 2.0	98 ± 2.8	109 ± 1.5	108 ± 2.9	12 ± 0.6	9 ± 1.1
	33/25	76 ± 1.4	80 ± 1.0	81 ± 0.5	83 ± 1.0	4 ± 1.2	3 ± 0.5
	37/29	79 ± 0.0	78 ± 2.6	87 ± 0.3	87 ± 4.5	7 ± 1.0	9 ± 0.2
$LSD_{(0.05)}$		6		7		3	

changed climate scenarios have not yet incorporated such effects (Horie et al. 1994; Jin et al. 1994).

Globally, irrigated indica rices are much more important than upland types. About 70% of world rice production comes from the irrigated lands and less than 5% from the uplands. Therefore, any significant interaction effect observed for indica rices may have extensive effects on food production.

Implications for Rice Productivity and Future Research

Benefits of increased atmospheric CO_2 to rice production will most likely be limited in a warmer world. Although elevated CO_2 partly mitigates acute heat stress in heat-susceptible rice genotypes, elevated CO_2 does not compensate for, and may instead exacerbate, the adverse chronic effects of high temperature. Because of the significant interactions between CO_2 and temperature, climate change impacts will be location-specific. With warmer present-day temperatures, the tropics will have a much greater probability of adverse effects of a warmer, higher CO_2 environment than temperate regions.

Impacts of global warming will also depend on whether warming results in greater temperature variability as well as greater average temperatures. For example, higher average temperature alone would most likely benefit rice production in temperate regions through reduced frequency of cold stress. If both average temperatures and temperature variability increase, however, the combined effect is likely to be negative. If average temperatures rise in the tropics, with or without a change in temperature variation, the impact will probably be negative.

The range of rice genetic variability for tolerance to acute heat stress and for response to elevated CO_2 suggests that germplasm improvement may partly mitigate adverse impacts of climate change. Future research should confirm and develop this potential.

There are three general strategies to prevent or mitigate potential adverse impacts of climate change on agricultural production and food security: improve adaptation of crops to changed climates, control emissions of greenhouse gases, and reduce demand for agricultural products through human population control. This chapter, and others in this volume, have discussed data that show a limited potential to pursue the first of these general strategies. Specifically, we should establish breeding programs to improve heat tolerance and improve the efficiency of CO_2 use as well as promote cultivation of heat-tolerant crops. In addition, farmers are likely to adjust their cropping calendars and to shift crop-production areas toward the poles with little need for research support. The contributions on methane emissions address agricultural contributions to atmospheric greenhouse gas emissions, but agricultural scientists generally do not address issues related to nonagricultural sources of greenhouse gases (e.g., control of fossil fuel combustion or reversing trends in deforestation). Nor do agricultural scientists often address the need for human population control. As evidence accumulates in support of the global warming hypothesis, and as research shows a limit for continued increases in agricultural production in a changed climate, agricultural scientists should begin to speak out on the need to pursue nonagricultural strategies to solve the problems of climate change.

Acknowledgment. Although the research described in this chapter has been funded wholly or in part by the US Environmental Protection Agency under cooperative agreement number 817425 to the International Rice Research Institute, it has not been subject to the Agency's review and therefore does not necessarily reflect the views of the Agency, and no official endorsement should be inferred.

Simulation Modeling

Effect of Anticipated Change in Global Environment on Rice Yields in Japan

T. Horie, M.J. Kropff,[2] H.G. Centeno,[2] H. Nakagawa,[1]
J. Nakano,[3] H.Y. Kim,[1] and M. Ohnishi[1]

Almost all rice in Japan is produced in rice fields flooded with water supplied through well-developed waterways. Therefore, unlike most other countries in the world, water plays a negligible role as a yield-reducing factor in Japanese rice production. The major climatic factors that influence rice production in Japan are temperature, solar radiation, and strong winds and heavy rainfalls associated with typhoons. Although Japan as a whole loses a certain amount of rice every year because of typhoons, the location and timing of typhoon damages are unpredictable. Except for typhoons, temperature and solar radiation are the main factors that produce spatial and yearly variations of rice yield in Japan. Cool summer temperatures in northern Japan (Hokkaido and Tohoku districts) can cause severe yield reductions and may significantly impact on national rice production (as occurred in 1993). Horie (1987) and Horie et al. (1992) quantitatively explained that yield variation in Japan was based on temperature and solar radiation using the simulation model SIMRIW (SImulation Model for Rice-Weather relationships).

Models of the dynamics of the atmosphere of the earth, known as general circulation models or GCMs, predict that with increasing levels of greenhouse gases in the atmosphere, global climate change is likely (Hansen et al. 1984). Because predicted climate change may have enormous effects on Japanese rice production, it is vitally important to assess the impacts on regional rice yield to provide a basis for counter measures (e.g., cultivar improvements, alterations of cropping seasons, and cultivation technologies). For this purpose, Horie (1993) modified SIMRIW to include processes describing: direct effects of atmospheric CO_2 on growth, and high temperature-induced spikelet sterility of rice. The modified model was then used to assess impacts of doubled CO_2 and climate change on rice yield in three representative rice-producing prefectures in Japan. Preliminary analysis indicated that doubled CO_2 and the associated climate change had different effects on rice yield in different locations, but the study was not sufficiently detailed to clarify the overall effect on rice production in the whole country. The objectives of this study are to validate the simulation model SIMRIW using rice growth and yield data obtained under widely different

[1] Faculty of Agriculture, Kyoto University, Kyoto 606, Japan
[2] International Rice Research Institute, P.O. Box 933, 1099 Manila, Philippines
[3] Faculty of Agriculture, Tottori University, Tottori, Japan

S. Peng et al. (Eds.) Climate Change and Rice

environmental conditions and to assess, by use of SIMRIW, the effects of elevated CO_2 and climate change on rice yields in the major rice-producing areas of Japan.

The Model

The details of SIMRIW have already been reported (Horie 1987; Horie et al. 1992). This brief description emphasizes the derivation of components that govern the direct effects of atmospheric CO_2 on growth and temperature-induced spikelet sterility of rice.

Basic Structure

SIMRIW consists of three major parts that describe the processes of ontogenetic crop development, biomass accumulation, and yield formation. Ontogenetic development of rice from emergence to heading is represented by a continuous variable, the developmental index (DVI). The value of DVI is defined to be zero at emergence, 1.0 at heading, and 2.0 at maturity. The value of DVI is computed daily by integrating the developmental rate (DVR) with respect to time. The DVR is a nonlinear function of daily mean temperature and day length (Horie and Nakagawa 1990).

Dry-matter accumulation is simulated based on the principle that dry weight of the crop at any moment is proportional to the absorbed solar radiation accumulated up to that moment (Monteith 1977). This process of biomass accumulation is characterized by one crop parameter, the solar radiation conversion efficiency (Cs). The Cs is assumed to be constant up to heading (DVI=1), and thereafter to decrease as a function of DVI to simulate maturation and senescence processes. Leaf-area growth rate, which governs radiation interception, is modeled as a unique function of temperature. This approach contrasts with traditional models that simulate leaf dry-weight accumulation and then calculate leaf area by multiplying leaf dry-weight by specific leaf area. The SIMRIW models leaf-area growth independent of leaf dry weight, thereby accounting for their largely independent natures (Horie et al. 1979).

Grain yield is simulated in SIMRIW from calculated total biomass by multiplying by harvest index. Harvest index is a function of DVI and spikelet sterility. The harvest index-DVI relationship makes the yield formation process dynamic, and simulates premature cessation of growth when crops encounter autumn coolness. Spikelet sterility is a function of cooling degree-days (Uchijima 1976) during the period when spikelets are sensitive to cool temperature ($0.75<DVI<1.2$).

The SIMRIW estimates the climatic potential yield of a given cultivar of irrigated rice under optimal cultivation technologies. Horie (1987) showed a close linear relationship between the simulated potential yield (Yp) and actual yield (Ya) across respective locations in Japan and in the United States. The relationship can be represented by:

$$Ya = KYp, \tag{1}$$

where K is an index of overall technology level of rice culture.

Modeling Effects of Carbon Dioxide and High Temperature on Rice

To determine the parameters for elevated CO_2 and high-temperature effects on rice for SIMRIW, temperature × CO_2 experiments were conducted at Kyoto University for three cropping seasons (1990, 1991, and 1992) using temperature-gradient tunnels (TGTs) with and without elevated CO_2. The TGT is a newly developed experimental system that can impose long-term temperature × CO_2 treatments on crops under seminatural environmental conditions (Horie et al. 1991).

Long-term temperature × CO_2 treatment revealed that nearly doubled daytime CO_2 concentration has negligible effects on nitrogen uptake and leaf-area development, but enhances dry-matter production. The observed insensitivity of leaf area to CO_2 is consistent with observations of Imai et al. (1985) and Baker et al. (1992b). The experimental results with TGT and those of previous workers (Imai et al. 1985; Baker et al. 1992b) indicate that doubled CO_2 concentration increases the dry weight of rice by 24% mainly through the enhancement of radiation conversion efficiency, not through enhanced light interception. These experimental data also showed that temperature did not consistently affect growth enhancement by elevated CO_2. From these findings and from an analysis of the rice canopy photosynthesis versus atmospheric CO_2 relationship by Baker et al. (1990a), Horie (1993) obtained an equation to describe the effects of CO_2 concentration (C_a) on radiation-conversion efficiency (Cs) of rice:

$$Cs = Co\left\{Rm(Ca-330)/\left[(Ca-330)+Kc\right]+1\right\}, \tag{2}$$

where Co is the conversion efficiency at 330 μl/l CO_2, Rm+1 is the asymptotic limit of the relative response to CO_2, and Kc is the Michaelis-Menton constant (estimated values are Rm=0.54 and Kc=370 μl/l). Equation (2) is identical to the one obtained by Allen et al. (1987) for the response of soybean seed and biomass yield to CO_2.

Elevated CO_2 enhanced panicle dry weight to a similar level as crop dry weight under temperature conditions near ambient in Kyoto (about 27 °C), but this effect of CO_2 decreased sharply with increase in temperature. The decline of panicle dry weight with temperature resulted from an increase in numbers of unfertile grains. Rice spikelets have the highest sensitivity to high temperature at anthesis, and are liable to be sterile because of a failure in pollination when temperatures exceed 35 °C during flowering (Satake and Yoshida 1978; Matsui and Horie 1992). Because rice spikelets usually flower during the day, daily maximum temperature is considered to be more closely related to high temperature-induced spikelet sterility of rice than the average temperature. When spikelet fertility is plotted against average daily maximum temperature over a 10-day period close to heading for 'Akihikari' rice grown under elevated and ambient CO_2 conditions in TGTs (Fig. 1), we see that CO_2 concentration has no effect on the relationship between temperature and spikelet fertility. The relation shown in Fig. 1 may be approximated by (Horie 1993):

$$\delta = 100/\{1 + \exp[0.853(\mathrm{Tm} - 36.6)]\}, \tag{3}$$

where δ is fertility percentage, and Tm is average daily maximum temperature during flowering. Equations (2) and (3) were incorporated into SIMRIW to simulate rice growth and yield under elevated CO_2 concentration and global warming conditions.

Sensitivity Analysis

A sensitivity analysis of SIMRIW to conditions in the aerial environment was made by examining responses of simulated yield to daily mean temperature, solar radiation, and CO_2 concentration, under constant conditions of those environments over the entire growth season (Fig. 2). In this analysis, the cultivar Nipponbare was used and the diurnal range of temperature was set at 8 °C. Under constant environmental conditions, the optimum mean temperature for simulated yield was 22–23 °C. Below 22 °C, yield decreased sharply because of the increase of sterile spikelets from cold temperature damage. As temperature increased above the optimum, yield declined more or less linearly up to about

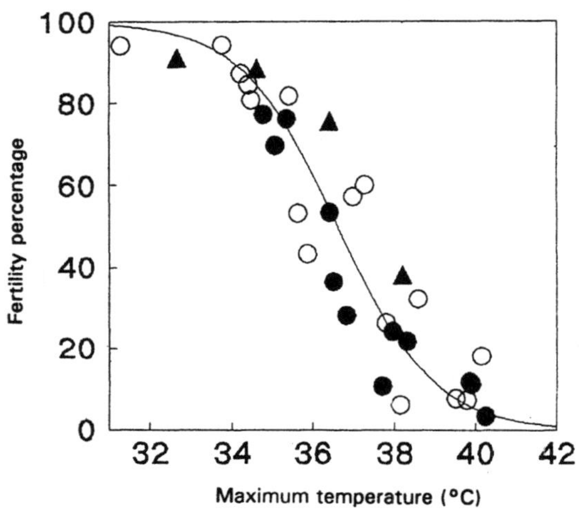

Fig. 1. Relationship between average daily maximum temperature during flowering period and fertility percentage of spikelets for Akihikari rice acclimated to different CO_2 concentrations. (Horie 1993). ○ 350 μl/l; ▲ 690 μl/l; ● 840 μl/l

Fig. 2. Simulated yield responses of Nipponbare rice to daily mean temperature, solar radiation, and CO_2 concentration under constant environmental conditions. Day length was set at 12 h in all cases. — 350 μl/l; ··· 700 μl/l

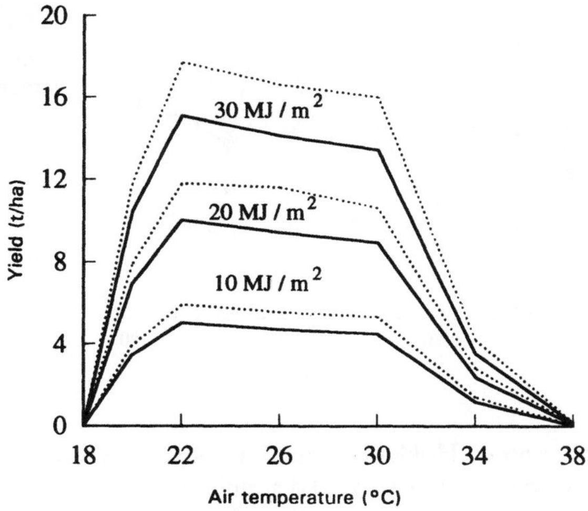

30 °C, above which it declined sharply. The linear decline of yield with increasing temperature from 22 to 30 °C results from shortened total crop duration because of acceleration of phenological development. Sharp decline of yield above 30 °C results from spikelet sterility caused by high-temperature damage.

The overall response pattern of simulated yields to temperature is similar to the results of Munakata (1976) in which the effect of temperature on yields was statistically analyzed using data from long-term field experiments from various regions in Japan. The linear decline of simulated yield over a temperature range between 22 and 30 °C is similar to the results of Yoshida and Parao (1976) in which a mathematical analysis was made of the effect of climate on rice using experimental data from IRRI.

Simulated rice yields were proportional to solar radiation over the entire temperature range, reflecting the axiom that biomass production, and therefore potential yield, is proportional to absorbed solar radiation. This response agrees with the results presented by Yoshida and Parao (1976). The model predicts that doubled CO_2 in the atmosphere alone increases rice yield by 24% under each temperature and radiation condition.

Validation

Prior to application of SIMRIW for the impact assessment of global environment change on Japanese rice production, the validity of the model was examined by using past weather and measured yield data in different rice-producing areas of Japan. For this validation, we used reported yield data for 1979–1990 from five prefectures: Hokkaido, Miyagi, Gunma, Aichi, and Miyazaki. We used daily weather data for 1979–1990 from one weather station for each prefecture:

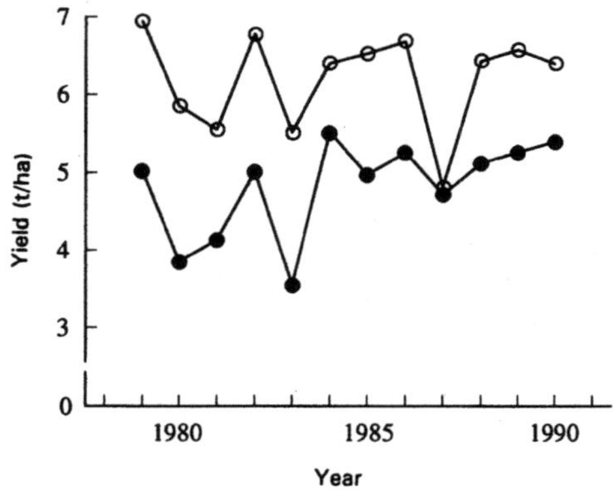

Fig. 3. Yearly changes in reported rice yields at Sapporo, Hokkaido, and those simulated by SIMRIW. ● actual; ○ simulated

Sapporo, Hokkaido; Sendai, Miyagi; Maebashi, Gunma; Nagoya,Aichi; and Miyazaki, Miyazaki. To simulate past rice yields, we used the leading cultivars for each prefecture: Ishikari for Hokkaido, Sasanishiki for Miyagi, Koshihikari for Gunma, Nipponbare for Aichi, and Mizuho for Miyazaki.

Figure 3 shows actual and simulated yearly variation of rice yield in Hokkaido for 1979–1990. Because the model gives climatic potential yield, simulated yields were much higher than the actual yields. The difference between simulated and actual yields decreased with time, indicating an advancement of rice-production technology. Therefore, the technological coefficient (K) that converts simulated yields to actual yields obtained by farmers is a function of year. By assuming a linear increase in rice-production technology by year, the simulated yield (Yp) may be converted to the yield obtained by farmers (Ya) by:

$$Ya = \left[b_0 + b_1(i-1)\right] Yp, \tag{4}$$

where i represents year since 1979, and b_0 and b_1 are the coefficients of the linear regression between K and time.

A multiple regression analysis was performed between the actual yield (Ya) and simulated yield (Yp) using Eq. (4) for each prefecture. Positive values of b_1 were found, with the highest value in Gunma (b_1=0.026) and the lowest in Aichi (b_1=0.005). This suggests that the increase in rice yield because of technological advancement is 2.6%/year in Gunma and 0.5%/year in Aichi.

Using Eq. (4) with coefficients for each prefecture, simulated yields were converted to yields predicted from weather conditions and cultivation technology level for a given year. The converted yields that were obtained, were plotted against actual yields obtained by farmers for each year for each prefecture (Fig. 4). SIMRIW explained 69% of the year-to-year variation of the rice yields in the five prefectures. Considering that rice yields vary not only in response to temperature and solar radiation, but in response to typhoon, pest, and disease damages, SIMRIW explained satisfactorily the regional yield variations in Japan from weather conditions.

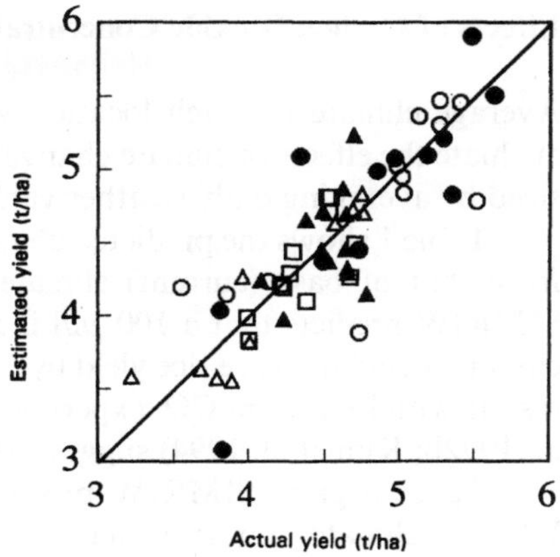

Fig. 4. Comparison between reported rice yields and those simulated by SIMRIW for five representative prefectures in 1979–1990. ○ Hokkaido; ● Miyagi; △ Gunma; ▲ Aichi; □ Miyazaki

Likely Effects of Doubled Carbon Dioxide on Rice Production in Japan

Scenarios for Doubled Carbon Dioxide Climate

To assess the impacts of doubled CO_2 climates on regional rice yields in Japan, we used the agroecological zones (AEZ) proposed by Ozawa (1962), which classified the Japanese islands into 14 zones based on climate and land use. Daily weather data during 12 years from 1979 to 1990 from nine representative weather stations were used as base climates. Besides this base climate, five climate scenarios were adopted for future environments: 450 μl/l CO_2 with no change in temperature or radiation; 450 μl/l CO_2 with a 2°C temperature rise; 700 μl/l CO_2 with the climate predicted by the GISS model; 700 μl/l CO_2 with GFDL-predicted climate; and 700 μl/l CO_2 with UKMO-predicted climate. These GCM climate scenarios were supplied by the data support section within the Scientific Computing Division of the National Center for Atmosphere Research (NCAR).

Among climate scenarios predicted by the three GCMs under a doubled CO_2 concentration, temperature rise is greatest in UKMO, moderate in GFDL, and smallest in GISS. Although the GFDL model predicts reductions of solar radiation under doubled CO_2 in most parts of Japan in most seasons, GISS and UKMO predict increases (particularly in the latter).

Future climate conditions were created by adding those monthly temperature changes in each scenario to current daily maximum and minimum temperatures of the same month, and by multiplying relative changes in monthly solar radiation by current daily solar radiations.

Effects of Carbon Dioxide Concentration and Climate Change

Average climate for each location was used as the baseline against which to evaluate the effects of climate change on rice yields. Average climate was synthesized by averaging daily weather values over the 12 years for each location.

Table 1 shows the predicted change in rice yield under each climate scenario from that of base (current) climate for the nine locations investigated. The SIMRIW predicts that a 100 μl/l increase in CO_2 concentration under current climate would increase rice yield by 7–8% in all locations. A comparison of these results with long-term CO_2 experimental data on rice (Imai et al. 1985; Baker et al. 1992b; Kim et al. 1994) suggests that this prediction is reasonable.

Figure 5 gives SIMRIW predictions of the effects of GFDL, GISS and UKMO climates on rice yields in Japan (as a relative yield change from the present). In AEZ X and XIII, where we did not have actual weather data, relative yield changes under the respective scenarios were interpolated from those in adjacent AEZs. Although the predicted effects of doubled CO_2 climate on Japanese rice yield differed quantitatively among climate scenarios, the directions of the effects were similar. Under all scenarios, SIMRIW predicted that climate change would have moderately positive effects on rice yield in northern and north-central Japan, and negative effects in south-central and southwestern Japan. The greatest negative effect would be in AEZ XI (Tokai district).

The positive effects of doubled CO_2 and global warming on rice yield in northern Japan are predicted because the temperatures suggested by the GCMs are not high enough to cause drastic yield reduction. The beneficial direct effects of doubling CO_2 more than offset the negative effects of warming. Negative effects of doubled CO_2 and global warming on simulated rice yields in south-central and southwestern Japan result because the temperatures predicted by GCMs are high enough to shorten rice-growth duration and to cause extensive spikelet sterility. Because temperatures above 35 °C during flowering cause spikelet sterility (Satake and Yoshida 1978; Matsui and Horie 1992), the results shown in Fig. 6 indicate that daily maximum temperatures under doubled CO_2 climates frequently exceeded 35 °C during flowering in south-central (AEZ XI) and southwestern Japan (AEZ XII, XIII, and XIV). Indeed, the SIMRIW predicts more than 30% yield reduction in AEZ XI under the GFDL and UKMO climate scenarios.

Probability Analysis on Effects of Doubled Carbon Dioxide Climate

Probability analyses were made of the effects of doubled CO_2 and climate change on rice yield for representative locations, using daily weather data from 1979 to 1990 at Sapporo (AEZ III), Sendai (AEZ V), Nagoya (AEZ XI), and Miyazaki (AEZ XIV) as base climates, and the described climate-change scenarios. Figure 6 gives the results of the probability analysis by SIMRIW. The simulated rice

Table 1. Predicted changes (%) in current rice yields at various locations under different CO_2 and climate conditions (predicted by SIMRIW model)

CO_2 (μl/l)	Temperature	Sapporo	Akita	Sendai	Maebashi	Toyama	Nagoya	Hiroshima	Kohchi	Miyazaki
450	+0 °C	+6.7	+8.0	+7.8	+7.7	+7.7	+7.7	+7.7	+7.7	+7.8
450	+2 °C	+0.7	+6.2	+3.8	+5.2	+2.4	–6.8	+1.9	+3.1	+7.3
700	GFDL	+0.0	+15.8	+10.3	+10.2	–8.0	–32.0	–18.7	–12.8	+6.7
700	GISS	+16.6	+29.8	+23.4	+25.7	+21.6	+0.5	–15.0	–26.9	+4.0
700	UKMO	+11.5	+27.6	+21.1	+16.7	+16.9	–39.2	–45.4	–34.8	–7.0

The prediction was based on experimental data on 'Akihikari' for high temperature effect on spikelet sterility.

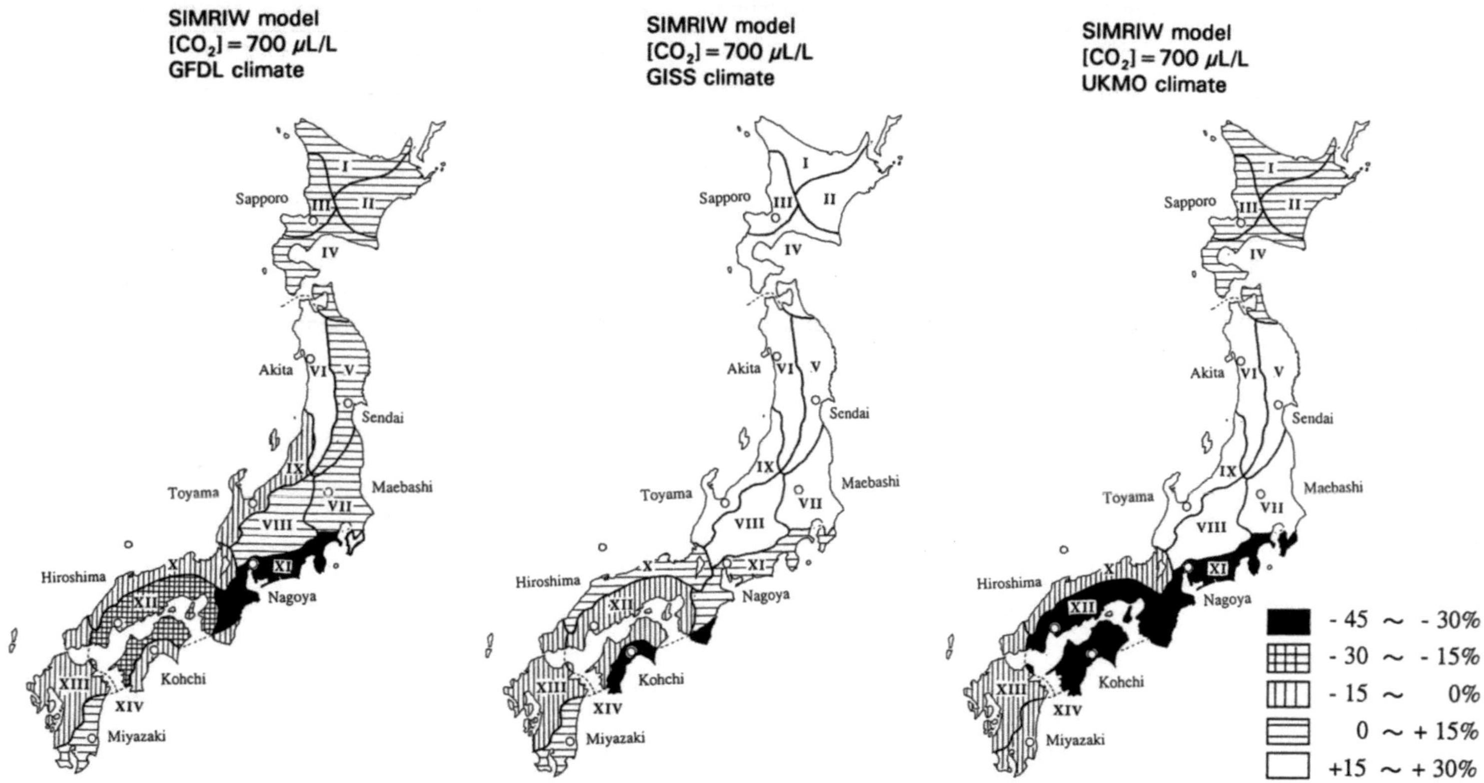

Fig. 5. Likely effects of doubled CO_2 and climate change predicted by three GCMs on regional rice yield in Japan (prediction by SIMRIW). The prediction was based on experimental data on ‘Akihikari’ rice for high temperature effect on spikelet sterility.

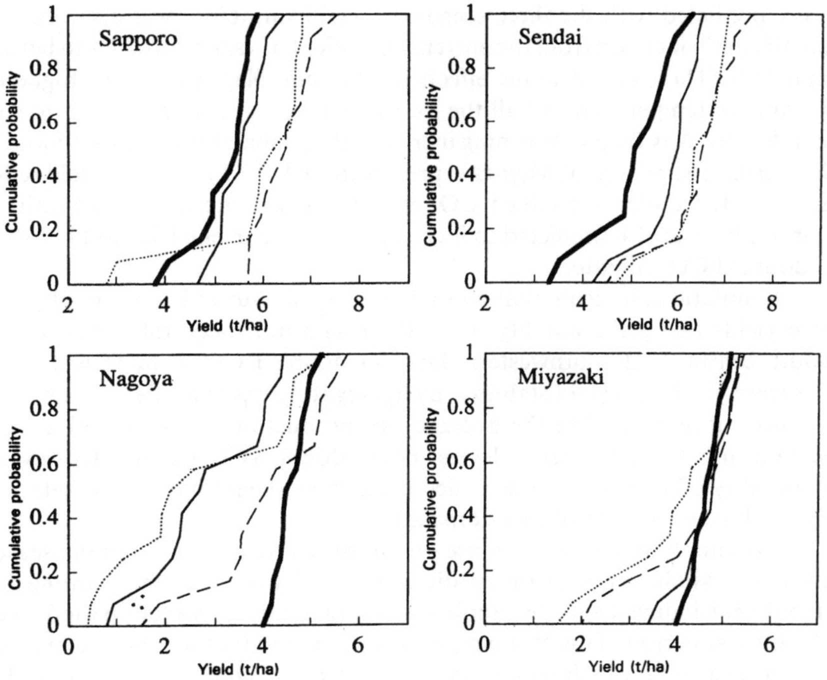

Fig. 6. Cumulative distribution functions for rice yield at four representative locations in Japan under three scenarios of global climate change (prediction by SIMRIW). — present; — GFDL; ---Giss; ··· UKMO

yields for 12 years under the three different climate scenarios were plotted as cumulative probability distributions.

At Sapporo in Hokkaido (AEZ I to IV), the average yield and its coefficient of variation (CV) under the current climate are 5.27 t/ha and 9.7%, respectively. The relative increases in average yield under doubled CO_2 climates of the GFDL, GISS, and UKMO were predicted to be 6, 23, and 15%, respectively. The GISS scenario gave the largest yields of the three GCMs because it predicts the smallest temperature rise and increased solar radiation levels. Because cool-summer damage is rare under the doubled CO_2 scenarios of GFDL and GISS, yield variability was also reduced. The temperature rise predicted by the UKMO scenario is so large that it causes heat-induced spikelet sterility, even in Hokkaido during some years. In general, a doubled CO_2 climate will substantially increase the average yield and reduce the yield variability in Hokkaido.

The largest positive effect of doubled CO_2 climates are expected in Tohoku district (AEZ V and VI). At Sendai, the predicted yield increase is 14–26%, depending on climate scenarios. Doubled CO_2 climates will reduce the yield variability in Tohoku by reducing the CV from the present 15% to less than 10%.

The most catastrophic effect of a doubled CO_2 climate was predicted at Nagoya, Aichi prefecture (AEZ XI), where average yield reductions of 8 and 38%

were predicted with the three scenarios. Yield variability was also predicted to significantly increase from the current coefficient of variation of 7% to between 27 and 61%. These catastrophic effects are because Nagoya has the highest daily maximum temperatures of all the locations investigated under current summer conditions. Any further warming increases the probability of heat-induced spikelet sterility. Similarly, at Miyazaki prefecture (AEZ XIV), 0–13% yield reduction is predicted under a doubled CO_2 climate, depending on the scenario. Yield variability was also predicted to increase from the current 4.7% to 11–26% under a doubled CO_2 climate.

Simulations indicate that a doubled CO_2 climate will substantially increase rice yields and yield stability in northern and north-central Japan, but in the south-central and southwestern Japan a doubled CO_2 climate will markedly decrease yields and yield stability. By aggregating regional effects over the whole country, it appears that the average rice production of Japan will not change significantly from the current levels under a doubled CO_2 climate. However, yield variability is likely to increase, reflecting increased heat-induced spikelet sterility in south-central and southwestern Japan.

This analysis assumes the use of current cultivars and cropping seasons. It may be possible to reduce the predicted instability by adjusting planting dates to avoid heat-induced spikelet sterility by allowing the flowering period to escape the hottest period. This strategy, however, may reduce average yields, because the hottest season in Japan is also associated with the highest solar radiation. Grain formation and grain filling in rice has been found to be strongly influenced by solar radiation levels during the reproductive period. Further studies are needed to quantitatively assess to what extent the predicted effects of a doubled CO_2 climate on rice yield can be mitigated by using different cropping seasons and cultivars.

Assessing Impacts of Climate Change on Rice Production: Strategies for Adaptation in Southern China

Z. Jin, D. Ge, H. Chen, and X. Zheng

Since 1949, great progress has been made in Chinese food production. China has only about 7% of the cultivated land in the world but supports more than one-fifth of the global population. Still, many constraints, such as population growth, sharp decreases in cultivated land, lack of water resources, pollution, and frequent natural disasters, make it difficult for China to increase its food production.

Recently, scientists have suggested that the addition of greenhouse gases to the atmosphere will alter global climate by increasing temperatures and changing rainfall and other climate patterns (IPCC 1990). The combined effects of climate change and the physiological effects of CO_2 on crops may result in a net increase in crop yields in some cases, especially in the high latitude regions. In low latitudes, agriculture could be negatively affected (Parry et al. 1988).

The objective of this study was to characterize the direction, magnitude, and degree of uncertainty of the potential impacts of global climate change on rice production in southern China. This study is important because: China is the largest rice producer and consumer in the world, southern China plays an important role in national rice production (more than 90% of both area planted and production), and southern China appears to be more vulnerable to climate change than the northern regions of China (Jin et al. 1990).

The study was based on simulations with the CERES-rice model, using actual daily weather data (about 30 years) from nine sites in seven provinces of southern China (Fig. 1) and regional climate-change scenarios that were created from the outputs of general circulation models (GCMs). Rice growth was simulated for both rainfed and automatic irrigation conditions to represent both mountain rice without irrigation and plains rice with an extensive irrigation system. Changes in rice yield, growth duration, evapotranspiration, and water applied for irrigation were estimated under baseline and climate-change conditions. The physiological effects of CO_2 were also analyzed under each scenario.

Jiangsu Academy of Agricultural Sciences, Nanjing 210014, China

S. Peng et al. (Eds.) Climate Change and Rice

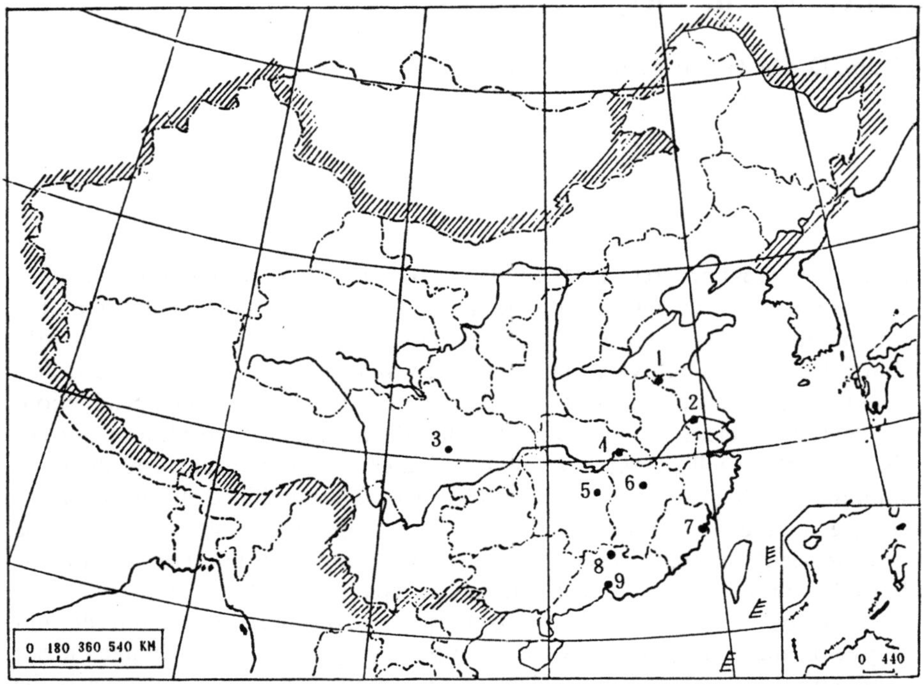

Fig. 1. Distribution of the nine sites in southern China. *1* Xuzhou; *2* Nanjing; *3* Chengdu; *4* Wuhan; *5* Changsha; *6* Nanchang; *7* Fuzhou; *8* Shaoguan; *9* Guangzhou

Methods

Climate Data

Local daily climate data from nine sites (Nanjing, Xuzhou, Nanchang, Wuhan, Changsha, Chengdu, Fuzhou, Shaoguan, and Guangzhou) for a period of 20–30 years (about 1957–1986) were taken from the *Monthly Report of Chinese Meteorological Record on the Surface* (China National Meteorological Bureau) and the *Daily Solar Radiation Record of China* (Beijing Meteorological Bureau). Daily solar radiation at Xuzhou was not available and was calculated according to Gao and Lu (1982). Weather data for the calibration and validation experiments were collected from the local meteorological stations.

Climate-Change Scenarios

Effects of doubled CO_2 on climate from GCM outputs were simulated to be fixed differences in mean monthly temperatures, and proportional changes in precipitation and solar radiation. These GCM-generated climates were combined with

Table 1. Annual temperature changes (deg °C) and precipitation differences (%) projected by the GISS, GFDL and UKMO Climate change scenarios for sites in Southern China

Site	Temperature differences (°C)			Precipitation differences (%)		
	GISS	GFDL	UKMO	GISS	GFDL	UKMO
Nanjing	4.1	4.6	6.5	43.7	9.3	13.1
Xuzhou	4.1	4.7	6.5	45.2	4.6	14.0
Nanchang	3.5	4.4	5.7	17.8	34.9	0.5
Wuhan	6.0	4.4	6.5	–9.1	35.0	12.3
Changsha	6.0	4.4	5.7	–7.3	30.3	–0.6
Chengdu	4.7	3.2	6.7	–20.4	89.1	34.0
Fuzhou	3.5	4.2	5.7	18.6	–11.4	1.6
Shaoguan	6.0	4.0	3.3	–7.5	6.8	26.2
Guangzhou	3.5	4.0	3.4	79.9	9.3	31.8

daily local historic climate data to generate climate-change scenarios for differrent sites. Three GCMs were used: Goddard Institute for Space Studies (GISS) (Hansen et al. 1983); Geophysical Fluid Dynamics Laboratory (GFDL) (Manabe and Wetherald 1987); and United Kingdom Meteorological Office (UKMO) (Wilson and Mitchell 1987).

Climate-change scenarios were created by combining GCM outputs and daily local climate, and therefore include local geographic variation within a grid box. In this study, the average difference of monthly GCM outputs between doubled and ambient (or their ratios) in each grid were used to modify the local historic climate data to create the climate-change scenarios. These climate-change scenarios maintain the geographic local climate variation found in the historic base (observed) period; therefore, the climate-change scenarios for different sites within a grid are different. Table 1 presents annual mean temperature and precipitation changes generated by the GISS, GFDL, and UKMO scenarios for different locations in southern China.

CERES-Rice Crop Model

We used the CERES-Rice model (Ritchie et al. 1987; IBSNAT 1989; Godwin et al. 1990; Jones et al. 1990). The model simulates crop growth, development, and yields under different management and climatic conditions. The CERES-rice model was chosen because: it simulates the effects of major factors (e.g., climate and soil management) on rice growth, development, and yield, it simulates direct effects of CO_2 on crop photosynthesis and evapotranspiration, the genetic coefficients that characterize different cultivars can be calibrated easily for experimental data, the model has been available and documented for several years and has been tested under a wide range of soil and climatic conditions.

Crop data on local rice cultivars (sowing date, soil type, sowing depth, transplanting density, row spacing, maturity date, biomass, and grain yields) were taken from local field experiments near the nine sites. Soil data included soil type, albedo, organic matter, texture, structure, and bulk density. Representative soil types and profiles were chosen according to the *Soil Atlas of China* (Institute of Soil Science, Academic Sinica 1986) and *Chinese Soils* (Xiong Yi and Li Qingkui 1987).

Calibration

Eight genetic coefficients define a rice cultivar in CERES-rice. These coefficients characterize quantitatively how a cultivar responds to its environment. Genetic coefficients for the rice cultivars used in this study were estimated by comparing simulated and experimental results. Field observations included growth duration, dry-matter accumulation, and yield. Genetic coefficients of local cultivars were estimated by trial and error; so that, the simulated yield and maturation date were as close as possible to observed values.

Simulation Scenarios

The three GCM climate-change scenarios were used for each site and each simulation was done under rainfed and irrigated conditions. Simulations of irrigated conditions assumed automatic irrigation to avoid any water stress. Because climate-change scenarios involve doubling of atmospheric CO_2 (Allen et al. 1987; Peart et al. 1989; Smith and Tirpak 1989; Rosenzweig 1990), the physiological CO_2 effects on crop yield and water use were also included in some simulations.

In addition, the study evaluated possible adaptive strategies to rice management under future climate. Scenarios were created by altering current agricultural practices to maximize yield under the conditions of climate change.

Multiple Crop Index

The Multiple Crop Index can be used to evaluate different cropping systems in China (Gao et al. 1987). Possible changes in the index were analyzed by using simulated rice-growth output under different conditions of climate change. Gao et al. (1987) defined the rice-growing season as the number of days from safe sowing date to the safe heading date plus 40 days for japonica rices, and the safe heading date plus 30 days for indica rices. The safe sowing date is the date when mean daily temperature has a 80% probability of being above 10 °C. The safe heading period is the date when mean daily temperature over 3 consecutive days is ≥20 °C.

Based on this index, we can compute the lengths of rice-growing seasons for the current climates and for the three GCM climate-change scenarios, and then evaluate the changes in rice cropping systems in the future. According to the China National Rice Research Institute (1988), the index of more than 10 °C accumulated temperature (ΣT) for different rice-based cropping systems are: single rice cropping ΣT=2000–4500 °C, double rice cropping ΣT=4500–7000 °C (with ΣT=5300 °C the northern limit for double rice), and triple rice cropping ΣT>7000 °C. These thermal values can also be used to estimate change in the rice cropping systems under different climate-change scenarios.

Simulation Outputs

The results of the study should not be considered as predictions, but as indications of the possible directions of global change in the sites modeled.

Impact of Climate Change on Rice Yield

Doubled CO_2 would influence rice yield directly through physiological processes and indirectly through climate. Two cases are discussed separately, climate-change effects, namely changes in temperature, precipitation, and solar radiation, and the combined effects of climate change with physiological effects of CO_2 enrichment.

Climate-Change Effects

Simulated rice yields decreased under the three scenarios considered compared with the baseline yields for all locations. Under the GISS scenario, rainfed yields decreased 10–78% from the baseline yield (Table 2). Under the GFDL scenario, yields decreased 6–33% and under the UKMO scenario yields decreased 7–35%. Yield decreases result from high temperatures, which shorten the rice-growing cycle and cause water stress in some regions. More severe yield decreases under the GISS scenario were caused mostly by precipitation decreases during the growing season. For example, July rainfall at Chengdu declined 90% compared with current rainfall under the GISS scenario. The growing season in the changed climate at Chengdu had less than 40 mm of total rainfall and a 78% yield reduction (Jin et al. 1990). Although the enhancement of respiration by high temperature may also cause a decrease in rice yields, it was not taken into account in the CERES-rice model.

Automatic irrigation simulations under the three climate-change scenarios also resulted in decreases in rice yields (Table 2). Yields declined 15–33% under the GISS scenario, 14–37% under the GFDL scenario, and 19–33% under the UKMO scenario. Therefore, irrigation does not fully offset the negative effects of increased temperature on rice yields. The primary effect of high temperature was

Table 2. Simulated yield changes under GCM climate-change scenarios compared with base yields. The direct effects of CO_2 are not included[a]

Site	Rainfed						Irrigated					
	GISS		GFDL		UKMO		GISS		GFDL		UKMO	
	Y%	SD%	Y%	SD%	Y%	SD%	Y%	SD%	Y%	SD%	Y%	SD%
Nanjing	–19.8	10.1	–33.2	10.0	–32.6	10.2	–16.8	3.9	–20.1	3.9	–20.2	4.0
Xuzhou	–10.2	8.0	–31.1	7.6	–22.9	7.7	–15.0	2.8	–18.0	2.7	–19.2	2.7
Nanchang	–23.4	12.3	–4.8	13.6	–34.8	11.8	–23.7	4.1	–18.8	3.8	–32.8	3.8
Wuhan	–27.1	8.8	–13.9	9.1	–27.1	8.9	–22.3	5.3	–25.6	5.3	–26.2	5.3
Changsha	–26.7	11.7	–6.1	12.6	–11.3	12.5	–32.7	5.4	–37.0	5.2	–31.4	5.4
Chengdu	–78.1	4.8	–13.6	2.4	–27.7	3.2	–31.5	2.4	–14.2	2.2	–20.1	2.3
Fuzhou	–21.0	14.6	–19.3	6.1	–17.8	6.1	–27.9	5.3	–24.7	5.4	–27.3	5.0
Shaoguan	–33.3	9.0	–21.8	9.4	–7.9	9.7	–25.3	5.1	–29.6	5.4	–26.1	4.5
Guangzhou	–44.8	11.5	–26.7	4.1	–24.8	3.8	–15.9	3.5	–20.9	3.9	–22.0	3.6
Mean	–31.4	9.3	–21.1	7.4	–21.1	7.5	–24.1	4.2	–21.0	4.2	–24.4	4.2

[a] Y yield, SD standard deviation.

shortening of rice-growth duration, particularly grain-filling. Nevertheless, the negative effects of climate change on rice yields are partly offset by full irrigation in the sites where rainfall during the rice-growing season greatly decreases under the climate-change scenarios. For example, at Chengdu and Guangzhou with the GISS scenario, rainfed yields decreased 78 and 45%, respectively; whereas, irrigated yields declined only 32 and 16%. Yield variability, as estimated by standard deviations of yield changes, was also much smaller for irrigated than for rainfed rice yields (Table 2).Therefore,one strategy to prepare for climate change would be to improve the irrigation systems in such regions.

Climate Change Combined with Physiological Effects of Carbon Dioxide

Simulated effects of climate change combined with physiological effects of CO_2 on yields and water use are shown in Table 3. Under the GISS scenario, rainfed rice yields increased at the northern (Xuzhou and Nanjing) and eastern (Fuzhou and Nanchang) sites. At these sites, scenario precipitation does not limit rice production. In Wuhan, Changsha, and Shaoguan, direct effects of CO_2 ameliorate, to a certain degree, the negative effect of increased temperatures on rainfed rice yields. In Chengdu and Guangzhou, however, because of large decreases in precipitation during the rice-growing season, yields remain much lower than the baseline even when direct effects of CO_2 are included. Under the GFDL and UKMO scenarios, direct effects of CO_2 largely compensated for negative impacts of climate change on rice yields in many, but not all, sites.

In the irrigated simulations, rice yields declined in the southern region (Nanchang, Changsha, Wuhan, and Shaoguan) under the three climate-change scenarios (Table 3). Yield declines result from the 4–7 °C regional annual temperature increases projected by the GCMs in that region. In contrast, the direct CO_2 effects increased simulated yields of irrigated rice at Xuzhou under all three climate-change scenarios. Increased temperatures may benefit rice production in the northern areas of the studied region. As in the case of climate change alone, yield variability was greater under rainfed than under irrigated conditions (Table 2).

Impact of Climate Change on Rice Growth Duration

Although CERES-rice does not simulate direct effects of CO_2 on duration of rice growth, increased temperatures under all climate-change scenarios decreased the duration of rice growth at all sites. Mean rice-growth duration decreased more in the northern areas (Xuzhou, Nanjing, and Chengdu) than in the southern areas. Because current temperatures in the northern sites are relatively low, rice has a long growth period. Long growth durations are more sensitive than short ones to temperature increases. In contrast, current temperatures at the southern sites already causes the rice-growth duration to be short. Further temperature increase does not have a very large effect on growth duration.

Table 3. Simulated yield changes (%) under GCM climate-change scenarios compared with base yields. The direct effects of CO_2 are included in the yield simulations [a]

Site	Rainfed						Irrigated					
	GISS		GFDL		UKMO		GISS		GFDL		UKMO	
	Y%	SD%	Y%	SD%	Y%	SD%	Y%	SD%	Y%	SD%	Y%	SD%
Nanjing	3.9	11.4	–12.8	10.9	–12.4	11.2	2.7	4.2	–1.9	4.2	–1.8	4.3
Xuzhou	19.6	14.5	–8.6	8.1	2.9	8.4	6.1	3.0	1.7	3.0	1.1	2.9
Nanchang	1.5	14.2	38.6	15.5	–12.7	12.7	0.0	4.9	4.4	4.2	–11.4	4.3
Wuhan	–8.7	9.5	10.8	10.1	–6.4	9.8	–2.9	5.6	–5.2	5.7	–4.9	5.8
Changsha	–2.0	12.8	27.3	13.7	20.6	13.9	–11.5	5.7	–14.7	5.5	–9.3	5.5
Chengdu	–71.8	2.3	3.8	2.7	–2.7	3.7	–6.2	3.2	3.0	2.6	3.0	2.7
Fuzhou	5.2	6.8	7.1	6.8	9.5	6.5	–4.0	5.7	0.2	5.7	–3.0	5.4
Shaoguan	–13.5	9.5	5.4	9.9	19.4	10.2	–3.0	5.1	–6.6	5.6	–4.5	4.2
Guangzhou	–31.7	4.9	–5.5	3.9	–5.9	4.0	3.3	4.2	–2.1	4.2	–4.8	4.0
Mean	–10.8	9.5	7.3	9.1	1.4	8.9	–1.7	4.6	–2.4	4.5	–4.0	4.3

[a] Y yield, SD standard deviation

Impact of Climate Change on Irrigation

An adequate water supply is one of the most important factors in rice production. The amount of irrigation water needed for rice depends on climate (rainfall, temperature, solar radiation) and crop and soil characteristics, depth of water table, and topography. Among these factors rainfall is the most important. Simulations were done with automatic irrigation assuming 100% efficiency of application and water availability.

Under the GISS scenario, demand for irrigation water rose in seven of the nine sites, with an overall mean increase of 103%. Climate change increased irrigation demand sixfold at Chengdu and twofold at Guangzhou. These increases were caused by large decreases in summer rainfall and increases in temperature and solar radiation.

Under the GFDL and UKMO scenarios, simulated changes in irrigation demands were mixed. Irrigation demands changed at most sites according to changes in predicted precipitation, but total change was small. Mean irrigation demand across all sites decreased 1% with the GFDL and increased 5% with the UKMO scenario. All three scenarios agreed that irrigation demand would increase in Chengdu and Xuzhou, and decrease in Fuzhou.

Possible Adaptation Strategies to Climate Change

Crop simulation with the physiological effects of CO_2 under the GISS scenario, was used to test four strategies for adaptation to climate change (Table 4): new rice cultivars, different planting dates (tested in 10-day steps using present cultivars), changes in both cultivars and planting dates, and improvements in irrigation systems.

Table 4. Simulated percentage changes in yield under different management conditions under the GISS scenario compared with base yields. The direct effects of CO_2 were included in the climate-change scenario simulation [a]

	CC%	CC+C%	CC+PD%	CC+PD+C%
Nanjing	2.7	–2.0	14.3	0.0
Xuzhou	6.1	10.3	10.1	11.3
Nanchang	–6.9	39.6	2.6	42.8
Wuhan	–2.9	6.4	–0.7	21.0
Changsha	–11.5	9.7	–10.1	24.7
Chengdu[b]	–71.8	–54.4	–	–41.3
Fuzhou	–4.0	28.8	–2.4	29.8
Shaoguan	–3.0	–1.2	5.0	2.8
Guangzhou[b]	–31.7	–47.2	–	–35.6
Mean	–13.7	–1.1	2.7	6.2

[a] CC climate change effect (GISS scenario), CC+C effects of climate change plus change in the cultivar, CC+PD effects of climate change plus changing sowing date, and CC+PD+C combined effect of climate change, changes in the sowing date and cultivar.
[b] Rainfed, the other sites were irrigated.

An upland rice cultivar (UPL Ri-5) was simulated in the sites where precipitation declined dramatically under scenario conditions (Table 4). With the new cultivar, yield at Chengdu increased compared with that of the original baseline cultivar, but still declined 54% as compared with the no climate change simulations. At Guangzhou, however, introducing the same cultivar did not improve yields under scenario conditions.

Cultivar IR43 was simulated in the other seven locations. In five sites (Nanchang, Fuzhou, Xuzhou, Changsha, and Wuhan), IR43 yielded more than the original cultivar under conditions of climate change. At the other two sites (Nanjing and Shaoguan), change of cultivar did not improve modeled yields. A change in planting date improved rice yields at the northern sites (Xuzhou and Nanjing) but not at the more southern sites (Table 4). When both cultivars and planting dates were changed, rice yields increased by 3–43% at six of the seven locations tested (only at Nanjing did yields not increase).

Impacts of Climate Change on Chinese Rice-Based Cropping Systems

Under all doubled CO_2 scenarios, temperatures and the accumulated temperature index increased. These higher temperatures, may extend the season for rice growth but increased temperatures shorten the life cycle of rice and reduce yields. The combined effects of prolonged growing season and shortened growth duration would shift the northern limits of various rice-based cropping systems toward higher latitudes. As a result, the Multiple Cropping Index in China would increase, and varieties and management systems would have to be adjusted to the new conditions.

The current climatic classification for rice production in China (Gao et al. 1987), would no longer apply in a changed climate. Under the GISS scenario, maturity dates of rice in China advance an average of 19 days, the rice-growing season averages 45 days longer, and the accumulated temperature index increases by an average of 1523 °C. The northern limits for double and triple-rice cropping would shift north (Fig. 2).

According to the GISS scenario, thermal conditions at Beijing would be more favorable for rice than the current thermal conditions at Nanjing, and it would be possible to grow rice after wheat in Beijing. Thermal conditions at Shenyang would be more favorable than current conditions at Beijing. Without the current low temperature problems, Shenyang could produce three crops in 2 years. In Harbin, the most northern area of China, the accumulated temperature would reach 3696 °C with a rice-growing season of 159 days, and it would be safe to grow a single crop of early japonica rice with a late maturity date.

Similar results would be obtained with the GFDL scenario, because temperature increases are similar to those under the GISS scenario. The UKMO scenario gives the most extreme results. In the UKMO scenario, the northern limit for a double crop of rice would shift to the Shenyang region, and the limit for a triple crop of rice would move to a line from Jinan to Zhengzhou. In southern China, rice could be grown throughout the year.

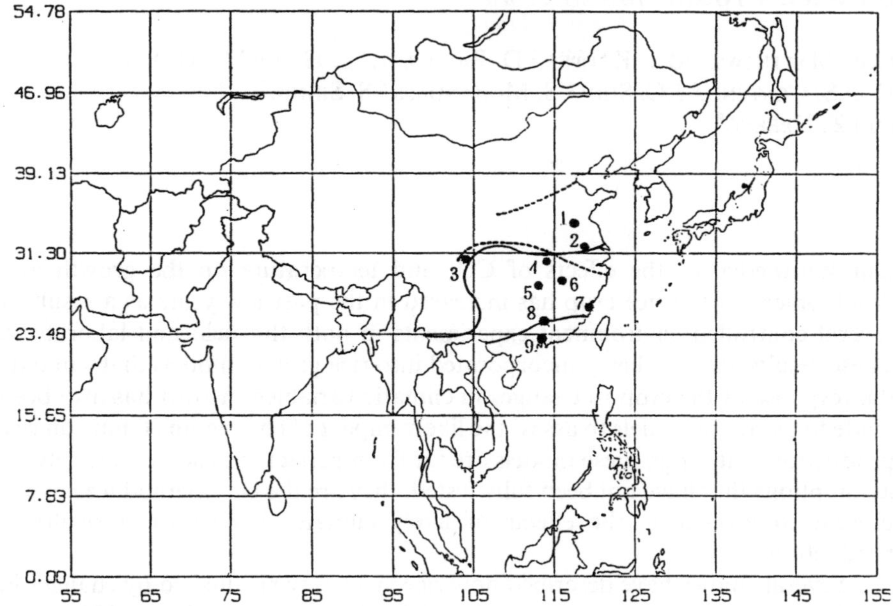

Fig. 2. The northern limits for double rice crops (*upper broken line*) and triple rice crops (*lower broken line*) under the GISS climate-change scenario. *Corresponding solid lines* are the northern limits of the current climate. (Gao et al. 1987). *1* Xuzhou; *2* Nanjing; *3* Chengdu; *4* Wuhan; *5* Changsha; *6* Nanchang; *7* Fuzhou; *8* Shaoguan; *9* Guangzhou

Current patterns of rice varieties used throughout China would shift, and the so-called south indica and north japonica varieties would change. Varieties tolerant of high temperatures would need to be grown as far north as the line extending from Jinan to Zhengzhou.

A doubled CO_2 climate would result in tremendous changes in rice production in China. Effective strategies for adapting to climate change must be developed. These strategies should include improvement of crop breeding to develop more heat- and drought-resistant rice varieties, production of chemical fertilizers, and irrigation systems. We may also need new pest- and disease-management systems. These strategies imply an increased need for agricultural inputs.

Acknowledgments. The authors express their appreciation to Dr. C. Rosenzweig, Prof. L.Z. Gao, Dr. J.T. Ritchie, and Dr. T.Y. Chou for their suggestions, encouragement, assistance, collaboration, and advice in this work.

The authors also thank the following units for providing agronomic data for this study: China National Rice Research Institute, the National Academy of Agricultural Sciences, Guangdong Academy of Agricultural Sciences, Jiangxi Agricultural University, Jiangsu Agricultural Bureau, Jiangsu Academy of Agricultural Sciences, Shaoguan Prefecture Institute of Agricultural Sciences in Guangdong province, and Lixiahe Prefecture of Agricultural Sciences in Jiangsu province. This project was supported by the National Science Foundation of China. Most software, including GCM outputs and the CERES-rice model, were provided by the United States Environmental Protection Agency.

Modeling the Impact of Climate Change on Rice Production in Asia

R.B. Matthews,[1] M.J. Kropff,[2] D. Bachelet,[3] T. Horie,[4] M.H. Lee,[5] H.G.S. Centeno,[6] J.C. Shin,[5] S. Mohandass,[6] S. Singh,[7] and Z. Defeng[8]

Our knowledge of the effects of CO_2 and temperature on the growth and development of the rice crop has increased in the past few years as a result of several controlled environment experiments in both the field and laboratory. These results are now being incorporated into rice growth models that simulate the response of the crop to changes in climatic variables. A start has also been made to use these models to assess the likely impact of changes in climate on rice production. Although these models are far from perfect, and sometimes rely on assumptions that have not been fully tested, they are the best method we have at present to investigate the effects of likely climate changes on agricultural production.

Several studies have described how rice yield may be affected by changes in the climate. Yoshino et al. (1988) predicted that lowland rice yields could increase in Japan by about 9% following a doubling of CO_2 and subsequent climatic changes as predicted by the Goddard Institute of Space Studies (GISS) general circulation model (Hansen et al. 1988). Jansen (1990) simulated potential yield, water-use efficiency, and nitrogen-limited yield of rice for combinations of three temperatures and three CO_2 scenarios for the years 2020 and 2100 using weather data from seven sites in Asia. He concluded that the effect would be a reduction in yield in the high temperature scenarios and an increase in the low temperature scenarios. Similar results were obtained by Penning de Vries (1993). These effects were related to predicted increases in photosynthesis at higher CO_2 and a reduction in the length of the growing season at higher temperatures. However, their models did not account for the decrease in phenological development rates

[1] Research Institute for Agrobiology and Soil Fertility, Bornsesteeg 65, 6700 AA Wageningen, Netherlands
[2] International Rice Research Institute, P.O. Box 933, 1099 Manila, Philippines
[3] Formerly, United States Environmental Protection Agency, Environmental Research Laboratory, 200 SW 35th Street, Corvallis, OR 97333, USA. Current address: c/o 29 rue Jeanne Marvig, 31055 Toulouse, Cédex, France
[4] Faculty of Agriculture, Kyoto University, Kyoto 606, Japan
[5] Crop Experiment Station, Seodundong 209, Suweon, South Korea
[6] Tamil Nadu Rice Research Institute, Adutherai, Tamil Nadu 612101, India
[7] Universiti Pertanian Malaysia, 43400 UPM Serdang, Selangor, Malaysia
[8] China National Rice Research Institute, Hangzhou, 310006 Zhejiang Province, China

S. Peng et al. (Eds.) Climate Change and Rice

at supraoptimal temperatures. Rosenzweig et al. (1993), working with collaborators from 22 countries, predicted reductions in yield at low latitudes and increases at high latitudes for a number of major crops. Leemans and Solomon (1993), using a very simple model and long-term monthly average climatic data in a world-wide study, predicted a yield increase of 0.4% for the current rice-growing environments, but little change in the areas sown because of the sharp temperature and moisture gradient along the northern border of the primary distribution of rice in eastern Asia. From these various estimates, it is clear that a wide range of predictions has been made on the likely effect of climate change on the production of rice. These studies contain many uncertainties, partly because of uncertainties in the predictions of the GCMs themselves, partly from the use of limited sites for which historical weather data are available, and partly from the quality of the crop-simulation models, especially for rainfed conditions (Bachelet et al. 1993).

In 1989, the US EPA Environmental Research Laboratory and IRRI initiated a major research project to investigate relationships between climate change and rice production. One component of this project was the integration of knowledge about rice responses to temperature and CO_2 into existing crop simulation models. This simulation component was executed by IRRI in collaboration with four teams of the Simulation and Systems Analysis for Rice Production (SARP) network (a collaborative network of scientists in national research stations in Asia, two institutes in the Netherlands, and IRRI), the Crop Science Laboratory at Kyoto University in Japan, and the US EPA Environmental Research Laboratory in Oregon, United States. The purpose of this chapter is to describe the model used and some of the preliminary results obtained from the study. Further details of the study are given in Matthews et al. (1994).

The ORYZA1 Model

The study used the ORYZA1 rice potential production model developed jointly at IRRI, Wageningen, and the collaborating institutes of the SARP network (Kropff et al. 1993b). ORYZA1 assumes that the potential yield of a crop is determined only by varietal characteristics, the seasonal pattern of temperature, radiation and daylength, and the ambient CO_2 concentration. The model has limitations: it assumes that weeds, diseases, and insect pests are absent, that water and fertilizers are abundant, that there are no adverse soil conditions, and that no extreme weather events (e.g., typhoons) occur. Although this may not be completely realistic, this approach has been found to explain a large part of the current yearly variation in rice yields in Japan because of weather (Horie 1994) and enables an understanding of the effect of climate on the underlying processes that influence crop yield.

The model simulates the processes of photosynthesis, respiration, dry matter partitioning, and leaf-area growth on a daily basis and uses values of solar radiation and minimum and maximum temperatures as inputs. Phenological events (e.g., dates of panicle initiation, 50% flowering, and maturity) are simulated using temperature and photoperiod as driving variables. A detailed description of ORYZA1 is given in Kropff et al. (1993b).

Dry Matter Production

Total daily canopy photosynthesis is calculated from the daily incoming radiation, temperature, leaf area index, and canopy nitrogen level by integrating instantaneous rates of leaf CO_2 assimilation at different levels in the canopy and at different times of day. After subtraction of respiration requirements, net daily growth is obtained. Dry matter produced is then partitioned among the various plant organs.

The relationship between photosynthesis rate and incident light level is commonly described by the light-response curve, a negative exponential relation that is characterized by two parameters: P_{max} the (asymptotic) maximum rate of photosynthesis attainable under high light conditions, and ε light-use efficiency at very low levels of light (Goudriaan 1982). There is a gradual decline in ε as temperature increases, mainly because of the increased affinity of the carboxylating enzyme rubisco for O_2 in relation to that for CO_2 (Ehleringer and Pearcy 1983). There is little effect of temperature on P_{max} between 20 and 37 °C, but outside these limits, photosynthetic rate declines rapidly (Penning de Vries et al. 1989).

The effect of CO_2 on the photosynthetic parameters was included according to the function derived by Jansen (1990) for the effect on ε, and a new relationship was derived from unpublished data collected at IRRI (Weerakoon et al.) for the effect of CO_2 on P_{max}. This relationship was derived by correcting the data for differences in the N concentration of the leaf. The relationship derived by Jansen from data of Akita (1980) for ε (g CO_2/MJ) is:

$$\varepsilon = \varepsilon^* \left[1 - \exp(-0.00305\, C - 0.222)\right] / \left[1 - \exp(-0.00305 \times 340 - 0.222)\right], \quad (1)$$

where ε^* is the initial light-use efficiency at standard values of 340 $\mu l\ l^{-1}$ CO_2 and 25 °C, and C is the atmospheric CO_2 concentration ($\mu l\ l^{-1}$). For P_{max} (g CO_2 m^{-2} h^{-1}) the following relationship was derived:

$$P_{max} = P^*_{max} \left(49.57 / 34.26 \left\{1 - \exp\left[-0.208\,(C - 60) / 49.57\right]\right\}\right), \quad (2)$$

where P^*_{max} is the value of P_{max} at 340 $\mu l\ l^{-1}$ CO_2 and 25 °C.

In the process of maintenance respiration, part of the carbohydrates produced by photosynthesis are converted back into CO_2 and water to provide the energy for maintaining existing biomass. Although this process consumes between 15 and 30% of the carbohydrates produced by a crop in a growing season (Penning de Vries et al. 1989), it is poorly understood at the biochemical level. In the model, we use an adapted version of the simple approach developed by Penning de Vries and van Laar (1982), in which maintenance requirements are approximately proportional to the dry weights of the plant organs to be maintained. The effect of temperature on maintenance respiration is described by:

$$R_m = R^*_m \, 2^{[(T_{av}-25)/10]}, \tag{3}$$

where R_m is the actual rate of maintenance respiration (g CO_2 m^{-2} d^{-1}), R^*_m is the maintenance respiration rate at 25 °C (g CO_2 m^{-2} d^{-1}), and T_{av} is the daily mean temperature (°C). The rate of maintenance respiration increases proportionately more as temperature increases.

Any carbohydrates in excess of maintenance requirements are available for conversion into structural plant material, during which a relatively constant fraction of assimilate is lost because of growth respiration. We have assumed that there is no direct effect of either temperature or CO_2 on this fraction.

After the daily increment in dry matter has been calculated, this is partitioned according to fixed ratios (depending on the stage of development of the crop) into that required for growth of the leaves, stems, roots, and panicles. We have assumed that there is no direct effect of temperature and CO_2 on these partitioning ratios.

Phenology

Because many physiological and morphological processes change with the phenological stage of the plant, an accurate description of phenological development is essential. We divided the life cycle of the rice crop into four phenological phases: basic vegetative phase (BVP), from sowing to the start of the photoperiod-sensitive phase; photoperiod-sensitive phase (PSP), during which daylength influences development rate (the PSP lasts until panicle initiation (PI); panicle formation phase (PFP), starting at the switch from the vegetative to the reproductive phase at PI, and lasting until 50% flowering; and grain filling phase (GFP), from 50% flowering to physiological maturity. The model assumes that there is a variety-specific constant number of developmental days required for completion of each phase.

The original version of ORYZA1 used a linear relationship between the daily mean temperature above a base temperature and the rate of development toward a given phenological event. A plateau was reached at about 30 °C, beyond which development rate did not change (Kropff et al. 1993b). However, data for many crops indicate that at temperatures above an optimum, the development rate decreases (Mohamed et al. 1988; Hammer et al. 1989; Hodges 1991). Although comprehensive data are lacking in rice for temperatures greater than 33 °C, recent data collected at IRRI suggest that there is a similar response (Manalo et al. 1994). Therefore, for the higher temperatures encountered in climate-change scenarios, we modified the temperature versus phenology function.

The new temperature versus phenology function assumes that the rate of progression through each phase is linearly related to daily mean temperature above a base temperature (T_{base}) up to an optimum temperature (T_{opt}), beyond which the rate decreases, again linearly, until a maximum temperature (T_{high}) is reached (Kiniry et al. 1991). For temperatures below the base temperature or above the maximum temperature, development rate is zero. This bilinear response is generally observed only when daily temperatures are constant (Mohamed et al. 1988). If temperatures fluctuate between a minimum and a maximum value, the response becomes curvilinear, particularly near each cardinal temperature. To account for this, we superimposed a sine wave representing the diurnal fluctuation in air temperature onto the bilinear response (Matthews and Hunt 1994). We also assume that there is no effect of CO_2 concentration on development rate.

Spikelet Fertility

In rice, spikelet fertility is influenced adversely by extreme temperatures immediately before and during flowering (Yoshida 1981). A reduction in spikelet fertility reduces the number of grains that can form, so that even if carbohydrate production in the grain-filling period is unaffected, yields are reduced. For this reason, ORYZA1 computes numbers of spikelets formed and the fraction of spikelets that form grains.

In experiments at IRRI, we found a good relationship between spikelet number at flowering and total crop growth from PI to 50% flowering. This relationship holds across wet and dry seasons, across nitrogen application levels from 0 to 285 kg/ha, across planting densities from 25 to 125 plants m^{-2}, and for severe drought stress. Effects of solar radiation, temperature, nitrogen, competition, and water on spikelet formation, therefore, can seemingly be integrated by their effects on crop growth over the panicle-formation period. We call the slope of this relationship the spikelet formation factor (γ). For a given variety, the relationship is remarkably consistent, although there are differences between genotypes. For example, γ has a value of about 65 spikelets/g of shoot dry matter for IR72, but ranges from 45 to 70 spikelets/g in other genotypes.

The model tracks the amount of growth from panicle initiation, and then calculates the number of spikelets at flowering (S_f m^{-2}) as the product of this growth and γ:

$$S_f = \left(\sum_{i=P}^{F} G_i \right) \gamma, \tag{4}$$

where P and F are the dates of panicle initiation and 50% flowering, respectively, and G_i is the increase in crop dry weight on day i (g m^{-2} day^{-1}). Therefore, at flowering, a certain number of spikelets have been produced, of which only a certain fraction may develop into fertile grains.

Effects of extreme temperatures on γ are simulated using relations developed by Uchijima (1976) for low temperature and by Horie (1994) for high temperature. Data of Horie (1994) suggest that there is no effect of CO_2 concentration on the shape of the high temperature relationship; this is also assumed for the low temperature relationship. The combined temperature response of spikelet fraction to form grain, $f(T_m)$ has a nearly linear increase from 0 at 17 °C to 0.9 at 22 °C, a plateau at 0.9 from 22 to 30 °C, a nearly linear decrease from 30 °C to 0.1 at 35 °C, and an exponential decay from 35 °C to 0 at 40 °C. Maximum yield (Y_{max}, g m^{-2}) is therefore given by:

$$Y_{max} = S_f f(T_m)(G_{max} \times 10^3), \tag{5}$$

where G_{max} is the grain size (mg grain^{-1}) and is assumed to be constant for a variety (Yoshida 1981).

Actual grain yield (Y, g m^{-2}) depends on the amount of assimilate produced from flowering to maturity, plus any translocated assimilate from vegetative tissues, provided Y does not exceed Y_{max}:

$$Y = \min\left[Y_{max},\ \sum_{i=F}^{M}(G_i + T_i)\right], \tag{6}$$

where F and M are the dates of 50% flowering and maturity, respectively, and G_i and T_i (g m^{-2} day^{-1}) are crop weight increase and amount of assimilate translocated, respectively, on day i. The model stops simulation when $Y = Y_{max}$, or when maturity is reached, whichever occurs first.

Overall Response of Grain Yield to Temperature and Carbon Dioxide

Incorporation of these relationships into ORYZA1 gives an overall response of grain yield to temperature and CO_2 (Fig. 1). Temperature responses of spikelet fertility constrains the upper and lower boundaries of the total response. The highest yields are obtained at 20–22 °C. Yields decline as temperatures increase further because of a shortening of crop duration. Between 28 and 32 °C there is a leveling of this response as crop duration reaches its minimum and then increases again because of a decrease in development rate at supraoptimal temperatures, before high spikelet sterility again reduces yields. At all temperatures, increased CO_2 concentration increases the yield. The model does not simulate any interactive effect between temperature and CO_2. The exact change in yield expected at a given site under a changed climate depends, therefore, on the current temperature and the magnitude of the temperature change predicted for the site.

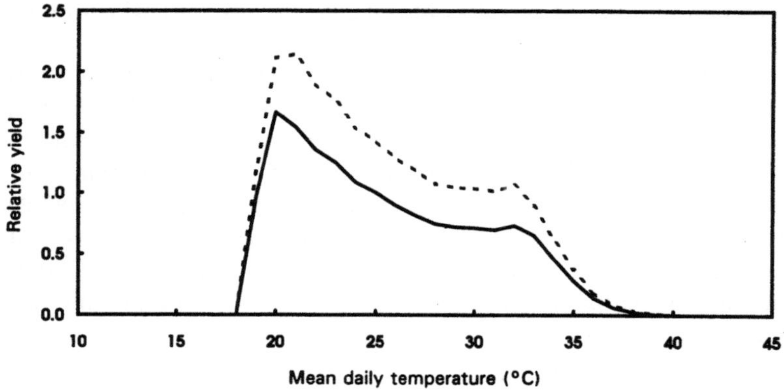

Fig. 1. Overall effect of temperature on grain yield predicted by the ORYZA1 model at current (340 μl l^{-1}) and doubled CO_2 concentrations.– –340 μl l^{-1}; --- 680 μl l^{-1} (base 25 °C, 340 μl l^{-1}).

Input Data

Weather Data

Daily weather data from 70 stations from eleven countries in Asia were used in the analysis (Table 1). For most of the stations, more than 10 years of data were available, giving a total of about 780 years of weather data. To enable extrapolation over a wider area, weather stations were classified according to agroecological zone (AEZ, Harris and Goebel 1987). Simulated yields were averaged for each weather station to represent the yield potential for the whole zone.

Simulations were made for both fixed-change scenarios and for scenarios predicted by three general circulation models (GCMs) for a doubled-CO_2 climate. A total of 15 different scenarios were simulated. In the fixed-change scenarios, temperature was increased by 1, 2, or 4 °C above current values for each site, and CO_2 concentration was increased from 340 to 510 or 680 μl l^{-1} (1.5 or 2 times the current CO_2 level). Temperature and CO_2 were changed independently and in various combinations.The three GCM scenarios used were those of the General Fluid Dynamics Laboratory model (GFDL, Hansen et al. 1988), the Goddard Institute of Space Studies model (GISS, Manabe and Wetherald 1987), and the United Kingdom Meteorological Office model (UKMO, Wilson and Mitchell 1987). Details of each GCM are shown in Table 2.

For both fixed-change and GCM scenarios the historic daily temperature values at each site were adjusted by the fixed-change increment or the predicted change. Solar radiation was not altered from the current values in the fixed-change scenarios. In the three GCM scenarios, predicted changes in solar radiation were also used to adjust observed values at each site. We assumed that the pattern of day-to-day temperature and solar radiation in each scenario would be

Table 1. Details of the countries used in the study, including the agroecological zone (AEZ), number of weather stations, total years of weather data, number of planting seasons, and planting dates used in the simulations. Because of unavailability of weather data in AEZ 8 in China and in India, simulated changes from AEZ 8 in Japan were used for these regions

Country	AEZ	No. of sites	Total years	Seasons	Planting dates (Julian date)
Bangladesh	3	4	43	3	105, 160, 350
China	5	2	18	1	136
	6	2	17	1	105
	7	6	53	2	91, 180
	8	–	–	–	–
India	1	6	65	1	161
	2	2	9	1	213
	5	–	–	–	–
	6	1	3	1	191
	8	–	–	–	–
Indonesia	3	7	49	2	270, 80
Japan	8	9	108	1	96–139[a]
Malaysia	3	3	30	2	228, 80
Myanmar	2	4	26	2	140, 290
Philippines	3	9	148	3	166, 4, 100
South Korea	6	11	158	1	120
Taiwan	7	1	8	2	350, 120
Thailand	2	3	44	2	180, 350
Total	–	70	779	–	–

[a] Planting dates vary according to latitude in Japan.

Table 2. Details of the three General Circulation Models used in the study

GCM	Year[a]	Base CO_2 conc. (µl/l)	Resolution (°)	Change in global average: Temp. (°C)	Change in global average: Precip. (%)	Reference
GFDL	1988	300	4.4 × 7.5	4.0	8	Hansen et al. (1988)
GISS	1982	315	7.8 × 10	4.2	11	Manabe and Wetherald (1987)
UKMO	1986	323	5.0 × 7.5	5.2	15	Wilson and Mitchell (1987)

[a] Refers to the year in which the GCM simulation run was made.

the same as for the present observed climate, and that only the daily mean values would change by the amount predicted in the scenario.

Dates of sowing and transplanting were, in general, supplied by the collaborating institutions. Where this information was not given, transplanting dates were obtained from IRRI (1991), and the date of sowing in the seedbed was assumed to be 25 days prior to this. For sites where more than one rice crop was grown per year, simulations were made for all crops.

Genotype Characteristics

Indica varieties are usually grown in the tropical regions (Bangladesh, India, Indonesia, Malaysia, Myanmar, Philippines, southern China, Taiwan, and Thailand in our study) and japonica varieties in the northern latitudes (Japan, northern China, and South Korea). We used genetic coefficients for the indica variety IR64 and the japonica variety Ishikawi. IR64 is early maturing and weakly photoperiod-sensitive, whereas Ishikawi is a very early maturing, photoperiod-insensitive variety, developed from selections that have been made over 100 years of rice breeding in the Hokkaido region (Yoshino et al. 1988).

Results and Discussion

Fixed-change scenarios enable an evaluation of the separate effects of temperature and CO_2 on potential yields at each site. Figure 2A shows the effect of increasing CO_2 level to 510 or 680 $\mu l\ l^{-1}$. At all sites, increasing CO_2 alone increased simulated yields. Comparison of regression coefficients of the simulated current yield against the simulated yield under each scenario indicated that the average yield increase was 23.6% for 510 $\mu l/l$ and 36.8% for 680 $\mu l/l$. Increments in temperature decreased simulated yields (Fig. 2B). Regression analysis indicated yield declines of 6.7, 14.1 and 29.4% for temperature increases of 1, 2 and 4 °C, respectively (a yield decrease of 6.7 and 7.4% for each 1°C increase in temperature). These values are similar to the 7–8% yield decrease per 1 °C increase measured by Baker et al. (1992a). The decrease in correlation coefficient (r^2) with increased temperature reflects an increase in variability of yields associated with temperature increases (Fig. 2B). This impact on yield variabilty will have some significance for risk assessment studies. Figure 3 shows the overall effect of combinations of fixed changes in temperature and CO_2 levels on potential yields, averaged across all sites and years. Similar patterns have also been found using other models (Bachelet et al. 1993).

Although fixed-change scenarios provide a 'sensitivity-analysis' of the effect of temperature and CO_2 on potential rice yields, scenarios predicted by the GCMs represent the best estimates we have at present of the likely changes in climate caused by an increase in CO_2 levels. Linking these predictions to crop simulation models, therefore, provides some idea of the effect that climate change will have on rice productivity. Table 3 shows the estimated changes in overall rice production for the individual AEZs and countries in the study, and for the whole region. In this analysis, the current production of each country is adjusted by the predicted fractional change in potential yield for each of the three GCM scenarios. Predicted change is the mean of all sites and planting seasons for an AEZ within the country. It is, therefore, assumed that proportional changes at the potential production level are the same at suboptimal levels of management, and that the area of production does not change. Model simulations predict a 6.5%

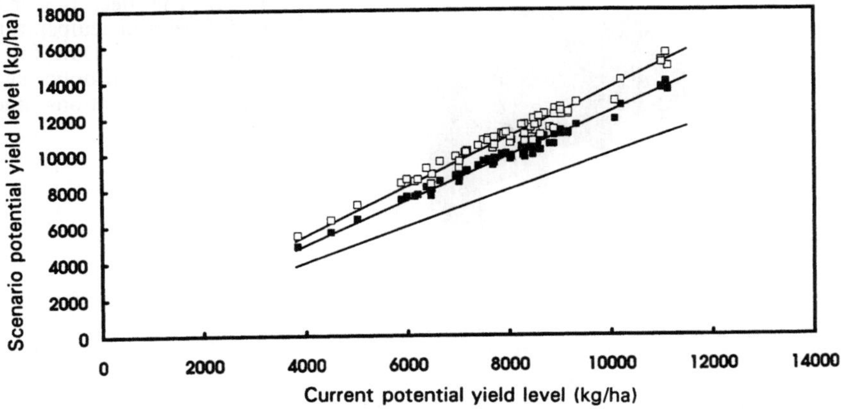

A

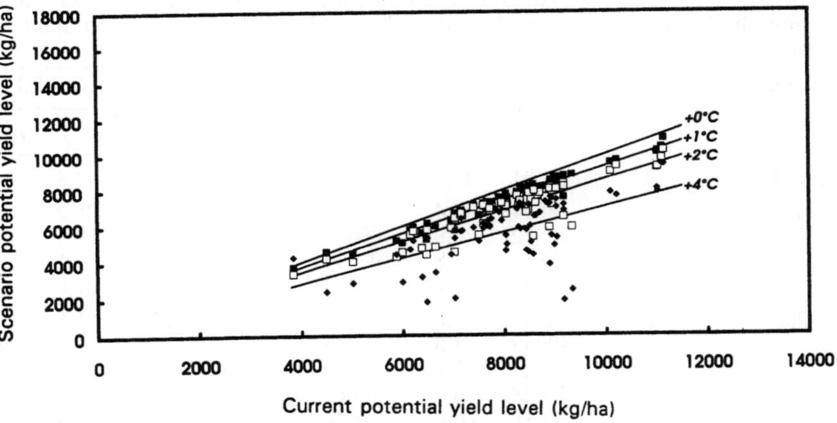

B

Fig. 2. **A** The relationship between predicted current yields and predicted yields at two levels of increased CO_2. Temperature at each site is unaltered from current values. Regression lines are forced through the origin (510 $\mu l\ l^{-1}$: $y = 1.23\ x$, $r^2 = 0.982$; 680 $\mu l\ l^{-1}$: $y = 1.368\ x$, $r^2 = 0.962$). - - - - 340 $\mu l\ l^{-1}$, ■ 510 $\mu l\ l^{-1}$, □ 680 $\mu l\ l^{-1}$. **B** The relationship between predicted current yields and predicted yields at three temperature increments (1, 2, and 4 °C above current values). The CO_2 level at each site is set to 340$\mu l/l$. Regression lines are forced through the origin (+1 °C $y = 0.933\ x$, $r^2 = 0.954$; +2 °C $y = 0.859\ x$, $r^2 = 0.795$; +4 °C $y = 0.706\ x$, $r^2 = 0.400$)

increase in rice production under the GFDL scenario, a 4.4% decrease under the GISS scenario, and a 5.6% decrease under the UKMO scenario.

There were, however, substantial differences in predicted yield changes between countries. Averaged across all three GCM scenarios, the mean change predicted in total production for China was –4.2%, which agrees closely with the –7.4% predicted by Zhou (1991). Similarly, decreases in production were predicted for Thailand under the GISS and UKMO scenarios. Changes in production in both of these countries is likely to have serious repercussions on regional trading patterns as China is the major importer of rice in the region (43% of total

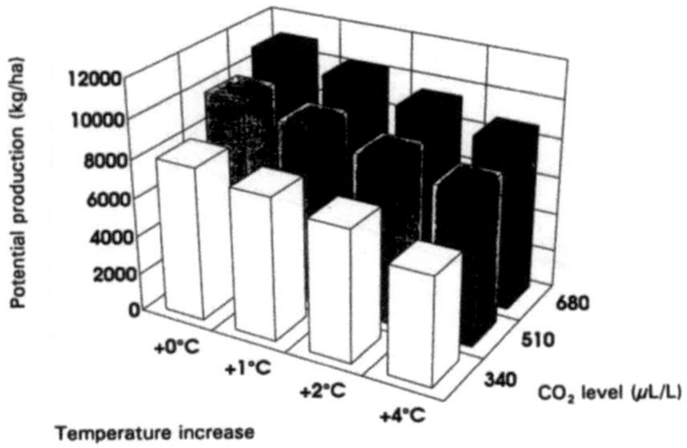

Fig. 3. The effects of increasing CO_2 level and temperature on potential yield. Data for main planting season only; mean of all weather stations and all years

Table 3. Estimated changes in total rice production for each country and in the region under the three GCM scenarios. Current actual production figures in each AEZ on a country basis (IRRI 1991) are adjusted by the simulated changes in total annual production (see text for explanation)

Country	AEZ	Current (t × 10³)	GFDL		GISS		UKMO	
			(%)	(t × 10³)	(%)	(t × 10³)	(%)	(t× 10³)
Bangladesh	3	27691	14.2	31621	–5.0	26298	–2.8	26919
China	5	8854	–7.4	8201	0.3	8881	–25.2	6619
	6	79872	0.8	80484	–21.7	62514	–19.5	64334
	7	91828	5.8	97196	5.8	97135	3.1	94695
	8	2361	–6.4	2209	–14.2	2026	–27.6	1710
India	1	32807	4.6	34305	–10.8	29272	–5.5	31017
	2	49949	1.8	50849	–2.9	48493	–7.9	46002
	5	227	–7.4	210	0.3	228	–25.2	170
	6	26628	5.4	28069	3.2	27480	–1.3	26287
	8	1011	–6.4	946	–14.2	867	–27.6	732
Indonesia	3	44726	23.3	55155	9.0	48748	5.9	47387
Japan	8	12005	–6.4	11231	–14.2	10300	–27.6	8696
Malaysia	3	1744	24.6	2173	17.6	2050	26.8	2211
Myanmar	2	13807	21.5	16776	–10.5	12356	1.2	13974
Philippines	3	9459	14.1	10797	–11.8	8340	–4.7	9018
South Korea	6	8192	–13.6	7078	–5.3	7755	–21.9	6401
Taiwan	7	2798	11.8	3128	12.8	3156	28.0	3583
Thailand	2	20177	9.3	22044	–4.7	19230	–0.9	19989
Total		434136		462472		415129		409743
% change				6.5		–4.4		–5.6

regional imported rice; IRRI 1991), while Thailand is one of the major rice exporting countries of the region (87% of total regional exported rice). Similarly, in Bangladesh, the predicted decline in production by both crop models in the GISS and UKMO scenarios is likely to increase the country's need to import even more rice than at present. On the other hand, countries such as Indonesia,

Malaysia, and Taiwan, and parts of India and China, are all predicted to benefit from any change in climate. There are, however, some striking discrepancies in predicted changes between the three scenarios. In AEZ 6 of China, as one example, a +0.8% increase in yields was predicted for the GFDL scenario, but a –21.7% decrease for the GISS scenario. These large fluctuations are mainly due to the sensitivity of spikelet sterility to temperatures in the region of 33 °C, where a difference of 1 °C can result in a modest yield increase becoming a large yield decrease. In many areas, therefore, it seems that the accuracy of any prediction of changes in rice production depends on the exact nature of changes in the climate there; until there is some consensus in predictions of climate change for an area, therefore, accurate prediction of potential production changes is difficult.

The use of simulation models to predict the likely effects of climate change on crop production is, of necessity, an evolving science. As both general circulation models and crop simulation models become more sophisticated, as more high-quality historical weather data for a larger number of sites become available, and as better physiological data become available to model rice responses to climate change variables, predictions will become more accurate. Some predictions have already been revised. Horie (1991), for example, calculated that there would be an overall decline in rice production in Japan under predicted climate-change scenarios, but Horie et al. (1994) now conclude that, although there will be a shift in the rice-producing regions within the country, the overall rice production of Japan will not be appreciably altered.

This study, therefore, can be seen as part of this evolutionary process, and complementary to previous studies. A number of studies that use models of differing complexity and assumptions but reach similar conclusions may help to reduce the general reluctance by many governments to take action to mitigate both the rate of change of climate and the detrimental effects these changes may have.

Acknowledgment. The information in this chapter has been funded wholly or in part by the US Environmental Protection Agency under cooperative agreement number 817426 to the International Rice Research Institute and contract number 68–C8–0006 to ManTech Environmental Technology Inc. It has been subject to the Agency's peer and administrative review. It has been approved for publication as an EPA document.

Global Warming and Rice Arthropod Communities

K.L. HEONG,[1] Y.H. SONG,[2] S. PIMSAMARN,[3] R. ZHANG,[4] and S.D. BAE[5]

The greenhouse effect on earth has been a widely debated issue over the last 10 years. There is now consensus among scientists that the lower atmosphere and the earth's surface will warm with virtual certainty. Near the earth's surface, the global average warming will lie between +1.5 °C and 4.5 °C, with a "best guess" of 2.5 °C (IPCC 1990). The surface warming at high latitudes will probably be greater than the global average in winter, but smaller than in summer and in the tropics. Surface warming and seasonal variations are probably less. However, regional temperature changes can be much larger and some regions even with negative changes (Schneider 1993). Even more certain are effects of temperature changes on climate variability such as frequency and magnitude of precipitation patterns and temperature extremes.

Rice ecosystems are expected to respond to global warming changes in several fundamental ways. In particular, the natural limits and composition of arthropods in rice ecosystems may change significantly. This expectation is based on the assumption that climate defines the boundaries of arthropod species. Rapid climate change would result in redistribution in abundance. Predictions of the impact of these changes are often based on the association between present distributions and climatic factors and on the anticipated expansion of species range as these factors change. Evolution is usually assumed to be unimportant because species may not be able to evolve rapidly enough to adapt to climate changes (Peters and Darling 1985). Fossil studies suggest that this might explain why many lineages display evolutionary stasis during periods of climate change (Travis and Futuyma 1993). However, fossil studies cannot detect changes in physiological traits, which are probably more important in climatic adaptation. Physiological and behavioral changes may occur without morphological changes. Therefore, evolutionary adaptations to global warming in many arthropods may occur but may not be detectable (Hoffman and Blows 1993). In fact, these adaptations may have occured in environments that are frequently under temperature extremes.

[1] International Rice Research Institute, P.O. Box 933, 1099 Manila, Philippines
[2] Gyeong-Sang University, Chinju, South Korea
[3] Khon Kaen University, Khon Kaen 40002, Thailand
[4] Zhongshan University, Guangzhou, China
[5] Yeong-Nam Crop Experiment Station, 1085 Naidong, Milyang, South Korea.

S. Peng et al. (Eds.) Climate Change and Rice

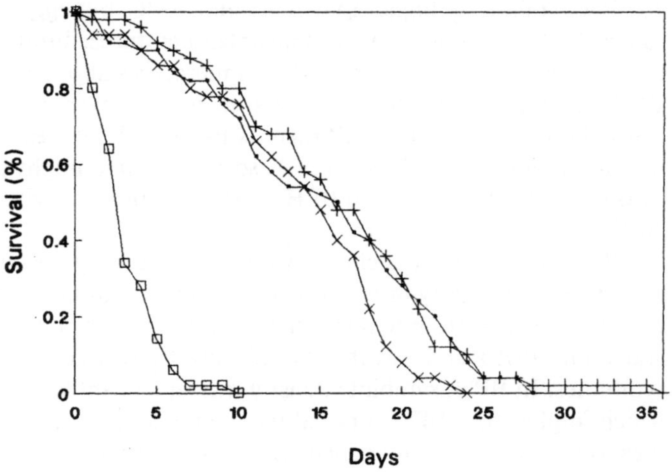

Fig. 1. Daily survival rates of *Nilaparvata lugens* adult females at different temperature regimes. ■ 25 °C, + 30 °C, × 35 °C; □ 40 °C

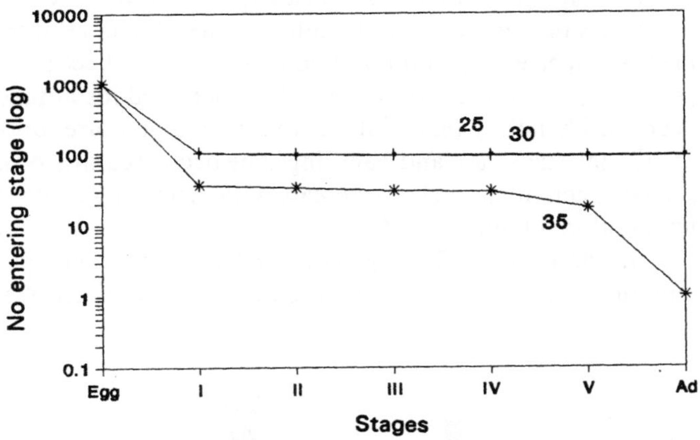

Fig. 2. Survivorship curves of *Nilaparvata lugens* egg reared in different temperature conditions

Direct Effects of Temperature on Rice Arthropod Biology

Climatic factors limit the distribution and abundance of animals. Temperature, for example, can affect any stage of the life cycle and therefore limit distribution and abundance through its effects on survival, reproduction, and development. In life table studies carried out at IRRI, adult survival of the brown planthopper (BPH) (*Nilaparvata lugens*) remained almost unchanged between 25 and 35 °C, but was drastically reduced at 40 °C (Fig. 1). Oviposition of females at 35 and

40 °C was relatively higher than at 25 and 30 °C, but egg survival was markedly reduced at 35 °C (Fig. 2). At the higher temperatures, durations of preoviposition periods were also reduced. Clearly, temperatures above 35 °C are likely to limit BPH development based on these studies. However, in temperature environments between 25 and 30 °C, there is likely to be increased fitness because of increased oviposition. Based on these studies, one might conclude that global warming is likely to increase BPH abundance in areas with temperatures below 30 °C.

Symbiotes are intracellular yeast-like microorganisms that exist in some delphacids. They play important roles in nutrition, metabolism, and reproduction of the hosts. High temperatures can destroy or inhibit their activity (Noda and Saito 1979) and insect survival may be related to the survival of these symbiotes. In BPH, symbiote density was significantly reduced at 35 °C (Fig. 3) which implies that BPH survival may depend on symbiote survival. However, these observations are not conclusive and do not show a cause-effect relationship.

In another pest species, survival of the different stages of the rice leaffolder (*Cnaphalocrocis medinalis*) was greatly affected at 35 °C (Table 1). Adults emerging from pupae reared at 35 °C were unable to lay eggs. The upper temperature threshold for survival of this species appears to lie between 30 and 35 °C.

Shifts in climate can differentially affect the development rates of pest and predator species. In addition, temperatures can affect predator search. The egg predator (*Cyrtorhinus lividipennis*) had increased instantaneous attack rates and decreased handling times with increasing temperatures until 32 °C (Table 2). At 35 °C the attack rate and handling time decreased drastically. This implies that predator activity is likely to increase with increasing temperatures up to a critical temperature of about 35 °C.

In addition to affecting biological characters, global warming may cause temporal asynchrony between interacting populations. Although natural selec-

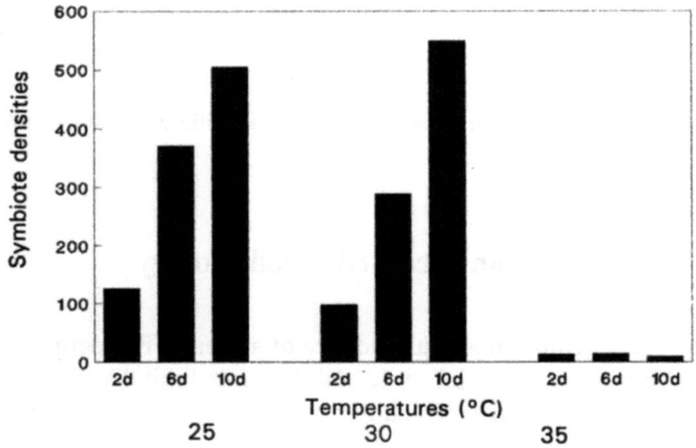

Fig. 3. Symbiote densities at different nymphal stages of *Nilaparvata lugens* reared at different temperatures

Table 1. Survival rates to the subsequent growth stage (% ± 95% confidence level) for *Cnaphalocrocis medinalis* in different temperature regimes

Temperature regime (°C)	Generation	Egg	1st Instar	3rd Instar	Pupa
25	G1	90.1 ± 11.3	32.1 ± 6.7	26.4 ± 8.3	24.3 ± 7.2
	G2	88.5 ± 9.8	33.7 ± 6.6	25.6 ± 6.5	23.5 ± 6.1
	G3	81.0 ± 4.4	29.1 ± 5.9	16.8 ± 4.6	15.5 ± 4.4
30	G1	94.0 ± 4.8	61.7 ± 16.8	44.4 ± 20.7	40.4 ± 20.1
	G2	88.3 ± 8.9	51.5 ± 6.9	36.4 ± 9.5	30.8 ± 11.2
	G3	87.2 ± 4.2	44.0 ± 9.5	31.5 ± 7.8	28.6 ± 6.6
35	G1	47.9 ± 13.0	19.7 ± 6.0	9.4 ± 3.0	6.5 ± 3.0
	G2	48.8 ± 15.0	23.5 ± 9.0	11.3 ± 6.0	9.8 ± 4.9
	G3	56.2 ± 12.0	26.7 ± 11.3	16.0 ± 6.6	13.0 ± 4.5
40	G1	2.0 ± 2.0	–	–	–
	G2	–	–	–	–
	G3	–	–	–	–

Table 2. Parameter estimates of functional response equations for *Cyrtorhinus lividipennis* female adults feeding on brown planthopper eggs at different temperature regimes

Temperature (°C)	Parameter estimates[a]		Asymptotic SEs		F
	a	Th	a	Th	
20	0.19	0.66	0.03	0.13	321
23	0.30	0.48	0.03	0.06	477
26	0.39	0.44	0.05	0.07	284
29	0.47	0.41	0.04	0.04	477
32	0.51	0.36	0.04	0.04	780
35	0.16	0.07	0.05	0.38	196

[a] a searching efficiency, Th handling time or time (in hours) between successful searches, and F is F.value of the F-test.

tion will tend to increase synchrony between hosts and parasitoids, asynchrony may occur if host and parasitoid respond differentially to changes in weather patterns (Hassell et al. 1993). Asynchrony may introduce a partial refuge effect that can reduce parasitism. However, the lack of temporal coincidence between searching parasitoids and hosts can contribute to stability.

Potential Adaptive Responses

Adaptation of arthropods to global warming can occur either by evolution of enhanced tolerance or by evasion (Hoffman and Blows 1993). Increased tolerance may involve traits that prevent mortality as well as traits that enhance fitness. Evasion may occur when the insects migrate, alter behavior, or switch resources to avoid the stressful conditions. These two traits may, in practice, interact.

Tolerance to High Temperatures

Using artificial selection experiments in the laboratory, individuals tolerant to the stressful conditions may be selected. These types of experiments have been successfully carried out with several insect species (White et al. 1970; Stephanon and Alahiotis 1983; Quintana and Prevosti 1990; Huey et al. 1991). Genetic variation to heat tolerance can also be documented with techniques other than artificial selection. For example, copepods from a power station were found to have higher heat tolerance than those from other sites (Bradley 1978, 1981). *Drosophila melanogaster* was found to be more tolerant to environmental extremes than four *melanogaster* subgroup species restricted to tropical Africa (*D. erecta, D. mauritiana, D. tessieri,* and *D. yakuba*) (Parsons 1983). This was documented using a standard method to quantify tolerance.

We used probit analysis to quantify tolerance to high temperatures (Domingo and Heong 1992). Batches of test insects were placed in the chamber set at the lethal temperature, and mortality was periodically recorded. The median lethal exposure time (LT_{50}) can then be calculated using the statistical procedure developed by Finney (1978). The LT_{50} is a quantitative expression of

Table 3. Median effective dose (LT_{50}) of *Nilaparvata lugens* and selected predators in two locations

	Location[a]	LT_{50} (h)	Fiducial limits	Slope
Nilaparvata lugens				
1st Instar	IRRI	23.4	18.3 – 35.4	5.0
	KKU	57.2	44.4 – 82.2	0.8
3rd Instar	IRRI	22.8	16.6 – 33.1	2.6
	KKU	144.6	87.9 – 396.2	0.7
5th Instar	IRRI	27.3	25.2 – 30.2	4.6
Brachypterous females	IRRI	16.9	14.8 – 17.1	5.6
	KKU	285.9	129.4 – 2038.4	0.6
Macropterous females	IRRI	47.3	38.6 – 72.1	3.4
	KKU	140.2	86.5 – 372.1	0.7
Cyrtorhinus lividipennis				
	IRRI	2.8	2.5 – 3.0	11.0
Tytthus chinensis				
	KKU	92.9	60.9 – 182.6	0.6
Pardosa pseudoannulata				
	IRRI	>280.0	–	–

[a] IRRI International Rice Research Institute, Los Baños, Philippines, KKU Khon Kaen University, Thailand.

the tolerance of the particular species (or population) to the lethal temperature. Species or populations may be compared with the relative potency ratios, expressed as the ratio of LT_{50} values.

Populations of *N. lugens* in Khon Kaen, Thailand, were more tolerant to 40 °C than those in Los Baños, Philippines (Table 3). The first-instar nymphs were 2.4 times more tolerant, whereas the brachypterous (adult, nonmigratory) females were 17 times more tolerant. The shallow slopes of the probit lines obtained from the Khon Kaen populations indicate heterogeneity in tolerance. Climatically, the two sites, Khon Kaen and Los Baños, differ significantly in temperature ranges and this might have contributed to the higher tolerance in BPH.

There were also significant differences in 40 °C tolerance between the different instars. The macropterous (adult, migratory) females were two times more tolerant than the first-instar nymphs. These differences contradict the conclusions of Noda and Saito (1979) that planthopper mortality at high temperatures was caused by the mortality of the symbiotes. Tolerance to heat in BPH is probably a physiological adaptation.

The wolf spider (*Pardosa pseudoannulata*) an important predator of BPH, was extremely tolerant to 40 °C. It would seem likely that the wolf spider might remain an important predator under higher temperature conditions. The BPH egg predator (*Cyrtorhinus lividipennis*) had a narrow range of tolerance. Because the LT_{50} was about eight times lower than BPH first-instar nymphs, range dissociation (Peters 1991) between these two species might occur as a result of global warming. Differential response to temperature changes may imply that global warming can reduce egg predation. However, another mirid predator was common in Khon Kaen, where temperatures often exceed 40 °C between April and July each year.This predator (*Tytthus chinensis*) is at least 33 times more tolerant to 40 °C than *C. lividipennis* in the Philippines. Using 45 °C as the lethal temperature, *C. lividipennis* and *T. chinensis* from Khon Kaen were both equally tolerant (LT_{50}=4.0 and 3.7, respectively). These findings imply that in Khon Kaen, both mirid egg predators of BPH are well adapted to the high temperature environment that often occurs in the area. It is also likely that heat-tolerant populations of pests and related predators may now exist in other areas that often experience high temperature extremes.

These studies on high-temperature tolerance of selected pest and predator species in the rice ecosystem suggest that there are diverse ways arthropods can evolve when faced with stresses likely to arise from global warming. However, whether all species or populations will show such adaptive responses is not clear. Predicting evolutionary changes becomes a difficult task if each species or population responds in a different way. It is of interest to test for general responses. The genes or mechanisms responsible could then be assessed to determine the distribution of tolerant populations, and the species adaptation potentials could be assessed.

Evasion of Climatic Stress

Climate change might indirectly cause evolutionary changes by affecting food availability. When this occurs, polyphagous species may switch resources.The rice crop may become less palatable because of increased production of phenolic compounds induced by climate change. Increased CO_2, reduction in soil nitrogen, and increased temperatures and UV radiation, tend to reduce tissue nitrogen, which is detrimental to many herbivores (Ayres 1993). Differential developments in other host plants could also cause polyphagous pests to switch onto rice plants. Therefore, meaningful predictions would require that we know the initial states of host plants and their physiological responses to climate change.

Insect may also evade climate-change stress through changes in life cycle. Insect populations from environments with higher temperatures may have higher fecundity and shorter growthstage durations to increase fitness. For example, in BPH, egg duration was 10.4 days at 25 °C and 7.9 days at 27–28 °C (Mochida and Okada 1979).

Behavioral responses of arthropods to climatic stresses are largely unexplored. For example, insects may retreat to refugia where temperatures are cooler. Another behavioral response is migration. Rice planthoppers (*N. lugens*

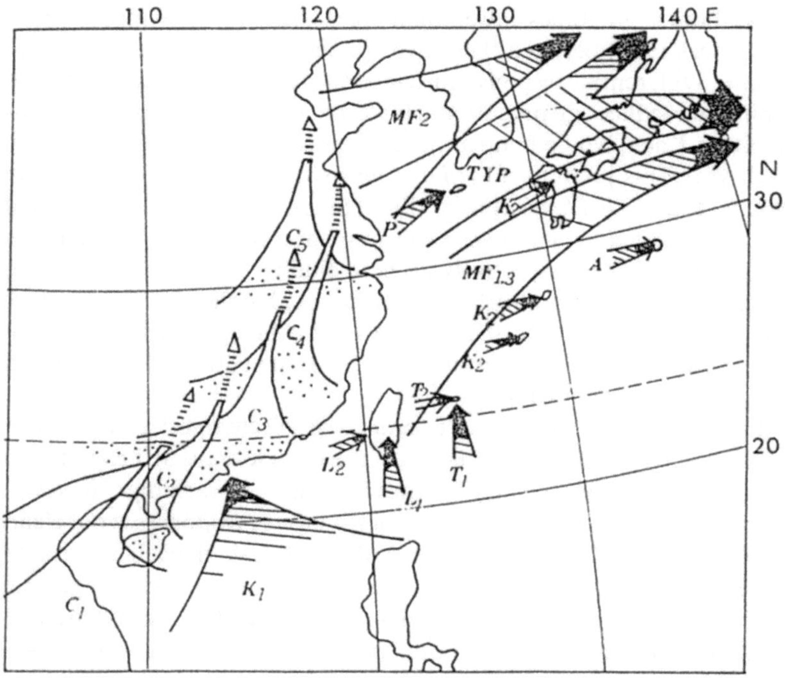

Fig. 4. Northward migration routes of *Nilaparvata lugens* and *Sogatella furcifera* in East Asia. (Kisimoto 1987)

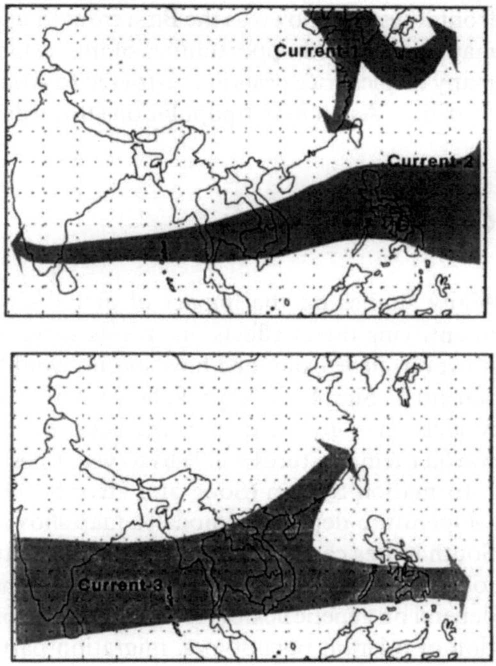

Fig. 5. Possible migration routes of insects in Asia determined from wind-pattern analysis using weather charts at 850 mB for 1992 released by WMO

and *Sogatella furcifera*), both of tropical origin, migrate annually into Japan and Korea (Kisimoto 1976). The migration probably involves two main steps (Fig. 4; Kisimoto 1987). First, in March and April, insects migrate from the permanent breeding areas in the tropics into subtropical regions. Then, in June and July, they migrate from subtropical to temperate regions. The axis of the low-level jet stream appearing in the East China Sea at this time coincides with the route of long-distance hopper migration (Seino et al. 1987; Heong and Sogawa 1994). Using World Meteorological Organization weather charts at 850 mb (about 3000-m altitude), a computer program was designed to predict immigration into Japan (Watanabe et al. 1988). Similar migration routes for Southeast Asia can be outlined using a modified program (Fig. 5).

Migration has important implications to the spread of high-temperature tolerant populations. Our preliminary study suggests that BPH populations native to an environment with higher temperatures also has higher tolerance. With wind assisted migration, tolerant populations can be easily distributed. Range shifts from the effects of global warming are unlikely to affect the survival of these populations. Predators (e.g., *C. lividipennis* and *Tytthus chinensis*) also migrate with the rice planthoppers (Kisimoto 1991). Therefore, range dissociations between predator-prey relationships may or may not be adversely affected by global warming (Heong and Domingo 1992).

Because distribution of these species depends on wind patterns, the effect of global warming on wind patterns will be important. For example, if the weather frontal system known as the Baiu front (Seino et al. 1987) shifts northward, there may be increased opportunities of migration into Japan and Korea. At present, many tropical rice pests are restricted to Asia. Changes in wind patterns may also distribute Asian insect populations towards Africa.

Conclusions

Many studies on the impact of global climate change have been devoted to quantifying direct effects on species survival under the expected stresses. Few, however, have examined how species would interact, evolve, or adapt to global warming. Simplistic approaches, such as those based on temperature-driven models, may give erroneous predictions. For example, they may conclude that warmer temperatures will lead to wider pest distribution. Using a Geographical Information System (Song and Heong 1993) and applying predictions from a temperature-dependent model, a map showing an expanded BPH distribution in South Korea can be produced. However, actual arthropod community responses to global warming cannot be so simple. The impact of global warming should depend on genetic flexibility of the populations (Kareiva et al. 1993), the distribution of tolerant phenotypes, migration patterns, and competition. Community and evolutionary dynamics will play a constraining or amplifying role, and it would be difficult to predict outcomes. A general trend may be obtained by comparing the food web structures of rice-arthropod populations (Cohen et al. 1994) in environments with high temperature extremes.

Monitoring the distribution and spread of stress tolerance in arthropod species may be useful to provide information for predictions. For this purpose, more portable field techniques must be established. The lethal median time (LT_{50}) can provide a precise statistic for comparison. However, the method requires a growth chamber set at the lethal temperatures and periodic recordings. Alternatively, a mortality-time response curve derived using single batches of test insects in a chamber with a preset temperature may suffice for comparisons. With the availability of portable oven units, a standardized method may be developed and used to identify regions where populations adapted to temperature extremes now occur.

These methods often use extremely high temperatures that may not be experienced by the insects. Exposure times are often too long. Studies on the effects of high temperatures on fitness characters would be most useful. Nonlethal fitness effects may be the critical factors that determine pest distribution and abundance. To provide adequate estimates of population fitness, it may be necessary to carry out studies for several generations rather than one because populations surviving the first generation may sometimes produce sterile adults.

Laboratory-determined heritabilities may not reflect field values because environment influences genetic expression (Hoffman and Blows 1993). For example, laboratory studies on insecticide resistance suggest a polygenic basis for resistance, but field studies generally indicate a monogenic basis (Rousch and McKenzie 1987).

In conclusion, relevant data on possible responses of rice arthropod communities to global warming are limited. Any predictions made from these limited studies may be grossly in error. However, the studies do indicate the flexibility inherent within species as well as in the arthropod communities. Because species do not encounter global warming in isolation, community-level or ecosystem-level responses to the change must be investigated.

Acknowledgment. The authors are grateful to Mr. Isaias Domingo who provided the technical support and to Ms. Elena G. Genil who typed the manuscript. Although the research described in this chapter has been funded wholly or in part by the US Environmental Protection Agency under cooperative agreement number 817425 to the International Rice Research Institute, it has not been subject to the Agency's review and therefore does not necessarily reflect the views of the Agency, and no official endorsement should be inferred.

References

Abeles FB (1986) Plant chemiluminescence. Annu Rev Plant Physiol 37: 49–72
Abeles FB, Leather GR, Forrence LE (1978) Plant chemiluminescence. Plant Physiol 62: 696–698
Acharya CN (1935) Studies on the anaerobic decomposition of plant materials. II. Some factors influencing the anaerobic decomposition. Biochem J 29: 953–960
Acock B, Allen LH Jr (1985) Crop responses to elevated carbon dioxide concentrations. In: Strain BR, Cure JD (eds) Direct effects of increasing carbon dioxide on vegetation. DOE/ER-O238, United States Department of Energy, Washington, pp 53–98
Adachi N, Imai K (1991) Effects of CO_2 concentration and phosphorus nutrition on growth of rice. Jpn J Crop Sci 60 (extra issue 2): 93–94 (in Japanese)
Adams RM, Chang C, McCarl BA, Callaway JM (1992) The role of agriculture in climate change: a preliminary evaluation of emission-control strategies. In: Reilly JM, Anderson M (eds) Economic issues in global climate change: agriculture, forestry and natural resources. Westview Press, Boulder, pp 273–287
Adamse P, Britz SJ (1992) Spectral quality of two fluorescent UV sources during long-term use. Photochem Photobiol 56: 641–644
Aiyer PAS (1920) The gases of swamp rice soils V: methane-oxidizing bacterium for rice soils. Mem Dep Agric India 5: 175–180
Akita S (1980) Studies on the differences in photosynthesis and photorespiration among crops. II. The differential responses of photosynthesis, photorespiration and dry matter production to carbon dioxide concentration among species. Bull Natl Inst Agric Sci (Ser D) 31: 59–94
Alexander M (1977) Introduction to soil microbiology. John Wiley, New York
Allen LH Jr (1990) Plant responses to rising carbon dioxide and potential interactions with air pollutants. J Environ Qual 19: 15–34
Allen LH Jr (1992) Free air CO_2 enrichment field experiments: an historical overview. Crit Rev Plant Sci 11(2–3): 121–134
Allen LH Jr (1994) Carbon dioxide increase: direct impacts on crops and indirect effects mediated through anticipated climatic changes. In: Boote KJ, Bennett JM, Sinclair TR, Paulsen GM, (eds) Physiology and determination of crop yield. American Society of Agronomy, Crop Science Society of America, and Soil Science Society of America, Madison, WI (in press)
Allen LH Jr, Jones P, Jones JW (1985) Rising atmospheric CO_2 and evapotranspiration. In: Advances in evapotranspiration. Proc Nat Conf on Advances in Evapotranspiration. ASAE Publ 14–85. American Society of Agricultural Engineers, St. Joseph, MI, pp 14–27
Allen LH Jr, Boote KJ, Jones JW, Jones PH, Valle RR, Acock B, Rogers HH, Dahlman RC (1987) Response of vegetation to rising carbon dioxide: photosynthesis, biomass, and seed yield of soybean. Global Biogeochem Cycles 1: 1–14
Allen LH Jr, Baker JT, Boote KJ, Jones JW, Jones P (1988) Growth and yield of rice as affected by carbon dioxide concentration. In: Agronomy Abstracts. American Society of Agronomy, Madison, WI, p 12

Allen LH Jr, Drake BG, Rogers HH, Shinn JH (1992) Field techniques for exposure of plants and ecosystems to elevated CO_2 and other trace gases. Crit Rev Plant Sci 11(2–3): 85–119

Allen LH Jr, Baker JT, Albrecht SL, Boote KJ, Pan D, Vu JCV (1994) Carbon dioxide and temperature effects on rice. In: Peng et al. (eds) Climate change and rice. International Rice Research Institute, Los Baños, Philippines

Amthor JS (1991) Respiration in a future, high-CO_2 world. Plant Cell Environ 14: 13–20

Anastasi C, Dowding M, Simpson VJ (1992) Future CH_4 emissions from rice production. J Geophys Res 97: 7521–7525

Anderson IC, Poth M, Homstead J, Burdige D (1993) A comparison of NO and N_2O production by the autotrophic nitrifier *Nitrosomonas europaea* and the heterotrophic nitrifier *Alcaligenes faecalis*. Appl Environ Microbiol 59: 3525–3533

Andreae M, Fishman J, Garstang M, Goldammer J, Justice C, Levine J, Scholes R, Stocks B, Thompson A, vanWilgen B, STARE/TRACE-A/SAFARI-92 Science Team (1994) Biomass burning in the global environment: first results from the IGAC/BIBEX field campaign STARE/TRACE-A/SAFARI-92. In: Prinn R (ed) Global atmospheric-biospheric chemistry. Plenum Press, New York, pp 83–102

Anonymous (1987) LI-6200 technical reference. LI-COR, Lincoln, Nebraska

Archer D, Maier-Reimer E (1994) Effect of deep-sea sedimentary calcite preservation on atmospheric CO_2 concentration. Nature 367: 260–263

Arp WJ (1991) Effects of source-sink relations on photosynthetic acclimation to elevated CO Plant Cell Environ 14(8): 869–875

Arp WJ, Drake BG (1991) Increased photosynthetic capacity of *Scirpus olneyi* after four years of exposure to elevated CO_2. Plant Cell Environ 14(9): 1003–1006

Arya SP (1988) Introduction to micrometeorology. Academic Press, New York

Aselmann I, Crützen PJ (1990) A global inventory of wetland distribution and seasonality, net primary productivity, and estimated methane emissions. In: Bouwman AF (ed) Soils and the greenhouse effect. John Wiley, New York, pp 441–449

Ayres MP (1993) Plant defense, herbivory, and climate change. In: Kareiva PM, Kingsolver JG, Huey RB (eds) Biotic interactions and global change. Sinauer, Sunderland, MA, pp 75–94

Bachelet D, Neue H-U (1993) Methane emission from wetland rice areas of Asia. Chemosphere 26: 219–237

Bachelet D, Barnes PW, Brown D, Brown M (1991) Latitudinal and seasonal variation in calculated solar ultraviolet-B irradiance (280–320 nm) for rice-growing regions of Asia. Photochem Photobiol 54: 411–422

Bachelet D, van Sickle J, Gay CA (1993) The impacts of climate change on rice yield: evaluation of the efficacy of different modeling approaches. In: Penning de Vries F.W.T, Teng P, Metselaar K (eds) Systems approaches for agricultural development. Kluwer, Dordrecht, pp 145–174

Baker JT, Allen LH Jr (1993a) Contrasting crop species responses to CO_2 and temperature: Rice, soybean, and citrus. Vegetation 104/105: 239–260

Baker JT, Allen LH Jr (1993b) Effects of CO_2 and temperature on rice: a summary of five growing seasons. J Agric Meteorol 48: 575–582

Baker JT, Allen LH Jr (1994) Assessment of the impact of rising carbon dioxide and other potential climate changes on vegetation. Environ Pollut 83: 223–235

Baker JT, Allen LH Jr, Boote KJ, Rowland-Bamford AJ, Jones JW, Jones PH, Bowes G, Albrecht SL (1988) Response of rice to subambient and superambient carbon dioxide concentrations 1986–1987. Progress Report of Research submitted to US Department of Energy, Washington, from US Department of Agriculture, ARS, and Institute of Food and Agricultural Sciences, University of Florida, Gainesville

Baker JT, Allen LH Jr, Boote KJ, Jones P, Jones JW (1989) Response of soybean to air temperature and carbon dioxide concentration. Crop Sci 29: 98–105

Baker JT, Allen LH Jr, Boote KJ (1990a) Growth and yield responses of rice to subambient, ambient, and superambient carbon dioxide concentrations. J Agric Sci 115: 313–320

Baker JT, Allen LH Jr, Boote KJ, Jones P, Jones JW (1990b) Rice photosynthesis and evapotranspiration in subambient, ambient and superambient carbon dioxide concentrations. Agron J 82: 834–840

Baker JT, Allen LH Jr, Boote KJ, Jones JW, Jones P (1990c) Developmental responses of rice to photoperiod and carbon dioxide concentration. Agric For Meteorol 50: 201–210

Baker JT, Allen LH Jr, Boote KJ (1992a) Response of rice to carbon dioxide and temperature. Agric For Meteorol 60: 153–166

Baker JT, Allen LH Jr, Boote KJ (1992b) Temperature effects on rice at elevated CO_2 concentration. J Exp Bot 43: 959–964

Baker JM, Norman JM, Bland WL (1992c) Field-scale application of flux measurement by conditional sampling. Agric For Meteorol 62: 31–52

Baker JT, Laugel F, Boote KJ, Allen LH Jr (1992d) Effects of daytime carbon dioxide concentration on dark respiration in rice. Plant Cell Environ 15: 231–239

Baker JT, Albrecht SL, Pan Deyun, Allen LH Jr, Pickering NB, Boote KJ (1994a) Carbon dioxide and temperature effects on rice (*Oryza sativa* L., cv. IR-72). Soil Crop Sci Soc Fla Proc 53 (in press)

Baker JT, Allen LH Jr, Boote KJ (1994b) Potential climate change effects on rice: carbon dioxide and temperature. In: Rosenzweig C, Harper LA, Hollinger SE, Jones JW, Allen LH, Jr (eds) Proceedings of a symposium on climate change and international impacts. ASA Spec Publ. American Society of Agronomy, Crop Science Society of America, and Soil Science Society of America, Madison, WI (in press)

Baker JT, Allen LH Jr, Boote KJ, Pickering NB (1994c) Assessment of rice responses to global climate change: CO_2 and temperature. In: IGBP/global change and terrestrial ecosystems state-of-the-science symposium on terrestrial ecosystem response to elevated CO_2, 18–31 October 1993, S. Miniato, Italy (in press)

Banin A (1986) Global budget of N_2O: the role of soils and their change. Sci Total Environ 55: 27–38

Barker R, Herdt RW, Rose B (1985) The rice economy of Asia. Resources for the Future, Washington, DC

Barnes PW, Jordan PW, Gold WG, Flint SD, Caldwell MM (1988) Competition, morphology and canopy structure in wheat (*Triticum aestivum* L.) and wild oat (*Avena fatua* L.) exposed to enhanced ultraviolet-B radiation. Funct Ecol 2: 319–330

Barnes PW, Flint SD, Caldwell MM (1990) Morphological responses of crop and weed species of different growth forms to ultraviolet-B radiation. Am J Bot 77: 1354–1360

Barnes PW, Maggard S, Holman SR, Vergara BS (1993) Intraspecific variation in sensitivity to UV-B radiation in rice. Crop Sci 33: 1041–1046

Bartlett KB, Harriss RC, Sebacher DI (1985) Methane flux from coastal salt marshes. J Geophys Res 90: 5710–5720

Bartlett KB, Crill PM, Sebacher DI, Harriss RC, Wilson JO, Melack JM (1988) Methane fluxes from the central Amazonian floodplain. J Geophys Res 93: 1571–1582

Batjes NH (1992a) Methane. In: A review of soil factors and processes that control fluxes of heat, moisture and greenhouse gases, prepared by N.H. Batjes and E.M. Bridges, Technical Paper 23, International Soil Reference and Information Centre (ISRIC), Wageningen, pp 37–66

Batjes NH (1992b) The WISE database: general considerations on a possible structure and attribute-data. In: Batjes NH, Bridges EM, A review of soil factors and processes that control fluxes of heat, moisture and greenhouse gases. Tech Pap 23, International Soil Reference and Information Centre (ISRIC), Wageningen, pp 149–158

Batjes NH (1993a) World inventory of soil emission potentials: guidelines for soil profile selection and protocol for completing the WISE data entry sheets. Working Paper 93/02, International Soil Reference and Information Centre (ISRIC), Wageningen

Batjes NH (1993b) World inventory of soil emission potentials: profile database user's manual (version 1.0). Working Paper 93/04, International Soil Reference and Information Centre (ISRIC), Wageningen

Batjes NH, Bridges EM (eds) (1992) World inventory of soil emission potentials. WISE Report 2, International Soil Reference and Information Centre (ISRIC), Wageningen
Batjes NH, Bridges EM (1994) Potential emissions of radiatively active gases from soil to atmosphere with special reference to methane: development of a global database (WISE). J Geophys Res 99: 16479–16489
Beauchamp EG, Kidd GE, Thurtell G (1978) Ammonia volatilization from sewage sludge applied in the field. J Environ Qual 7: 141–146
Berner RA, Lasaga AC (1989) Modeling the geochemical carbon cycle. Sci Am 260: 54–61
Besford RT, Ludwig LJ, Withers AC (1990) The greenhouse effect: acclimation of tomato plants growing in high CO_2, photosynthesis and ribulose-1,5-bisphosphate of carboxylase protein. J Exp Bot 41: 925–931
Biggs RH, Kossuth SV (1978) Effects of ultraviolet-B radiation enhancement under field conditions on potatoes, tomatoes, corn, rice, southern peas, peanuts, squash, mustard and radish. In: UV-B biological and climatic effects research (BACER), Final Report, Environmental Protection Agency, Washington
Biggs RH, Webb PG (1986) Effects of enhanced ultraviolet-B radiation on yield and disease incidence and severity for wheat under field conditions. In: Worrest RC, Caldwell MM (eds) Stratospheric ozone reduction, solar ultraviolet radiation and plant life. Springer, Berlin Heidelberg New York, pp 303–311
Biggs RH, Webb PG, Garrad LA, Sinclair TR, West SH (1984) The effects of enhanced ultraviolet-B radiation on rice, wheat, corn, soybean, citrus, and duckweed. Year-3 Interim Report, Environmental Protection Agency Report 808075–03, US Environmental Protection Agency, Washington
Björn LO (1989) Computer programs for estimating ultraviolet radiation in daylight. In: Diffey BL (ed) Radiation measurement in photobiology. Academic Press, London, pp 161–189
Björn LO, Murphy TM (1985) Computer calculation of solar ultraviolet radiation at ground level. Physiol Vég 23: 555–561
Blake DR, Rowland FS (1988) Continuing worldwide increase in tropospheric methane 1978–1987. Science 239: 1129–1131
Bliss NB (1990) A hierarchy of soil databases for calibrating models of global climate change. In: Bouwman AF (ed) Soils and the greenhouse effect. John Wiley, New York, pp 523–529
Blumthaler M (1993) Solar UV measurements. In: Tevini M (ed) UV-B radiation and ozone depletion: effects on humans, animals, plants, microorganisms, and materials. Lewis Publishers, Boca Raton, pp 71–94
Boden AT, Sepanski RJ, Stoss FW (eds) (1992) Trends 91: a compendium of data on global change—highlights. Publ No ORNL/CDIAC 49. ESD Publ 3797 Oak Ridge, TN
Bohne K, Horn R, Baumgartl T (1993) Providing the parameters of unsaturated soil hydraulic functions on the basis of soil texture, bulk density and type of horizon. Z Pflanzenernähr Bodenkd 156: 229–233
Bolin B (1986) How much CO_2 will remain in the atmosphere? In: Bolin B, Böös BR, Jäger J, Warrick RA (eds) The greenhouse effect, climatic change and ecosystems. SCOPE 29. John Wiley, New York, pp 93–155
Bolin B, Böös BR, Jäger J, Warrick RA (eds) (1986) The greenhouse effect, climatic change and ecosystems. SCOPE 29. John Wiley, New York
Booker FL, Fiscus EL, Philbeck R, Miller JE, Heagle AS, Heck WW (1992a) A supplemental ultraviolet-B radiation system for open-top field chambers. J Environ Qual 21: 56–61
Booker FL, Miller JE, Fiscus EL (1992b) Effects of ozone and UV-B radiation on pigment, biomass and peroxidase activity in soybean. In: Berglund RL (ed) Tropospheric ozone II: effects, modeling and control. Air and Waste Management Association, Pittsburgh, PA, pp 489–503
Boote KJ, Pickering NB (1994) Modeling photosynthesis of row crop canopies. Hort Science 29: 13–24

Boote KJ, Jones JW, Singh P (1992) Modeling growth and yield of groundnut. In: Nigam SN (ed) Groundnut – a global perspective. Proc of Int Workshop. International Crops Research Institute for the Semi-Arid Tropics, Patencheru, India, pp 331–343

Bornman JF (1989) Target sites of UV-B radiation in photosynthesis of higher plants. J Photochem Photobiol B Biol 4: 145–158

Bouma J, van Lanen HAJ (1987) Transfer functions and threshold values: From soil characteristics to land qualities. In: Beek KJ, Burrough PA, McCormack DE (eds) Quantified land evaluation procedures. Publ 6, International Institute for Aerospace Survey and Earth Sciences (ITC), Enschede, pp 106–110

Bouwman AF (1990) Soils and the greenhouse effect. John Wiley, New York

Bouwman AF (1991) Agronomic aspects of wetland rice cultivation and associated methane emissions. Biogeochemistry 15: 65–88

Bouwman AF, van der Hoek KW (1993) NxOy (Natural) and NH_3. In: Bouwman AF (ed) Working group report of the third workshop of the global emissions inventory activity (GEIA), Amersfoort, 31 January-2 February 1993. RIVM Report No 481507002, Bilthoven, The Netherlands, pp 31–34

Bouwman AF, Fung I, Matthews E, John J (1993) Global analysis of the potential for N_2O production in natural soils. Global Biogeochem Cycles 7: 557–597

Braatz BV, Hogan KB (1991) Sustainable rice productivity and methane reduction research plan. Office of Air and Radiation, United States Environmental Protection Agency, Washington

Bradley BP (1978) Genetic and physiological adaptation of the copepod *Eurytemora affinis* to seasonal temperatures. Genetics 90: 193–205

Bradley BP (1981) Models for physiological and genetic adaptation to variable environments. In: Denno RF, Dingle H (eds) Insect life-history patterns: habitats and genetic variation. Springer, Berlin Heidelberg New York, pp 33–50

Bremner JM, Blackmer AM (1978) Nitrous oxide emission from soils during nitrification of fertilizer nitrogen. Science 199: 295–296

Bronson KF, Mosier AR (1991) Effect of encapsulated calcium carbide on dinitrogen, nitrous oxide, methane and carbon dioxide emissions in flooded rice. Biol Fertil Soils 3: 116–120

Bronson KF, Singh U (1994) Nitrous oxide emissions from flooded rice. In: Peng et al. (eds) Climate change and rice. International Rice Research Institute, Los Baños, Philippines

Bronson KF, Neue H-U, Singh U (1993) Nitrous oxide and methane fluxes in a rainfed fallow-irrigated rice system. Agronomy Abstracts, American Society of Agronomy, Madison, WI, p 242

Brouwer FM, Thomas AJ, Chadwick MJ (eds) (1991) Land use changes in Europe: processes of change, environmental transformations and future trends vol 18, The GeoJournal Library. Kluwer, Dordrecht

Brown, KW, Turner FT, Thomas JC, Deuel LE, Keener ME (1978) Water balance of flooded rice paddies. Agric Water Manage 1: 277–291

Bunce JA (1990) Short-and long-term inhibition of respiratory carbon dioxide efflux by elevated carbon dioxide. Ann Bot 65: 637–642

Buresh RJ, Woodhead T, Shepherd KD, Flordelis E, Cabangon R (1989) Nitrate accumulation and loss in mungbean lowland rice cropping system. Soil Sci Soc Am J 53: 477–482

Burrough PA (1993) Soil variability: a late 20th century view. Soils Fertil 56: 529–562

Businger JA, Oncley S (1990) Flux measurement with conditional sampling. J Atmos Oceanic Technol 7: 349–352

Butler JN (1982) Carbon dioxide equilibria and their applications. Addison-Wesley, Reading, MA

Butterbach-Bahl K (1993) Mechanisms of methane production and emission in ricefields PhD Thesis, Fraunhofer Institute for Atmospheric Environmental Research, 82467 Garmisch-Partenkirchen, Germany (in german)

Buttler IW, Riha SJ (1989) GAPS: a general purpose simulation model of the soil-plant-atmosphere system (Version 1.1 User's Manual). Department of Agronomy, Cornell University, Ithaca, NY

Caldwell MM (1968) Solar ultraviolet radiation as an ecological factor in alpine plants. Ecol Monogr 38: 243–268

Caldwell MM (1971) Solar UV irradiation and growth and development of higher plants. In Giese AC (ed) Photophysiology, vol VI. Current topics in photobiology and photochemistry Academic Press, New York, pp 131–177

Caldwell MM (1981) Plant response to solar ultraviolet radiation. In: Lange OL, Nobel PS, Osmond CB, Ziegler H (eds) Encyclopedia of plant physiology: physiological plant ecology. I. Responses to the physical environment vol 12A. Springer, Berlin Heidelberg New York, pp 169–181

Caldwell MM, Robberecht R, Flint SD (1983a) Internal filters: prospects for UV-acclimation in higher plants. Physiol Plant 58: 445–450

Caldwell MM, Gold WG, Harris G, Ashurst CW (1983b) A modulated lamp system for solar UV-B (280–320 nm). Supplementation studies in the field. Photochem Photobiol 37: 479–485

Caldwell MM, Camp LB, Warner CW, Flint SD (1986) Action spectra and their key role in assessing biological consequences of solar UV-B radiation change. In: Worrest RC, Caldwell MM (eds) Stratospheric ozone reduction, ultraviolet radiation and plant life. NATO ASI Ser, vol G8, Springer, Berlin Heidelberg New York, pp 87–111

Caldwell MM, Teramura AH, Tevini M (1989) The changing solar ultraviolet climate and the ecological consequences for higher plants. Trends Ecol Evol 4: 363–367

Campbell AK (1988) Chemiluminescence: principles and applications in biology and medicine. VCH, Weinheim, Germany

Campbell WJ, Allen LH Jr, Bowes G (1988) Effects of CO_2 concentration on rubisco activity, amount, and photosynthesis in soybean leaves. Plant Physiol 88: 1310–1316

Carns HR, Graham JH, Ravitz SJ (1978) Effects of UV-B radiation on selected leaf pathogenic fungi and on disease severity. EPA-IAG-D6–1068, BACER Program, Environmental Protection Agency, Washington

Cassman KG, Pingali PL (1995) Extrapolating trends from long-term experiments to farmers fields: the case of irrigated rice systems of Asia. In Barnett V, Payne R, Steiner R (eds) Agricultural sustainability in economic, environmental, and statistical terms. John Wiley, London (in press)

Cassman KG, De Datta SK, Olk DC, Alcantara J, Samson M, Descalsota J, Dizon M (1994) Yield decline and the nitrogen economy of long-term experiments on continuous, irrigated rice systems in the tropics. In: Lat R and Stewart RA (eds) Sustainable management of soils. Lewis Publisher, Chelsea, MI (in press)

Cen YP, Bornman JF (1993) The effect of exposure to enhanced UV-B radiation on the penetration of monochromatic and polychromatic UV-B radiation in leaves of *Brassica napus* Physiol Plant 87: 249–255

Cen YP, Björn LO (1994) Action spectra for enhancement of ultraweak luminescence by uv radiation (270–340nm) in leaves of *Brassica napus*. Photochem Photobiol B: Biol. 22: 125–129

Chemical and Engineering News (1986) The changing atmosphere. Nov 64, American Chemical Society, Washington, DC, pp 14–61

Chen D (1993) Population structure of *Pyricularia grisea* (Cooke) Sacc. at two screening sites in the Philippines and characterization of resistance genes. PhD Thesis, University of Philippines, Los Baños, Philippines

Chen DZ, Wang MX, Shangguan XJ, Huang J, Rasmussen RA, Khalil MKA (1993a) Methane emission from southwest China. Adv Earth Sci (in press)

Chen DZ, Wang MX, Shangguan XJ, Toelg M, Wassmann R, Xie XL, Wang WD, Shen RX, Wang YS, Papen H, Rennenberg H, Seiler W (1994) Methane emission from rice fields in Hunan (PR China) III: influence of water management. Biogeochemistry (in press)

Chen ZL, Li DB, Shao KS, Wang BJ (1993b) Features of CH_4 emission from rice fields in Beijing and Nanjing. Chemosphere 26: 239–245

Chen ZL, Shao KS, Li DB and Wang BJ (1993c) CH_4 emission from rice fields and its relation to agriculture management (pers. comm.) (in Chinese)
China Agriculture Yearbook (1986) Ministry of Agriculture, People's Republic of China, Agricultural Press, Beijing
China National Rice Research Institute (1988) A classification for rice planting in China. Zhejiang Scientific and Technical Publishing House, Hangzhou, China (in Chinese)
Chow WS, Strid A, Anderson JM (1992) Short-term treatment of pea plants with supplementary ultraviolet-B radiation: recovery time-courses of some photosynthetic functions and components. In: Murata N (ed) Research in photosynthesis vol IV. Kluwer Dordrecht, pp 361–364
Christy JR, McNider RT (1994) Satellite greenhouse signal. Nature 367: 325
Chumley FG, Valent B (1990) Genetic analysis of melanin-deficient, nonpathogenic mutants of *Magnaporthe grisea.* Mol Plant-Microbe Interact 3: 135–143
Cicerone RJ, Oremland RS (1988) Biogeochemical aspects on atmospheric methane. Global Biogeochem Cycles 2: 299–327
Cicerone RJ, Shetter JD (1981) Sources of atmospheric methane: measurements in rice paddies and a discussion. J Geophys Res 86: 7203–7209
Cicerone RJ, Shetter JD, Delwiche CC (1983) Seasonal variation of methane flux from a California rice paddy. J Geophys Res 88: 11022–11024
Cicerone RJ, Delwiche CC, Typer SC, Zimmermann PR (1992) Methane emissions from California rice paddies with varied treatments. Global Biogeochem Cycles 6(3): 233–248
Cohen JE, Schoenly KG, Heong KL, Justo H, Arida G, Barrion AT, Litsinger JA (1994) A food-web approach to evaluating the effect of insecticide spraying on insect pest population dynamics in a Philippine irrigated rice ecosystem. J Appl Ecol 31: 747–763
Conn EE, Stumpf PK (1972) Outlines of biochemistry. John Wiley, New York
Connell WE, Patrick WH Jr (1969) Reduction of sulfate to sulfide in waterlogged soil. Soil Sci Soc Am Proc 33: 711–715
Conrad R (1989) Control of methane production in terrestrial ecosystems. In: Andreae MO, Schimel DS (eds) Exchange of trace gases between terrestrial ecosystems and the atmosphere. John Wiley, Chichester, pp 39–58
Conrad R, Rothfuss F (1991) Methane oxidation in the soil surface layer of a flooded rice field and the effect of ammonium. Biol Fert Soils 12: 28–32
Conroy J, Hocking P (1993) Nitrogen nutrition of C_3 plants at elevated atmospheric CO_2 concentrations. Physiol Plant 89: 570–576
Correll DS, Johnston MS (1970) Manual of the vascular plants of Texas. Texas Research Foundation, Renner
Correll DL, Clark CO, Goldberg B, Goodrich VR, Hayes DR Jr, Klein WH, Schecher WD (1992) Spectral ultraviolet-B radiation fluxes at the earth's surface: long-term variations at 39°N, 77°W.J Geophys Res 97: 7579–7591
Cramer WP, Solomon AM (1993) Climatic classification and future global redistribution of agricultural land. Clim Res 3: 97–110
Crawford W (1984) Methane production in Minnesota peat lands. Appl Environ Microbiol 47: 1266–1271
Crawford MS, Chumley FG, Weaver CG, Valent B (1986) Characterization of the heterokaryotic and vegetative diploid phases of *Magnaporthe grisea.* Genetics 114: 1111–1129
Crill PM, Bartlett KB, Harris RC, Verry ES, Sebacher DI, Madzar L, Sanner W (1988) Methane flux from Minnesota peatlands. Global Biogeochem Cycles 2: 371–384
Crout JD, Symmank DG, Peterson GA (1965) Soil survey of Jefferson County, Texas, US Department of Agriculture, Soil Conservation Service in cooperation with the Texas Agricultural Experiment Station, Beaumont
Crozier CR, DeLaune RD, Patrick WH Jr (1994) Methane production in Mississippi Deltaic Plain wetland soils as a function of soil redox species. In: Lal R et al. (eds) Soil and global change. Lewis Publishers, Chelsea, MI (in press)
Crützen PJ (1976) Upper limits on atmospheric ozone reductions following increased application of fixed nitrogen to the soil. Geophys Res Lett 3: 169–172

Crützen PJ (1991) Methane's sinks and sources. Nature 350: 380–381
Cure JD (1985) Carbon dioxide doubling responses: a crop survey. In: Strain BR, Cure JD (eds) Direct effects of increasing carbon dioxide on vegetation. United States Department of Energy, Washington, pp 99–116
Cure JD, Acock B (1986) Crop responses to carbon dioxide doubling: a literature survey. Agric For Meteorol 38: 127–145
Curry RB, Peart RM, Jones JW, Boote KJ, Allen LH Jr (1990) Simulation as a tool for analyzing crop response to climate change. Trans Am Soc Agric Eng 33: 981–990
Curtis PS, Drake BG, Whigham DF (1989) Nitrogen and carbon dynamics in C_3 and C_4 estuarine marsh plants grown under elevated CO_2 in situ. Oecologia 78: 297–301
Dai AG, Wang MX, Shen RX, Schütz H, Rennenberg H, Seiler W, Wu HB (1991) Methane emission from a rice paddy in Hangzhou of China during growing season of late rice. Sci Atmos Sin 37: 102–110 (in Chinese with English Abstract).
Dai Q, Coronel VP, Vergara BS, Barnes PW, Quintos AT (1992) Ultraviolet-B radiation effects on growth and physiology of four rice cultivars. Crop Sci 32: 1269–1274
Davidson EA (1991) Fluxes of nitrous oxide and nitric oxide from terrestrial ecosystems. In Rogers JE, Whitman WB (eds) Microbial production and consumption of greenhouse gases: methane, nitrogen oxides, and halomethanes. American Society for Microbiology, Washington, pp 219–235
De Datta SK (1981) Principles and practices of rice production. John Wiley, New York
De Datta SK (1987) Advances in soil fertility research and nitrogen fertilizer management for lowland rice. In Efficiency of nitrogen fertilizer for rice. International Rice Research Institute, Los Baños, Philippines, pp 27–41
De Datta SK, Patrick WH Jr (eds) (1986) Nitrogen economy of flooded rice soils. Development in plant and soil sciences. Martin Nijhoff, Dordrecht
DeLaune RD, Smith EJ, Patrick WH Jr (1983) Methane release from Gulf Coast wetlands. Tellus 35B: 8–15
DeLucia EH, Sasek TW, Strain BR (1985) Photosynthetic inhibition after long-term exposure to elevated levels of carbon dioxide. Photosynth Res 7: 175–184
DeLuisi J, Wendell J, Kreiner F (1992) An examination of the spectral response characteristics of seven Robertson-Berger meters after long-term field use. Photochem Photobiol 56: 115–122
Delwiche CC, Cicerone RJ (1993) Factors affecting methane production under rice. Global Biogeochem Cycles 7: 143–155
Denier van der Gon HAC, Neue H-U, Lantin RS, Wassmann R, Alberto MCR, Aduna JB, Tan MJP (1992) Controlling factors of methane emission from rice fields. In: Batjes NH, Bridges EM (eds) World inventory of soil emission potentials. WISE Report 2, International Soil Reference and Information Centre (ISRIC), Netherlands, pp 81–92
Denmead OT (1983) Micrometeorological methods for measuring gaseous losses of nitrogen in the field. In: Freney JR, Simpson JR (eds) Gaseous loss of nitrogen from plant-soil systems. Martinus Nijhoff/Dr W Junk, The Hague, pp 133–157
Denmead OT (1993) Developments in flux measurements of greenhouse gases. J Agric Meteorol 48: 543–550
Denmead OT (1994) Measuring fluxes of CH_4 and N_2O between agricultural systems and the atmosphere. In: Minami K, Mosier A, Sass R (eds) CH_4 and N_2O: global emissions and controls from ricefields and other agricultural and industrial sources. National Institute of Agro-Environmental Sciences, Tsukuba, Japan, pp 209–234
Denmead OT, Raupach MR (1993) Methods for measuring atmospheric gas transport in agricultural and forest systems. In: Harper LA, Mosier AR, Duxbury JM, Rolston DE (eds) Agricultural ecosystem effects on trace gases and global climate change. ASA Spec Publ 55, American Society of Agronomy, Madison, WI, pp 19–43
Denmead OT, Freney JR, Simpson JR (1979) Studies of nitrous oxide emission from a grass sward. Soil Sci Soc Am J 43: 726–728

Denmead OT, Dunin FX, Wong SC, Greenwood EAN (1993) Measuring water use efficiency of Eucalyptus trees with chambers and micrometeorological techniques. J Hydrol 150: 649–664

Dlugokencky EJ, Masaire KA, Lang PM, Tans PP, Steele LP, Nisbet EG (1994) A dramatic decrease in the growth rate of atmospheric methane in the northern hemisphere during 1992. Geosphys Res Lett 21: 45–48

Domingo I, Heong KL (1992) Evaluating high temperature tolerance in the brown planthopper (BPH). Int Rice Res Newsl 17(3): 22

Dornbusch R, Poterba JM (1991) Global warming: economic policy responses. MIT Press, Cambridge, MA

Downs RJ, Hellmers H (1978) Controlled climate and plant research. World Meteorological Organization Technical Note 148, Academic Press, Geneva

Drake BG, Leadley PW, Arp WJ, Nassiry D, Curits PS (1989) An open-top chamber for field studies of elevated atmospheric CO_2 concentration on salt-marsh vegetation. Funct Ecol 3: 363–371

Drauschke G, Neumann W (1992) Investigation on influences of toxic substances on microbial methane production from cattle wastes. Zentralbl Mikrobiol 147: 308–318

Driessen PM (1986) Soil data, a matter of concern. In Annual Report, International Soil Museum [now International Soil Reference and Information Centre (ISRIC)], Wageningen, pp 43–47

Dumpert K, Knacker T (1985) A comparison of the effects of enhanced UV-B radiation on some crop plants exposed to greenhouse and field conditions. Biochem Physiol Pflanz 180: 599–612

Dunfield P, Knowles R, Dumont R, Moore TR (1993) Methane production and consumption in temperate and subarctic peat soils: response to temperature and pH. Soil Biol Biochem 25: 321–326

Duxbury JM, Harper LA, Mosier AR (1993) Contributions of agroecosystems to global climate change. In: Harper LA, Mosier AR, Duxbury JM, Rolston DE, Peterson GA, Baenziger PS, Luxmoore RJ, Kral DM, Bartels JM (eds) Agricultural ecosystem effects on trace gases and global climate change. ASA Spec Publ 55, American Society of Agronomy, Madison, WI, pp 1–18

Easterling WE, Rosenberg NJ, McKenney MS, Jones CA (1992) An introduction to the methodology, the region of study, and a historical analog of climate change. Agric For Meteorol 59: 3–15

Easterling WE, Crosson RP, Rosenberg NJ, McKenney MS, Katz LA, Lemon KM (1993) Agricultural impacts of and responses to climate change in the Missouri-Iowa-Nebraska-Kansas (MINK) region. Clim Change 24: 23–61

Egeh AO, Ingram KT, Zamora OB (1994) High temperature effects on nitrogen and carbohydrate accumulation in rice grains. Phillipp J Crop Sci (in press)

Ehleringer J, Pearcy RW (1983) Variation in quantum yield for CO_2 uptake among C_3 and C_4 plants. Plant Physiol 73: 555–559

Eichner MJ (1990) N_2O emissions from fertilized soils: summary of available data. J Environ Qual 19: 272–286

Ensminger PA (1993) Control of development in plants and fungi by far-UV radiation. Physiol Plant 88: 501–508

Esser G (1987) Sensitivity of global carbon pools and fluxes to human and potential climatic impacts. Tellus 39B: 245–260

Fan SM, Wofsy SC, Bakwin PS, Jacob DI, Anderson SM, Kebabian PL, McManus JB, Kolb CE (1992) Micrometeorological measurements of CH_4 and CO_2 exchange between the atmosphere an subarctic tundra. J Geophys Res 97: 16,627–16,643

FAO – Food and Agricultural Organization (1981) Report on the agroecological zones project. World Soil Resour Rep 48: 1–4

FAO – Food and Agricultural Organization (1991) The digitized soil map of the world. Notes. Vol 1: Africa (release 1.0). World Soil Res Rep 67/1, Food and Agricultural Organization, Rome

FAO-ISRIC – Food and Agricultural Organization – International Soil Reference and Information Centre (1989) FAO-ISRIC soil data base (SDB). World Soil Resour Rep 64, Food and Agricultural Organization, Rome
FAO-UNESCO – Food and Agricultural Organization – United Nations Educational, Scientific and Cultural Organisation (1971–1981) Soil map of the world, 1: 5 M, Vol 1–10, United Nations Educational, Scientific and Cultural Organisation, Paris
Farquhar GD, von Caemmerer S (1982) Modelling of photosynthetic response to environment. In: Lange OL, Nobel PS, Osmond CB, Ziegler H (eds) Encyclopedia of plant physiology, NS Vol 12B, Physiological Plant Ecology II. Springer, Berlin Heidelberg New York
Farquhar GD, von Caemmerer S, Berry JA (1980) A biochemical model of photosynthetic CO_2 assimilation in leaves of C_3 species. Planta 149: 78–90
Fillery IRP, Vlek PLG (1986) Reappraisal of the significance of ammonia volatilization as a N loss mechanism in flooded rice fields. In: De Datta SK, Patrick WH Jr (eds) Development in plant and soil sciences. Martin Nijhoff, Dordrecht, pp 79–98
Finney DJ (1978) Statistical method in biological assay. Charles Griffin, London
Firestone MK (1982) Biological denitrification. In: Stevenson FJ (ed) Nitrogen in agricultural soils. American Society of Agronomy, Madison, WI, pp 289–326
Fiscus EL, Booker FL (1993) UVB: a modified version of the Green/Björn and Murphy (1985) model. Ver. 3.0. PC and Macintosh applications. United States Department of Agriculture, Agricultural Research Service, Raleigh, NC
Fiscus EL, Miller JE, Booker FL (1994) Is UV-B a hazard to soybean photosynthesis and yield? Results of an ozone/UV-B interaction study and model predictions. In: Bigg RH Joyner MEB (eds) Stratospheric ozone depletion/UV-B radiation in the biosphere. NATO ASI Ser vol I 18. Springer, Berlin Heidelberg New York, pp 135–147
Flessa H, Fischer WR (1992) Plant-induced changes in the redox potential of rice rhizosphere. Plant Soil 143: 55–60
Frederick JE (1993) Ultraviolet sunlight reaching the earth's surface: a review of recent research. Photochem Photobiol 57: 175–178
Frederick JE, Snell HE (1990) Tropospheric influence on solar ultraviolet radiation: the role of clouds. J Climate 3: 373–381
Frederick JE, Weatherhead EC (1992) Temporal changes in surface ultraviolet radiation: a study of the Robertson-Berger meter and Dobson data records. Photochem Photobiol 56: 123–131
Frederick JE, Weatherhead EC, Haywood EK (1991) Long-term variations in ultraviolet sunlight reaching the biosphere: calculation for the past three decades. Photochem Photobiol 54: 781–788
Freney JR, Denmead OT, Watanabe I, Craswell ET (1981) Ammonia and nitrous oxide losses following applications of ammonium sulfate to flooded rice. Aust J Agric Res 32: 37–45
Freney JR, Trevitt ACF, De Datta SK, Obcemea WN, Real JG (1990) The interdependence of ammonia volatilization and denitrification as nitrogen loss processes in flooded rice fields in the Philippines. Biol Fert Soils 9: 31–36
Frenzel P, Rothfuss F, Conrad R (1992) Oxygen profiles and methane turnover in a flooded microcosm. Biol Fert Soils 14: 84–89
Fung I, John J, Lerner J, Matthews E, Prather M, Steele LP, Fraser PJ (1991) Three-dimensional model synthesis of the global methane cycle. J Geophys Res 96: 13033–13065
Galle B, Klemedtsson L, Griffith DW (1994) Application of an FTIR system for measurements of nitrous oxide fluxes using micrometeorological methods, an ultra-large chamber system and conventional field chambers. J Geophys Res 99: 16575–16583
Gao Guodong, Lu Yurong (1982) Radiation and heat balance on the ground surface in China. Scientific and Technological Publishing House, Beijing
Gao Liangzhi, Li Lin, Jin Zhiqing (1987) A classification for rice production in China. Agric For Meteorol 39: 55–65
Garcia JL, Raimbault M, Jacq V, Rinaudo G, Roger P (1974) Activities microbiennes dans les sols de rizières du Sénégal: relations avec les propriétés physicochimiques et influence de la rhizosphère. Rev Ecol Biol 11(2): 169–185

Gay LW, Greenberg RJ (1985) The AZET battery-powered Bowen ratio system. In: Proc 17th Conf on Agriculture and Forest Meteorology. American Meteorological Society, Boston, MA, pp 181–182
GEIA – Global Emission Inventory Activity (1993) In: Bouwman AF (ed) Rep 3rd workshop, Amersford, 31 January–2 February 1993. RIVM-Pep. 481507002, Bilthoven, The Netherlands
Genovesi AD, Magill CW (1976) Heterokaryosis and parasexuality in *Pyricularia oryzae* Carvar. Can J Microbiol 22: 531–536
George T, Ladha JK, Buresh RJ, Garrity DP (1993) Nitrate dynamics during the aerobic soil phase in lowland rice-based cropping systems. Soil Sci Soc Am J 57: 1526–1532
Gibson JH (1992) Criteria for status-and-trends monitoring of ultraviolet (UV) radiation. Natural Resource Ecology Laboratoy, Colorado State University, Fort Collins, CO
Gifford RM (1979) Growth and yield of carbon dioxide-enriched wheat under water-limited conditions. Aust J Plant Physiol 6: 367–378
Gleason JF, Bhartia PK, Herman JR, McPeters R, Newman P, Stolarski RS, Flyn L, Labow G, Larko D, Seftor C, Wellemeyer C, Komhyr WD, Miller AJ, Planew W (1993) Record low global ozone in 1992. Science 260: 523–526
Godwin DC, Singh U, Buresh RJ, De Datta SK (1990) Modeling of nitrogen dynamics in relation to rice growth and yield. In: Trans 14th Int Congr Soil Science, Kyoto, Japan, pp 320–325
Gold WG, Caldwell MM (1983) The effects of ultraviolet-B radiation on plant competition in terrestrial ecosystems. Physiol Plant 58: 435–444
Goudriaan J (1982) Some techniques in dynamic simulation. In: Penning de Vries FWT, van Laar HH (eds) Simulation of plant growth and crop production. Simulation Monographs, Pudoc, Wageningen, pp 66–84
Goudriaan J (1990) Atmospheric CO_2, global carbon fluxes and the biosphere. In: Rabbinge et al. (eds) Theoretical production ecology: reflections and prospects. Simulation Monographs 34, Pudoc, Wageningen, pp 17–40
Goudriaan J (1992) Biosphere structure, carbon sequestering potential and the atmospheric ^{14}C carbon record. J Exp Bot 43: 1111–1119
Goudriaan J, Ketner P (1984) A simulation study for the global carbon cycle, including man's impact on the biosphere. Clim Change 6: 167–192
Goudriaan J, Unsworth M H (1990) Implications of increasing carbon dioxide and climate change for agricultural productivity and water resources. In: Kimball BA, Rosenberg NJ, Allen LH Jr (eds) Impact of carbon dioxide, trace gases, and climate change on global agriculture. ASA Spec Publ No 353. American Society of Agronomy, Madison, WI
Green AES (1983) The penetration of ultraviolet radiation to the ground. Physiol Plant 58: 351–359
Green AES, Cross KR, Smith LA (1980) Improved analytic characterization of ultraviolet skylight. Photochem Photobiol 31: 59–65
Gregson K, Hector DJ, McGowan M (1987) A one-parameter model for the soil water characteristic. J Soil Sci 38: 483–486
Hall DO, Scurlock JMO (1991) Climate change and productivity of natural grasslands. Ann Bot 67: 49–55
Hammer GL, Vanderlip RL, Gibson G, Wade LJ, Henzell RG, Younger DR, Warren J, Dale AB (1989) Genotype-by-environment interaction in grain sorghum. II. Effects of temperature and photoperiod on ontogeny. Crop Sci 29: 376–384
Hansen J, Russell G, Rind D, Stone P, Lacis A, Lebedeff S, Ruedy R, Travis L (1983) Efficient three-dimensional global models for climate studies: Models I and II. Mon Weather Rev III (4): 609–662
Hansen J, Lacis A, Rind D, Russell G, Stone P, Fung I, Ruedy R, Lerner J (1984) Climate sensitivity: analysis of feedback mechanisms. In: Hansen J, Takahashi T (eds) Climate processes and climate sensitivity. Maurice Ewing Ser 5. American Geophysical Union, Washington, pp 130–163
Hansen J, Fung I, Lacis A, Lebedeff S, Rind D, Ruedy R, Russell G, Stone P (1988)

Global climate changes as forecast by the Goddard Institute for Space studies three-dimensional model. J Geophys Res 93: 9341–9364
Hanson RS (1980) Ecology and diversity of methylotrophic organisms. Adv Appl Microbiol 26: 3–40
Hargreaves KJ, Skiba U, Fowler D, Arah J, Wienhold F, Klemedtsson L, Galle B (1994) Measurements of nitrous oxide emission from fertilised grassland using micrometeorological techniques. J Geophys Res 99: 16569–16574
Harley PC, Weber JA, Gates DM (1985) Interactive effects of light, leaf temperature, CO_2, and O_2 on photosynthesis in soybean. Planta 165: 249–263
Harris HC, Goebel W (1987) Objectives, expectations and needs of the International Agricultural Research Centres. In: Bunting AH (ed) Agricultural environments: characterization, classification, and mapping. Commonwealth Agricultural Bureaux, Farnham Royal, UK, pp 5–10
Harrison WH (1914) The gases of swamp rice soils. II. Their utilization for the aeration of roots of the crop. Mem Dep Agric India 4: 1–17
Hassell MP, Godfray HCJ, Comins HN (1993) Effects of global change on the dynamics of insect host-parasitoid interactions. In: Kareiva PM, Kingsolver JG, Huey RB (eds) Biotic interactions and global change. Sinauer, Sunderland, MA, pp 402–423
Haun JR (1973) Visual quantification of wheat development. Agron J 65: 116–119
He J, Huang LK, Chow WS, Whitecross MI, Anderson JM (1993) Effects of supplemental ultraviolet-B radiation on rice and pea plants. Aust J Plant Physiol 20: 129–142
Heagle AS, Body DE, Heck WW (1973) An open top field chamber to assess the impact of air pollution on plants. J Environ Qual 2: 365–368
Heagle AS, Philbeck RB, Ferrell RE, Heck WW (1989) Design and performance of a large, field exposure chamber to measure effects of air quality on plants. J Environ Qual 18: 361–368
Henderson-Sellers A (1992) Decision support system prototypes for greenhouse advice. Ecol Model 6: 241–260
Heong KL, Domingo I (1992) Shifts in predator-prey ranges in response to global warming. Int Rice Res Newsl 17(6): 29–30
Heong KL, Sogawa K (1994) Management strategies for key insect pests or rice – critical issues. In: Teng PS, Heong KL, Moody K (eds) Advances in rice pest management. International Rice Research Institute, Los Baños, Philippines, pp 3–14.
Herold A (1980) Regulation of photosynthesis by sink activity: the missing link. New Phytol 86: 131–144
Hideg E (1993) On the spontaneous ultraweak light emission of plants. J Photochem Photobiol 18: 239–244
Hileman DR, Bhattacharya NC, Ghosh PP, Biswas PK, Allen LH Jr, Lewin KF, Hendry GR (1992a) Distribution of carbon dioxide within and above a cotton canopy growing in the FACE system. Crit Rev Plant Sci 11(2–3): 187–194
Hileman DR, Ghosh PP, Bhattacharya NC, Biswas PK, Allen LH Jr, Peresta G, Kim ball BA (1992b) A comparison of the uniformity of an elevated CO_2 environment in three different types of open-top chambers. Crit Rev Plant Sci 11(2–3): 195–202
Hodges T (1991) Temperature and water stress effects on phenology. In: Hodges T (ed) Predicting crop phenology CRC Press, Boca Raton, FL, pp 7–13
Hoffman AA, Blows MW (1993) Evolutionary genetics and climatic change: will animals adapt to global warming? In: Kareiva PM, Kingsolver JG, Huey RB (eds) Biotic interactions and global change. Sinauer, Sunderland, MA, pp 165–178
Hogan KB (ed) (1993) Current and future CH_4 emissions from natural sources (Rep to Congress). United States Environmental Protection Agency, Washington
Hogan KP, Smith AP, Ziska LH (1991) Potential effects of elevated CO_2 and changes in temperature on tropical plants. Plant Cell Environ 14: 763–778
Holdridge LR (1967) Life zone ecology. Tropical Science Center, San José, CA
Holzapfel-Pschorn A, Seiler W (1986) Methane emission during a cultivation period from an Italian rice paddy. J Geophys Res 91: 11803–11814
Holzapfel-Psch A, Conrad R, Seiler W (1985) Production, oxidation and emission of methane in rice paddies. FEMS Microbiol Ecol 31: 343–351

Holzapfel-Pschorn A, Conrad R, Seiler W (1986) Effects of vegetation on the emission of methane from submerged paddy soil. Plant Soil 92: 223–233
Hoogenboom G, Boote KJ, Jones JW, Getz RR (1990) Modeling the effect of temperature on peanut production in the Southeast. SE Climate Symposium 1990. Global Change: A Southern Perspective, Charleston, South Carolina
Horie T (1987) A model for evaluating climatic productivity and water balance of irrigated rice and its application to Southeast Asia. Southeast Asian Stud 25: 62–74
Horie T (1988) The effects of climatic variations on agriculture in Japan. 5. The effects on rice yields in Hokkaido. In: Parry ML, Carter TR, Konjin NT (eds) The impact of climatic variations on agriculture. Vol 1: Assessments in cool temperature and cold regions. Kluwer, Dordrecht, pp 809–826
Horie T (1991) Model analysis of the effect of climatic variation on rice yield in Japan. In: Geng S, Cady CW (eds) Climatic variation and change: implications for agriculture in the pacific rim. University of California, Davis, CA
Horie T (1993) Predicting the effects of climatic variation and effect of CO_2 on rice yield in Japan. J Agric Meteorol (Tokyo) 48: 567–574
Horie T (1994) Predicting the effects of increased temperature and elevated CO_2 on rice yields in Japan. In Matthews RB, Kropff MJ, Thiyagarajan TM, Kareem AA (eds) The development, testing and application of crop models simulating the potential production of rice. Proc Workshop held in Tamil Nadu, India, 25–28 January 1993. SARP Res Proc
Horie T, Nakagawa H (1990) Modelling and prediction of developmental processes in rice. I. Structure and method of parameter estimation of a model for simulating developmental processes toward heading. Jpn J Crop Sci 59: 687–695
Horie T, de Wit CT, Goudriaan J, Bensink J (1979) A formal template for the development of cucumber in its vegetative stage. Proc KNAW, Ser C, No 82: 433–479
Horie T, Nakano J, Nakagawa H, Wada K, Kim HY, Seo T (1991) Effects of elevated CO_2 and high temperature on growth and high temperature on growth and yield of rice. 1. Development of temperature gradient tunnels. Jpn J Crop Sci 60 (Extra Issue 2): 127–128
Horie T, Yajima M, Nakagawa H (1992) Yield forecasting. Agric Syst 40: 211–236
Horie T, Kropff MJ, Centeno HG, Nakagawa H, Kim HY, Onishi M (1994) Effect of anticipated global environment change on rice yield in Japan. In Peng et al. (eds) Climate change and rice. International Rice Research Institute, Los Baños, Phillippines
Horino T (1990) Mineral contents and the eating quality of rice grain. Jpn J Crop Sci 59: 605–611 (in Japanese)
Hotta I, Imai K (1994) Effects of atmospheric CO_2 concentration on photosynthesis, respiration and growth of young rice plants. Jpn J Crop Sci 63 (Extra Issue 1): 124–125 (in Japanese)
Houghton JT, Callander BA, Varney SK (eds) (1992) Climate change 1992. The supplementary report to the IPCC scientific assessment. Cambridge University Press, Cambridge
Huey RB, Partridge L, Fowler K (1991) Thermal sensitivity of *Drosophila melanogaster* responds rapidly to laboratory natural selection. Evolution 45: 751–756
Huke RE, Huke E, Woodhead T, Huang J (1993) Rice-wheat atlas of China. International Rice Research Institute, Los Baños, Philippines
Hutchinson GL, Livingston GP (1993) Use of chamber systems to measure trace gas fluxes. In: Harper LA, Mosier AR, Duxbury JM, Rolston DE (eds) Agricultural ecosystems effects on trace gases and global climate change. ASA Spec Publ 55. American Society of Agronomy, Madison, WI, pp 63–78
IAEA – International Atomic Energy Agency (1992) Manual of measurements of methane and nitrous oxide emissions from agriculture. Vienna
IBSNAT – International Benchmark Sites Network for Agrotechnology Transfer (1988) Experimental design and data collection procedures for IBSNAT (3rd edn). Technical Report 1. Department of Agronomy and Soils, University of Hawaii, Honolulu, Hawaii

IBSNAT – International Benchmark Sites Network for Agrotechnology Transfer (1989) Decision support system for agrotechnology transfer version 2.1 (DSSAT V2.1). Department Agronomy and Soil Soils, University of Hawaii, Honolulu, Hawaii

Idso SB (1990) Interactive effects of carbon dioxide and climate variables on plant growth. In: Kimball BA (ed) Impact of carbon dioxide, trace gases and climate change on global agriculture. ASA Spec Publ 53. American Society of Agronomy, Madison, WI, pp 61–69

Idso SB, Kimball BA, Anderson MG, Mauney JR (1987) Effects of atmospheric CO_2 enrichment on plant growth: the interactive role of air temperature. Agric Ecosyst Environ 20: 1–10

Imai K (1977) Studies on the effect of carbon dioxide concentration on growth and dry matter production of crop plants. Dissertation, University of Tokyo, Tokyo (in Japanese)

Imai K (1988) Carbon dioxide and crop production. Jpn J Crop Sci 57: 380–391 (in Japanese)

Imai K (1993) Global environmental change and field crops. Environ Control Biol 31: 129–137 (in Japanese)

Imai K, Coleman DF (1983) Elevated atmospheric partial pressure of carbon dioxide and dry matter production of konjak (*Amorphophallus conjac* K. Koch). Photosynth Res 4: 331–336

Imai K, Murata Y (1976) Effect of carbon dioxide concentration on growth and dry matter of crop plants. I. Effects on leaf area, dry matter, tillering, dry matter distribution ratio, and transpiration. Jpn J Crop Sci 45: 598–606 (in Japanese, English summary)

Imai K, Murata Y (1977) Effect of carbon dioxide concentration on growth and dry matter production of crop plants. II. Specific and varietal differences in the response of dry matter production. Jpn J Crop Sci 46: 291–297 (in Japanese, English summary)

Imai K, Murata Y (1978a) Effect of carbon dioxide concentration on growth and dry matter production of crop plants. III. Relationship between CO_2 concentration and nitrogen nutrition in some C_3 and C_4 species. Jpn J Crop Sci 47: 118–123 (in Japanese, English summary)

Imai K, Murata Y (1978b) Effect of carbon dioxide concentration on growth and dry matter production of crop plants. IV. After-effects of carbon dioxide-treatments on the apparent photosynthesis, dark respiration and dry matter production. Jpn J Crop Sci 47: 330–335 (in Japanese, English summary)

Imai K, Murata Y (1978c) Effect of carbon dioxide concentration on growth and dry matter production of crop plants. V. Analysis of after-effect of carbon dioxide-treatment on apparent photosynthesis. Jpn J Crop Sci 47: 587–595

Imai K, Murata Y (1979a) Effect of carbon dioxide concentration on growth and dry matter production of crop plants. VI. Effect of oxygen concentration on the carbon dioxide-dry matter production relationship in some C_3 and C_4 species. Jpn J Crop Sci 48: 58–65

Imai K, Murata Y (1979b) Effect of carbon dioxide concentration on growth and dry matter partitioning of crop plants. VII. Influence of light intensity and temperature on the effect of carbon dioxide environment in some C_3 and C_4 species. Jpn J Crop Sci 48: 409–417

Imai K, Nomura M (1992) Effects of phosphorus nutrition on gas exchanges of rice leaves at elevated atmospheric partial pressure of carbon dioxide. In: Proc Int Symp on Global Change (IGBP), Waseda University, Tokyo, pp 573–578

Imai K, Okamoto-Sato M (1991) Effects of temperature on CO_2 dependence of gas exchanges in C_3 and C_4 crop plants. Jpn J Crop Sci 60: 139–145

Imai K, Yoshida T, Takahashi M, Kanda A, Murata Y (1983) Effect of temeprature on dry matter production in young plants of cassava and rice. Jpn J Trop Agric 27: 107–113

Imai K, Coleman DF, Yanagisawa T (1984) Elevated atmospheric partial pressure of carbon dioxide and dry matter production of cassava (*Manihot esculenta* Crantz). Jpn J Crop Sci 53: 479–485

Imai K, Colman DF, Yanagisawa T (1985) Increase of atmospheric partial pressure of carbon dioxide and growth and yield or rice (*Oryza sativa* L.). Jpn J Crop Sci 54: 413–418

Imai K, Adachi N, Moss DN (1994) Effects of atmospheric CO_2 concentration, phosphorus nutrition and night temperature on growth and yield of rice. Environ Control Biol 32: 53–60

Ingram JSI (ed) (1993) Report of the IGBP-DIS/GCTE Global Soils Database Workshop, 8–9 October 1992, Silsoe

Insitute of Soil Science, Academic Sinica (1986) The soil atlas of China. Cartographic Publishing House, Beijing

Inubushi K, Hori K, Matsumoto S, Umebayashi M, Wada H (1989) Methane emission from the flooded rice soil to the atmosphere through rice plant. Jpn J Soil Sci Plant Nutr 60: 318–324 (in Japanese, English summary)

Inubushi K, Muramatsu Y, Taja D, Umebayashi M (1990a) Control of CH_4 emission from paddy soil. In: Proceedings 14th Int Cong Soil Science, Aug 12–18, Kyoto, Japan

Inubushi K, Taja D, Obata H, Umebayashi M (1990b) Suppression of methane emission from ricefield by heavy metal. Research related to the UNESCO's Man and the Biosphere Program in Japan 1989–1990

Inubushi K, Umebayashi M, Wada H (1990c) Methane emission from paddy fields. In: Proceedings 14th Int Cong Soil Science Aug 12–18, Kyoto, Japan pp 249–254

Inubushi K, Muramatsu Y, Umebayashi M (1992) Influence of percolation on methane emission from flooded paddy soil. Jpn J Soil Sci Plant Nutr 63: 184–189

IPCC – Intergovernmental Panel on Climate Change (1990) Introduction. In: Houghton JT, Jenkins GJ, Ephraums JJ (eds) Climate change: the IPCC scientific assessment. Cambridge University Press, Cambridge

IPCC – Intergovernmental Panel on Climate Change (1992) In: Houghton JT, Callandar BA, Varney SK (eds) Climate change 1992: the supplementary report to the IPCC scientific assessment. Cambridge University Press, Cambridge

IRRI – International Rice Research Institute (1989) IRRI towards 2000 and beyond. PO Box 933, Manila, Philippines

IRRI – International Rice Research Institute (1991) World Rice Statistics 1990. PO Box 933, Manila, Philippines

IRRI – International Rice Research Institute (1993) IRRI rice almanac. PO Box 933, Manila, Philippines

IRRI – International Rice Research Institute (1994) Program report for 1993. PO Box 933, Manila, Philippines

Jagger J (1985) Solar-UV actions on living cells. CBS, New York

Jagtap SS, Jones JW (1989) Evapotranspiration model for developing crops. Trans ASAE 32: 1342–1350

Jakobsen P, Patrick WH Jr, Williams BG (1981) Sulfide and methane formation in soils and sediments. Soil Sci 132: 279–287

Jansen DM (1990) Potential rice yields in future weather conditions in different parts of Asia. Neth J Agric Sci 38: 661–680

Japan Environmental Agency – US Environmental Protection Agency (1990) Methane emissions and opportunities for control. Workshop on Intergovernmental Panel on Climate Change (IPCC), Washington

Japanese Ministry of Agriculture, Forestry and Fisheries (various years) Crop statistics. MAFF, Tokyo (in Japanese)

Jenkinson DS, Adams DE, Wild A (1991) Model estimates of CO_2 emissions from soil in response to global warming. Nature 351: 304–306

Jermsawatdipong P, Murase J, Prabuddham P, Hasathon Y, Chinda M, Chaiwatana B, Khomthong N, Sattawatananon S, Naklang K, Watanabe A, Haraguchi H, Kimura M (1994) Methane emission from plots with differences in fertilizer application in Thai rice fields. Soil Sci Plant Nutr 40: 63–71

Jin Z, Chen H, Ge D, Fang J (1990) Using a hybrid between GCM outputs and historic weather data to generate China's regional climate scenarios with doubled CO_2, 1992 Chin Agrometeorol 13(5): 13–20 (in Chinese)

Jin Z, Ge D, Chen H, Zheng X (1994) Assessing impacts of climate change on rice production and strategies for adaptation in southern China. In: Peng et al. (eds) Climate change and rice. International Rice Research Institute, Los Baños, Philippines

Johnson DE, Hill TM, Ward GM, Johnson ME, Branine M E, Carmean BR, Lodman, DW (1993) Ruminants and other animals. In: Khalil MAK (ed) Atmospheric methane: sources and sinks, and role in global change. NATO-ASI Ser I, vol 13 Springer, Berlin Heidelberg New York, pp 199–229

Johnson FS, Mo T, Green AES (1976) Average latitudinal variation in ultraviolet radiation at the earth's surface. Photochem Photobiol 23: 179–188

Jones CA, Kiniry J (1986) CERES-Maize: a simulation model of maize growth and development. Texas A&M University Press, College Station, TX

Jones JW, Ritchie JT (1990) Crop growth models. In: Hoffman GJ, Howell TA, Solomon KH (eds) Management of farm irrigation systems. The American Society of Agricultural Engineers. St Joseph, MI

Jones JW, Boote KJ, Hoogenboom G, Jagtap SS, Wilkerson GG (1989) SOYGRO V 5.42: Soybean crop growth simulation model. User's guide. Florida Agric Exp Station Journal No 8304. Institute of Food and Agricultural Sciences, University of Florida, Gainesville

Jones JW, Jagtap SS, Hoogenboom G, Tsuji GY (1990) The structure and function of DSSAT. In: Part I. Proc IBSNAT Symp: Decision Support System for Agrotechnology Transfer, Las Vegas, Nevada, 16–18 Oct 1989. Department of Agronomy and Soil Science, College of Tropical Agriculture and Human Resources, University of Hawaii, Honolulu, Hawaii, pp 1–14

Jones P (1993) Is climate change occurring? Evidence from the insturmental record. In: Kaiser HM, Drennen TE (eds) Agricultural dimensions of global climate change. St Lucie Press, Delray Beach, FL

Jones P, Jones JW, Allen LH Jr, Mishoe JW (1984) Dynamic computer control of closed environment plant growth chambers. Design and verification. Trans ASAE 27: 879–888

Jones P, Allen LH Jr, Jones JW (1985a) Responses of soybean canopy photosynthesis and transpiration to whole-day temperature changes in different CO_2 environments. Agron J 77: 242–249

Jones P, Allen LH Jr, Jones JW, Valle R (1985b) Photosynthesis and transpiration responses of soybean canopies to short-term and long-term CO_2 treatments. Agron J 77: 119–126

Jones P, Jones JW, Allen LH Jr (1985c) Seasonal carbon and water balances of soybeans grown under stress treatments in sunlit chambers. Trans ASAE 28: 2021–2028

Jordan BR, He J, Chow WS, Anderson JM (1992) Changes in mRNA levels and polypeptide subunits of ribulose 1,5-bisphosphate carboxylase in response to supplementary ultraviolet-B radiation. Plant Cell Environ 15: 91–98

Jordan DB, Ogren WL (1984) The CO_2/O_2 specificity of ribulose 1,5-bisphosphate carboxylase/oxygenase. Dependance on ribulosebisphosphate concentration, pH and temperature. Planta 161: 308–313

Kaiser HM, Riha SJ, Rossiter DG, Wilks DS (1992) Agronomic and economic impacts of gradual global warming: a preliminary analysis of midwestern crop farming. In: Reilly JM, Anderson M (eds) Economic issues in global climate change: agriculture, forestry and natural resources. Westview Press, Boulder, pp 91–116

Kanda A, Imai K, Moriya T, Handa S, Terajima T (1984) A gas exchange measurement system and its application to rice and cassava leaves. Jpn J Crop Sci 53: 472–478

Kanemasu ET, Flitcroft ID, Shah TDH, Nie D, Thurtell GW, Kidd G, Simpson I, Lin M, Neue H-U, Bronson K (1994) Comparison of micrometeorological and chamber measurements of methane emissions from rice fields. In: Peng et al. (eds) Climate change and rice. International Rice Research Institute, Los Baños, Philippines

Kareiva PM, Kingsolve JG, Huey RB (eds) (1993) Biotic interactions and global change. Sinauer Associates, Sunderland, MA

Katyal JC, Carter MF, Vlek PLG (1988) Nitrification activity in submerged soils and its relation to denitrification loss. Biol Fert Soils 7: 16–22

Kawaguchi K, Kyuma K (1977) Paddy soils in tropical Asia: their material nature and fertility. The University Press of Hawaii, Honolulu, Hawaii
Keeling CD, Bacastow RB, Carter AF, Piper SC, Whorf TP, Heinmann M, Mook WG, Roeloffzen H (1989) A three dimensional model of atmospheric CO_2 transport based on observed winds: analysis of data. In: Peterson DH (ed) Aspects of climate variability in the Pacific and the Western Americas. Geophysical Monograph 55, American Geophysical Union, Washington, pp 165–236
Keeney DR, Sahrawat KL (1986) Nitrogen transformations in flooded rice soils. Fert Res 9: 15–38
Keerthisinghe DG, Freney JR, Mosier AR (1993) Effect of wax-coated calcium carbide and nitrapyrin on nitrogen loss and methane emission from dry-seeded flooded rice. Biol Fert Soils 16: 71–75
Kennedy BC, Sharp WE (1992) A validation study of the Robertson-Berger meter. Photochem Photobiol 56: 133–141
Kerr JB, McElroy CT (1993) Evidence for large upward trends of ultraviolet-B radiation linked to ozone depletion. Science 262: 1032–1034
Khalil MAK (1993) Methane emissions from rice fields. In: Van Amstel AR (ed) Proc int IPCC workshop on methane and nitrous oxide: methods in national emission inventories and options for control. National Institute of Public Health and Environmental Protection (RIVM), Bilthoven, The Netherlands, pp 239–244
Khalil MAK, Rasmusen RA (1987) Atmospheric methane: trends over the last 10,000 years. Atmos Environ 21: 2445–2452
Khalil MAK, Rasmussen RA (1992) The global sources of nitrous oxide. J Geophys Res 97: 14651–14660
Khalil MAK, Rasmussen RA (1993) Decreasing trend of methane: unpredictability of future concentrations. Chemosphere 26: 803–814
Khalil MAK, Rasmussen RA, Wang MX, Ren LX (1991) Methane emission from rice fields in China. Environ Sci Technol 25: 979–981
Kim HY, Horie T, Nakagawa H, Wada K, Seo T (1992) Effects of elevated CO_2 concentration and high temperature on growth and nitrogen use efficiency of rice. In: Proc 1st Asian crop science conference: crop production and improvement technology in Asia. Korean Society of Crop Science Guh Mog Publishing Co, Seoul, Korea, pp 205–211
Kim HY, Horie T, Nakagawa H, Wada K, Seo T (1994) Effects of elevated CO_2 concentration and high temperature on growth and yield of rice. Jpn J Crop Sci 63 (extra issue 1)
Kimball BA (1983) Carbon dioxide and agricultural yield: an assemblage and analysis of 430 prior observations. Agron J 75: 779–788
Kimura M (1992) Methane emission from paddy soils in Japan and Thailand. In: Batjes NH, Bridges EM (eds) World inventory of soil emission potentials. WISE Rep 2, International Soil Reference and Information Centre (ISRIC), Wageningen, Netherlands, pp 43–79
Kimura M, Wada H, Takai Y (1982) Effects of direct sowing cultivation on the rhizosphere of lowland rice. Soil Sci Plant Nutr 28: 173–182
Kimura M, Ando H, Haraguchi H (1991a) Estimation of potential CO_2 and CH_4 production in Japanese paddy fields. Environ Sci 4: 15–25
Kimura M, Miura Y, Watanabe A, Katoh T, Haraguchi H (1991b) Methane emission from rice field (Part 1). Effect of fertilization, growth stage and midsummer drainage: Pot experiment. Environ Sci 4: 265–271
Kimura M, Asai K, Watanabe A, Murase J, Kuwatsuka S (1992a) Suppression of methane fluxes from flooded paddy soil with rice plants by foliar spray of nitrogen fertilizers. Soil Sci Plant Nutr 38: 735–740
Kimura M, Miura Y, Watanabe A, Murase J, Kuwatsuka S (1992b) Methane production and its fate in rice fields. I. Effects of rice straw application and percolation rate on the leaching of methane and other soil components into the subsoil. Soil Sci Plant Nutr 38: 665–672

King GM, Adamsen PS (1992) Effects of temperature on methane consumption in a forest soil and in pure cultures of the methanotroph *Methylomonas rubra.* Appl Environ Microbiol 58: 2758–2763

Kiniry JR, Rosenthal WD, Jackson BS, Hoogenboom G (1991) Predicting leaf development of crop plants. In: Hodges T. (ed) Predicting crop phenology. CRC Press, Boca Raton, pp 29–42

Kisimoto R (1976) Synoptic weather conditions inducing long-distance immigration of planthoppers, *Sogatella furcifera* Horvath and *Nilaparvata lugens* Stål. Ecol Entomol 1: 95–109

Kisimoto R (1987) Ecology of planthopper migration. In: Wilson MR, Nault NR (eds) Proc 2nd international workshop on leafhoppers and planthoppers of economic importance. Commonwealth Institute of Entomology, London, pp 41–54

Kisimoto R (1991) Long distance migration of rice insects. In: Heinrichs EA, Miller TA (eds) Rice insects: management strategies. Springer, Berlin Heidelberg New York, pp 167–198

Kludze HK, DeLaune RD (1994) Methane emissions and growth of *Spartina patens* in response to soil redox intensity. Soil Sci Soc Am J 58: 1838–1845

Kludze HK, DeLaune RD (1995) Straw application effects on methane and oxygen exchange and growth in rice. Soil Sci Soc Am J 59: 287–292

Kludze HK, DeLaune RD, Patrick WH Jr (1993) Aerenchyma formation and methane and oxygen exchange in rice. Soil Sci Soc Am J 57: 386–391

Kludze HK, DeLaune RD, Patrick WH Jr (1994) A calorimetric method for assaying dissolved oxygen loss from container-grown rice roots. Agron J 86: 483–487

Knudson LL, Tibbitts TW, Edwards GE (1977) Measurement of ozone injury by determination of leaf chlorophyll concentration. Plant Physiol 60: 606–608

Koschorreck M, Conrad R (1993) Oxidation of atmospheric methane in soil: measurements in the field, in soil cores and in soil samples. Global Biogeochem Cycles 7: 109–121

Kramer GF, Norman HA, Krizek DT, Mirecki RM (1991) Influence of UV-B radiation on polyamines, lipid peroxidation and membrane lipids in cucumber. Phytochemistry 30: 2101–2108

Krizek DT, Kramer GF, Upadhyaya A, Mirecki RM (1993) UV-B response of cucumber seedlings grown under metal halide and high pressure sodium/deluxe lamps. Physiol Plant 88: 350–358

Kropff MJ, Centeno G, Bachelet D, Lee MH, Mohandass S, Horie T, Defeng S, Singh S, Penning de Vries FWT (1993a) Predicting the impact of CO_2 and temperature on rice production. International Rice Research Institute, Los Baños, Philippines (unpublished)

Kropff MJ, van Laar HH, ten Berge HFM (eds) (1993b) ORYZA1: a basic model for irrigated lowland rice production. Simulation and systems analysis for rice production (SARP) publication. International Rice Research Institute, Los Baños, Philippines

Krupa SV, Kickert RN (1989) The greenhouse effect: impacts of ultraviolet-B (UV-B) radiation, carbon dioxide (CO_2), and ozone (O_3) on vegetation. Environ Pollut 61: 263–393

Kuchler AW (1964) Manual to accompany the map of potential natural vegetation of the conterminous United States. Spec Publ 36, American Geographic Society, New York

Kumagai T, Sato T (1992) Inhibitory effects of increase in near-UV radiation on the growth of Japanese rice cultivars in a phytotron and recovery by exposure to visible radiation. Jpn J Breed 42: 545–552

Ladha JK, Tirol AC, Darby LG, Caldo G, Ventura W, Watanabe L (1986) Plant associated N_2 fixation (C_2H_2-reduction) by five rice varieties and relationship with plant growth characters as affected by straw incorporation. Soil Sci Plant Nutr 32: 91–106

Lave LB, Dowlatabadi H (1993) Climate change: the effects of personal beliefs and scientific uncertainty. Environ Sci Technol 27(10): 1962–1972

Leclerc MY, Thurtell GW (1990) Footprint prediction of scalar fluxes using a Markovian analysis. Boundary-Layer Meteorol 52: 247–258

Leemans R, Solomon AM (1993) Modeling the potential change in yield and distribution of the earth's crops under a warmed climate. Clim Res 3: 79–96
Lemon ER (ed) (1984) CO_2 and plants. American Association for the Advancement of Science Selected Symposium No 84. Westview Press, Boulder, CO
Lemon E, Stewart DW, Shawcroft RW (1971) The sun's work in a cornfield. Science 174: 371–378
Lenschow DH (1994) Micrometeorological techniques for measuring biosphere-atmosphere trace gas exchange. In: Matson P, Harriss R (eds) Methods in ecology: trace gases, Blackwell, Oxford (in press)
Leung H, Taga M (1988) *Magnaporthe grisea* (*Pyricularia* species), the blast fungus. Adv Plant Pathol 6: 175–188
Leung H, Borromeo ES, Bernardo MA, Notteghem JL (1988) Genetic analysis of virulence in the rice blast fungus *Magnaporthe grisea*. Phytopathology 78: 1227–1233
Leung H, Lehtinen U, Karjalainen R, Skinner D, Tooley P, Leong SA, Ellingboe A (1990) Transformation of the rice blast fungus *Magnaporthe grisea* to hygromycin B resistance. Curr Genet 17: 409–411
Lindau CW, Patrick WH Jr, DeLaune RD, Reddy KR (1990) Rate of accumulation and emission of N_2, N_2O and CH_4 from a flooded rice soil. Plant Soil 129: 269–276
Lindau CW, Bollich PK, DeLaune RD, Patrick WH Jr, Law VJ (1991) Effect of urea fertilizer and environmental factors on CH_4 emission from a Louisiana, USA, rice field. Plant Soil 136: 195–203
Lindau CW, Bollick PK, DeLaune RD, Mosier AR, Bronson KF (1993) Methane mitigation in flooded Louisiana rice fields. Biol Fert Soils 15: 174–178
Lovely DR, Klug MJ (1983) Sulfate reducers can outcompete methanogens at freshwater sulfate concentrations. Appl Environ Microbiol 45: 187–192
Lugo AE, Brown S (1986) Steady state terrestrial ecosystems and the global carbon cycle. Vegetation 68: 83–90
Lydon J, Teramura AH, Summer EG (1986) Effects of ultraviolet-B radiation on the growth and productivity of field growth soybean. In: Worrest RC, Caldwell MM (eds) Stratospheric ozone reduction, solar ultraviolet radiation and plant life. Springer, Berlin Heidelberg New York, pp 313–325
Ma Yi (1981) Biogas para la communidad rural en China popular. Interciencia 6: 75–76
Mackill DJ, Coffman WR, Rutger JN (1982) Pollen shedding and combining ability for high temperature tolerance in rice. Crop Sci 22: 730–733
Madronich S (1993) UV radiation in the natural and perturbed atmosphere. In: Tevini M (ed) UV-B radiation and ozone depletion: effects on humans, animals, plants, microorganisms, and materials. Lewis, Boca Raton, pp 17–69
Maene LM, De Vuyst A, Pradhan SB (1987) The importance of urea as a source of nitrogen for crop production in the Asian and Pacific region. Urea-Tech'87, 16–19 March, Kuala Lumpur, Malaysia
Manabe S, Wetherald RT (1987) Large-scale changes of soil wetness induced by an increase in atmospheric carbon dioxide. J Atmos Sci 44: 1211–1235
Manalo PA, Ingram KT, Pamplona RR, Egeh AO (1994) Atmospheric carbon dioxide and temperature effects on rice. Agric Ecosyst Environ 51: 339–347
Mariko S, Harazono Y, Owa N, Nouchi I (1991) Methane in flooded soil, water and the emission through rice plants to the atmosphere. Environ Exp Bot 31: 343–350
Martens CS, Berner RA (1974) Methane production in the interstitial water of sulfate-depleted marine sediments. Science 185: 1167–1169
Martin JK (1977) Factors influencing the loss of organic carbon from wheat roots. Soil Biochem 9: 1–7
Masscheleyn PH, DeLaune RD, Patrick WH Jr (1993) Methane and nitrous oxide emission from laboratory measurements of rice soil suspension: effect of soil oxidation-reduction status. Chemosphere 26: 251–260
Matson PA, Ojima DS (eds) (1990) Terrestrial biosphere exchange with global atmospheric chemistry. Global Change Rep 13, International Geosphere-Biosphere Programme, Stockholm

Matsui T, Horie T (1992) Effects of elevated CO_2 and high temperature on growth and yield of rice. II. Sensitive period and pollen germination rate in high temperature sterility of rice spikelets at flowering. Jpn J Crop Sci 61(extra issue 1): 148–149
Matthews E (1993) Global geographical databases for modelling trace gas fluxes. Int J Geographical Info Syst 7: 124–142
Matthews E, Fung I (1987) Methane emissions from natural wetlands: global distribution, area, and environmental characteristics. Global Biogeochem Cycles 1: 61–86
Matthews E, Fung I, Lerner J (1991) Methane emissions from rice cultivation: geographic and seasonal distribution of cultivated areas and emission. Global Biogeochem Cycles 9: 251–256
Matthews RB, Hunt LA (1994) A model describing the growth of cassava (*Manihot esculenta* L. Crantz). Field Crops Res 36: 69–84
Metthews RB, Kropff MJ, Bachelet D, van Laar HH (eds) (1994) The impact of global climate change on rice production in Asia: a simulation study. Report commissioned by United States Environmental Protection Agency, Corvallis, OR
Matthias AD, Blackmer AM, Bremner JM (1980) A simple chamber technique for field measurement of emissions of nitrous oxide from soils. J Environ Qual 9: 251–256
Mattson MD, Likens GE (1990) Air pressure and methane fluxes. Nature 347: 718–719
Mausbach MJ, Wilding LP (eds) (1991) Spatial variability of soils and landforms. SSSA Spec Publ 28, Soil Science Society of America, Madison, WI
McElroy MB, Wofsy FC (1986) Tropical forests: interactions with the atmosphere. In: Prance GT (ed) Tropical rain forests and the world atmosphere. Westview Press, Boulder
McFarland M, Kaye J (1992) Yearly review: chlorofluorocarbons and ozone. Photochem Photobiol 55: 911–929
McPeters RD, Krueger AJ, Bhartia PK, Herman JR, Oaks A, Ziuddin A, Cebula RP, Schlesinger BM, Swissler T, Taylor SL, Torres O, Wellemeyer CG (1993) Nimbus-7 total ozone mapping spectrometer (TOMS) data products users guide. NASA RP-1323, NSSDC, Goddard Space Flight Center, Greenbelt, MD
Mengel K, Kirby EA (1987) Principles of plant nutrition. 4th edn. International Potash Institute, Bern
Middleton EM, Teramura AH (1993) The role of flavonol glycosides and carotenoids in protecting soybean from ultraviolet-B damage. Plant Physiol 103: 741–752
Miller JE, Booker FL, Fiscus EL, Heagle AS, Pursley WA, Vozzo SF, Heck WW (1994) Ultraviolet-B radiation and ozone effects on growth, yield, and photosynthesis of soybean. J Environ Qual 23: 83–91
Minami K (1993) Methane from rice production. In: Van Amstel AR (ed) Proc international IPCC workshop on methane and nitrous oxide: methods in national emission inventories and options for control. National Institute of Public Health and Environmental protection (RIVM), Bilthoven, The Netherlands, pp 143–162
Minami K, Neue H-U (1994) Rice paddies as a methane source. Clim Change 27: 13–26
Minami K, Yagi K (1988) Method for measuring methane flux from rice paddies. Jpn J Soil Sci Plant Nutr 59: 458–463 (in Japanese, English summary).
Mirecki RM, Teramura AH (1984) Effects of ultraviolet-B irradiance on soybean. V. The dependence of plant sensitivity on the photosynthetic photon flux density during and after leaf expansion. Plant Physiol 74: 475–480
Miura Y, Watanabe A, Kimura M, Kuwatsuka S (1992a) Methane emission from rice field (part 2): main route of methane transfer through rice plant, and temperature and light effects on diurnal variation of methane emission. Environ Sci 5: 187–193
Miura Y, Watanabe A, Murase J, Kimura M (1992b) Methane emission and its fate in rice fields. II. Oxidation of methane and its coupled ferric oxide reduction in subsoil. Soil Sci Plant Nutr 38: 673–679
Mochida O, Okada T (1979) Taxonomy and biology of *Nilaparvata lugens* (Homoptera: Delphacidae). In: Brown planthopper: threat to rice production in Asia. International Rice Research Institute, Los Baños, Philippines, pp 21–43
Mohamed HA, Clark JA, Ong CK (1988) Genotypic differences in the temperature responses of tropical crops. I. Germination characteristics of groundnut (*Arachis hypogaea* L.) and pearl millet (*Pennisetum typhoides* S. & H.). J Exp Bot 39: 1121–1128

Monserud RA, Leemans R (1992) Comparing global vegetation maps with the Kappa statistics. Ecol Model 62: 275–293

Monteith JL (1977) Climate and efficiency of crop production in Britain. Philos Trans R Soc Lond B 281: 277–294

Morgan J (1968) The influence of added organic matter on certain processes occurring in anaerobically incubated soils. Proc 9th Int Congr Soil Science, Adelaide, Australia. Transactions 4: 699–707

Morrison JIL, Gifford RM (1984) Plant growth and water use with limited water supply in high CO_2 concentrations. I. Leaf area, water use and transpiration. Aust J Plant Physiol 11: 361–374

Mosier AR (1990) Gas flux measurement techniques with special reference to techniques suitable for measurements over large ecologically uniform area. In: Bouwman AF (ed) Soils and the greenhouse effect. John Wiley, New York, pp 289–301

Mosier AR, Mohanty SK, Bhadrachalam A, Chakravorti SP (1990) Evolution of dinitrogen and nitrous oxide from the soil to the atmosphere through rice plants. Biol Fert Soils 9: 61–67

Mott KA (1990) Sensing of atmospheric CO_2 by plants. Plant Cell Environ 13: 731–737

Msanya BM (1987) Soil resources inventories: characterization, quality and evaluation of their adequacy for specific land use. PhD Thesis, State University of Ghent, Belgium

Munakata K (1976) Effects of temperature and light on the reproductive growth and ripening of rice. In: Climate and rice. International Rice Research Institute, Los Baños, Philippines, pp 187–210

Murali NS, Teramura AH (1985) Effects of ultraviolet-B irradiance on soybean – VI. Influence of phosphorous nutrition on growth and flavonoid content. Physiol Plant 63: 413–416

Murali NS, Teramura AH (1986a) Intraspecific differences in *Cucumis sativus* sensitivity to ultraviolet-B radiation. Physiol Plant 68: 673–677

Murali NS, Teramura AH (1986b) Effects of supplemental ultraviolet-B radiation on the growth and physiology of field-grown soybean. Environ Exp Bot 26: 233–242.

Murase J, Kimura M (1994) Methane production and its fate in fields. VI. Anaerobic oxidation of methane in plow layer soil. Soil Sci Plant Nutr 40: 505–514

Murase J, Kimura M (1995) Methane production and its fate in paddy fields. IX. Methane flux distribution and decomposition of methane in the subsoil during the growth period of rice plants. Soil Sci Plant Nutr (in press)

Murase J, Kimura M, Kuwatsuka S (1993) Methane production and its fate in ricefields. III. Effects of percolation in rice field on methane flux distribution to the atmosphere and to subsoil. Soil Sci Plant Nutr 39: 63–70

Muzio LJ, Kramlich JC (1988) An artifact in the measurement of nitrous oxide from combustion sources. Geophys Res Lett 15: 1369–1372

Muzio LJ, Teague ME, Kramlich JC, Cole JA, McCarthy JM, Lyon RK (1989) Errors in grab sample measurements of N_2O from combustion sources. J Air Pollut Control Assoc 39: 287–293

Nagarajah S, Neue H-U, Alberto MCR (1989) Effect of Sesbania, Azolla and rice straw incorporation on the kinetics of NH_4, K, Fe, Mn, Zn and P in some flooded rice soils. Plant Soil 116: 37–48

National Research Council-NRC (1992) Policy implications of greenhouse warming: mitigation, adaptation, and the science base. National Academy Press, Washington

NASA-National Aeronautics and Space Administration (1988) Executive summary of the ozone trend panel. NASA, Washington

Neales TF, Incoll LD (1968) The control of leaf photosynthesis rate by the level of assimilate concentration in the leaf: a review of the hypothesis. Bot Rev 34: 107–125

Negash L, Björn LO (1986) Stomatal closure by ultraviolet radiation. Physiol Plant 66: 360–364

Neue H-U (1993) Methane emission from rice fields. BioScience 43: 466–474

Neue H-U, Roger PA (1993) Rice agriculture: factors controlling emission. In: Khalil MAK, Shearer M (eds) Global atmospheric methane: sources, sinks, and role in global

change. NATO ASI Ser I, Global Environmental Change Vol 13. Springer, Berlin Heidelberg NewYork, pp 254–298
Neue H-U, Sass R (1994) Trace gas emissions from rice fields. In: Prinn R (ed) Global atmospheric-biospheric chemistry. Plenum Press, New York, pp 119–148
Neue H-U, Scharpenseel HW (1984) Gaseous products of the decomposition of organic matter in submerged soils. In: Organic matter and rice. International Rice Research Institute, Los Baños, Philippines, pp 311–328
Neue H-U, Becker-Heidmann P, Scharpenseel HW (1990) Organic matter dynamics, soil properties, and cultural practices in rice lands and their relationship to methane production. In: Bouwman AF (ed) Soils and the greenhouse effect. John Wiley, New York, pp 457–466
Neue H-U, Quijano CC, Scharpenseel HW (1991) Conservation and enhancement of wetland soils. In: Evaluation for sustainable land management in the developing world. Vol 2: technical papers. IBSRAM Proc No 12(2), Bangkok, Thailand, pp 273–303
Neue H-U, Lantin R, Wassmann R, Aduna JB, Alberto CR, Andales MF (1994) Methane emission from rice soils of the Philippines. In: CH_4 and N_2O. National Institute of Agro-Environmental Sciences, Tsukuba, Japan, pp 55–63
Nishiyama I (1976) Effects of temperature on the vegetative growth of rice plants. In: Climate and rice. International Rice Research Institute, Los Baños, Philippines, pp 159–186
Noda H, Saito T (1979) Effects of high temperature on the development of *Laodelphax striatellus* (Homoptera: Delphacidae) and on its intracellular yeast like symbiotes. Appl Entomol Zool 14: 64–75
Nomura M, Imai K, Matsuda T (1993) Effects of atmospheric CO_2 concentration and phosphorus nutrition on the leaf structure of rice. Proc Kanto Branch, Crop Sci Soc Jpn 8: 49–50 (in Japanese)
Norgaard RB, Howarth RB (1992) Sustainability and intergenerational environmental rights: implications for benefit-cost analysis. In: Reilly JM, Anderson M (eds) Economic isues in global climate change: agriculture, forestry and natural resources. Westview Press, Boulder, pp 41–55
Nouchi I (1991) Effects of ultraviolet-B (UV-B) irradiation on growth of cucumber, radish and kidney bean plants. J Agric Meteorol 46: 205–214 (in Japanese, English summary).
Nouchi I (1993) Effects of ultraviolet-B radiation on growth of cucumber plants. J Agric Meteorol 48: 731–734
Nouchi I, Kobayashi K (1995) Effects of enhanced ultraviolet-B radiation with a modulated lamp control system on growth of 17 rice cultivars in the field. J Agric Meteorol 51: 11–20
Nouchi I, Kobayashi K, Hosono T (1993) Effects of ultraviolet-B radiation on growth of rice plants in the fields. In Proc of 15th Int Botanical Congress, Yokohama, Japan
Nouchi I, Mariko S, Aoki K (1990) Mechanism of methane transportation from the rhizosphere to the atmosphere through rice plants. Plant Physiol 94: 59–66
NRC—National Research Council (1992) Policy implications of greenhouse warming: mitigation, adaptation, and the science base. National Academy Press, Washington
Oldeman LR, Sombroek WG (1990) Use of the SOTER database to model global change. In: Trans 14th Int Congr Soil Science, vol 5, Kyoto, Japan, pp 136–141
Oldeman LR, van Engelen VWP (1993) A world soils and terrain digital database (SOTER): an improved assessment of land resources. Geoderma 60: 309–325
Olszyk DM, Tibbitts TW, Hertzberg WM (1980) Environment in open top field chambers utilized for air pollution studies. J Environ Qual 9: 610–615
Ordway DE (1969) An aerodynamicist's analysis of the Odum cylinder approach to net CO_2 exchange. Photosynthetica 3: 199–209
Oremland RS (1988) Biogeochemistry of methanogenic bacteria. In: Zehnder AJB (ed) Biology of anaerobic microorganisms. John Wiley, New York, pp 641–706
Oremland RS, Culbertson CW (1992) Importance of methane-oxidizing bacteria in the methane budget as revealed by the use of a specific inhibitor. Nature 356: 421–423
Oremland RS, Marsh LM, Polein S (1982) Methane production and simultaneous sulfate reduction in anoxic marsh sediments. Nature 296: 143–145

Orth AB, Teramura AH, Sisler HD (1990) Effects of ultraviolet-B radiation on fungal disease development in *Cucumis sativus*. Am J Bot 79: 1188–1192
Ou Sh (1985) Rice diseases. 2nd edn. Commonwealth Mycological Institute, Kew, Surrey, UK
Owens OH, Krizek DT (1980) Multiple effects of UV radiation (265–330 nm) on fungal spore emergence. Photochem Photobiol 32: 41–49
Ozawa Y (1962) Climatic zoning of Japan for land use. Chiri (Kokinshoin, Tokyo) 7: 300–364; 432–438 (in Japanese)
Pang IQ, Hays JB (1991) UV-B inducible and temperature-sensitive photoreaction of cyclobutane pyridine dimers in *Arabidopsis thaliana*. Plant Physiol 95: 536–543
Panikov NS, Semenov AM, Tarasov AA, Belyaev AS, Kravchenko IK, Smagina NS, Palejeva MV, Zelenev VV, Skupchenko KV (1993) Methane production and uptake in soils of the European part of the USSR. J Ecol Chem 1: 7–18
Papen H, Hellmann B, Papke H, Rennenberg H (1993) Emission of N-oxides from acid irrigated and limed soils of a coniferous forest in Bavaria. In: Oremland RS (ed) Biogeochemistry of global change. Chapman & Hall, New York, pp 245–260
Parashar D, Rai CJ, Gupta PK, Singh N (1990) Parameters affecting methane emission from paddy fields. Indian J Radio Space Phys 20: 12–17
Parrish JA, Jaenicke KR, Anderson RR (1982) Erythema and melanogenesis action spectra of normal human skin. Photochem Photobiol 36: 187–191
Parry M (1990) Climate change and world agriculture. Earthscan Publications, London
Parry ML, Carter TR, Konijn NT (1988) The impact of climatic variations on agriculture, vol 1. Assessments in cool temperate and cold regions, vol 2. Assessments in semiarid region. Kluwer, Dordrecht
Parsons PA (1983) The evolutionary biology of colonizing species. Cambridge University Press, New York
Patrick WH Jr (1982) Nitrogen transformations in submerged soils. In: Stevenson FJ (ed) Nitrogen in agricultural soils. American Society of Agronomy, Madison, WI, pp 449–465
Patrick WH Jr, Delaune RD (1977) Chemical and biological redox systems affecting nutrient availability in the coastal wetlands. Geosci Man 18: 131–137
Patrick WH Jr, Reddy CN (1978) Chemical changes in rice soil. Soils and Rice. IRRI. P.O. Box 933, Manila, Philippines, pp 361–380
Paulson CA (1970) The mathematical representation of wind speed and temperature profiles in the unstable atmospheric surface layer. J Appl Meteorol 9: 857–861
Peak MJ, Peak JG (1983) Use of action spectra for identitying molecular targets and mechanisms of action of solar ultraviolet light. Physiol Plant 58: 367–372
Peart RM, Jones JW, Curry RB, Boote K, Allen LH Jr (1989) Impact of climate change on crop yield in the Southeastern USA. In: Smith JB, Tirpak DA (eds) The potential effects of global climate change on the United States. Report to Congress. EPA-230-05-89-050. United States Environmental Protection Agency, Washington, pp 1–54
Penning de Vries FWT (1993) Rice production and climate change In: Penning de Vries FWT, Teng P, Metselaar K (eds) Systems approaches for agricultural development. Kluwer, Dordrecht, pp 175–192
Penning de Vries FWT, van Laar HH (eds) (1982) Simulation of plant growth and crop production. Simulation Monographs, Pudoc, Wageningen, Netherlands
Penning de Vries FWT, Jansen DM, ten Berge HFM, Bakema A (1989) Simulation of ecophysiological processes of growth in several annual crops. Simulation Monographs, Pudoc, Wageningen, Netherlands, and IRRI, Los Baños, Philippines
Peters RL (1991) Consequences of global warming for biological diversity. In: Wyman RL (ed) Global climate change and life on earth. Chapman and Hall, London, pp 99–118
Peters RL, Darling JDS (1985) The greenhouse effect and nature reserves. BioScience 35: 707–717
Pickering NB, Allen LH Jr, Albrecht SL, Jones P, Beaker JT (1994a) Environmental plant chambers: control and measurement using CR-10T dataloggers. In: Watson DG, Zazueta FS, Harrison TV (eds) Proc 5th Int Conf on Computers in Agriculture 1994. American Society of Agricultural Engineers, St Joseph, MI, pp 29–35

Pickering NB, Allen LH Jr, Baker JT, Boote KJ (1994b) Control and monitoring of rice experiments in closed environmental chambers using a distributed network of dataloggers. In: Peng et al. (eds) Climate change and rice. International Rice Research Institute, Los Baños, Philippines

Pickering NB, Jones JW, Boote KJ (1994c) Adapting SOYGRO V5.42 for prediction under climate change conditions. In: Rosenzweig C, Harper LA, Hollinger SE, Jones JW, Allen LH Jr (eds) Proc of a symposium on climate change and international impacts. American Society of Agronomy, Crop Science Society of America, and Soil Science Society of America, Madison, WI (in press)

Pingali PL (1992) Diversifying Asian rice-farming systems: a deterministic paradigm. In: Barghouti S, Garbux L, Umali D (eds) Trends in agricultural diversification: regional perspectives. World Bank Technical Paper No 180, The World Bank, Washington, pp 107–126

Pingali P (1993) Technological and policy options for sustaining rice productivity growth in Asia. Food and Fertilizer Technology Center, Extention Bulletin No 377, Taipei City, Taiwan

Pingali PL, Rosegrant MW (1993) Confronting the environmental consequences of the green revolution in Asia. International pre-conference on post green revolution agricultural development strategies in the Third World: what next? American Society of Agricultural Engineers, St. Joseph, MI

Pingali PL, Moya PF, Velasco LE (1990) The post-green revolution blues in Asian rice production – the diminished gap between experiment station and farmer yields. Social Sciences Divison Paper No 90–01, International Rice Research Institute, Los Baños, Philippines

Pittock AB (1991) The enhanced greenhouse effect and its agricultural impact. In: Geng S, Cady CS (eds) Proc of a workshop on climatic variation and change: implications for agriculture in the Pacific rim. The Public Service Research and Dissemination Program, University of California, Davis, CA

Pleijsier K (1989) Variability in soil data. In: Bouma J, Bregt AK (eds) Land qualities in space and time. Pudoc, Wageningen, Netherlands, pp 89–99

Porter JR (1993) AFRCWHEAT2: a model of the growth and development of wheat incorporating responses to water and nitrogen. Eur J Agron

Post WM, Pastor J, Zinke PJ, Stangenberger AG (1985) Global patterns of soil nitrogen storage. Nature 317: 613–616

Prinn R (1994) The interactive atmosphere: global atmospheric-biospheric chemistry. Ambio 23: 50–61

Prinn R, Hartley D (1992) Atmosphere, ocean and land: critical gaps in earth system models. In: Ojima D (ed) Modeling the earth system. UCAR, Boulder, pp 9–38

Quaite FE, Sutherland BM, Sutherland JC (1992) Action spectrum for DNA damage in alfalfa lowers predicted impact of ozone depletion. Nature 358: 576–578

Quintana A, Prevosti A (1990) Genetics and environmental factors in the resistance of *Drosophila subobscura* adults to high temperature schock. Theor Appl Genet 80: 847–851

Raimbault M (1975) Étude de l'influence inhibitrice de l'acétylene sur la formation biologique du méthane dans un sol de rizière. Ann Microbiol (Inst Pasteur) 126a: 217–258

Raimbault M, Rinaudo G, Garcia JL, Boureau M (1977) A device to study metabolic gases in the rice rhizosphere. Soil Biol Biochem 2: 193–196

Rasmussen RA, Khalil MAK (1983) Global production of methane by termites. Nature 301: 700–702

Reddy KR, Patrick WH Jr (1984) Nitrogen transformations and loss in flooded soils and sediments. CRC Crit Rev Environ Control 13: 273–309

Reddy KR, Patrick WH Jr (1986) Denitrification losses in flooded rice fields. Fert Res 9: 99–116

Reddy KR, Patrick WH Jr, Lindau CW (1986) Redox processes and nitrogen transformations in flooded soils In: Trans 13th Congr Int Soc of Soil Science, Hamburg, Germany, pp 729–738

Reekie EG, Bazzaz FA (1989) Competition and patterns of resource use among seedlings of five tropical trees grown at ambient and elevated CO_2. Oecologia 79: 212–222
Reilly JM, Anderson M (1992) Economic issues in global climate change: agriculture, forestry and natural resources. Westview Press, Boulder
Rennenberg H, Wassmann R, Paper H, Seiler W (1992) Trace gas exchange in rice cultivation. Ecol Bull 42: 164–173
Reynolds JF, Acock B (1985) Modeling approaches for evaluating vegetation response to carbon dioxide concentration. In: Strain BR, Cure JD (eds) Direct effects of increasing carbon dioxide on vegetation. DOE/ER-0283, United States Department of Energy, Carbon Dioxide Research Division, Washington
Ritchie JT (1985) A user-oriented model of the soil water balance in wheat. In: Fry E, Atkin TK (eds) Wheat growth and modeling. Plenum Press, New York
Ritchie JT, Otter S (1985) Description and performanc of CERES-Wheat: a user-oriented wheat yield model. Publ No ARS-38, United States Department of Agriculture, Agricultural Research Service, Raleigh, NC
Ritchie JT, Alocilja EC, Singh U, Uehara G (1987) IBSNAT and the CERES-Rice model. In: Weather and rice. International Rice Research Institute, Los Baños, Laguna, Philippines
Ritchie JT, Singh U, Godwin D, Hunt LA (1989) A user's guide to CERES-Maize-V2.10. International Fertilizer Development Center. Muscle Shoals, Alabama
Ritter JA, Barrick JDW, Sachse GW, Gregory GL, Woerner MA, Watson CE, Hill GF, Collins JE (1992) Airbone flux measurements of trace species in an Arctic boundary layer. J Geophys Res 97: 16601–16625
Robbins NS, Pharr DM (1988) Effect of restricted root growth on carbohydrate metabolism and whole plant growth of *Cucumis sativus* L. Plant Physiol 87: 409–413
Roberton AM, Wolfe RS (1970) ATP pools in methanobacterium. J Bacteria 102: 43–51
Rodhe H (1990) A comparison of the contribution of various gases to the greenhouse effect. Science 248: 1217–1219
Rogers HH, Bingham GE, Cure JD, Smith JM, Surano KA (1983a) Responses of selected plant species to elevated carbon dioxide in the field. J Environ Qual 12: 569–574
Rogers HH, Heck WW, Heagle AS (1983b) A field technique for the study of plant responses to elevated carbon dioxide concentrations. J Air Pollut Control Assoc 33: 42–44
Rosegrant M, Binswanger H (1994) Markets in tradable water rights: potentials for efficiency gains in developing-country irrigation. World Dev 22: 1613–1625
Rosengrant M, Svendson M (1993) Irrigation investment and management in Asia: trends, priorities, and policy directions. Food Policy 18(1): 13–22
Rosenzweig C (1989) Potential effects of climate change on agricultural production in the great plains – a simulation study. In: Smith JB (ed) The potential effects of global climate change on the United States: appendix C – agriculture. United States Environmental Protection Agency, Washington
Rosenzweig C (1990) Crop response to climate change in the Southern Great Plains: a simulation study. Prof Geog 42(1): 20–37
Rosenzweig C, Iglesias A (eds) (1994) Implications of climate change for international agriculture: crop modeling study. United States Environmental Protection Agency. Washington, EPA Publication Number EPA 230-B-94-003
Rosenzweig C, Parry ML (1994) Potential impact of climate change on world food supply. Nature 367: 133–138
Rosenzweig C, Parry M, Fisher G, Frohberg K (1993) Climate change and world food supply. Research Report No 3, Environmental Change Unit, University of Oxford, Oxford
Rosswall T, Woodmansee RG, Risserand PG (eds) (1988) Scales and global change: spatial and temporal variability in biospheric and geospheric processes. SCOPE 35. John Wiley, New York
Rotmans J, den Elzen MGJ, Krol MS, Swart RJ, van der Woerd HJ (1992). Stabilizing atmospheric concentrations: towards international methane control. AMBIO 21: 404–413

Rotty RM, Marland G (1986) Fossil fuel combustion: recent amounts, patterns, and trends of CO_2. In: Trabalka JR, Reichle DE (eds) The changing carbon cycle: a global analysis. Springer, Berlin Heidelberg New York, pp 474–490

Roulet NT, Matthews E (1993) Methane emissions from natural wetlands. In: Van Amstel AR, (ed) Proc of international IPCC workshop on methane and nitrous oxide: methods in national emission inventories and options for control. National Institute of Public Health and Environmental Protection (RIVM), Bilthoven, Netherlands, pp 163–170

Rousch RT, McKenzie RT (1987) Ecological genetics of insecticide and acaricide resistance. Annu Rev Entomol 32: 361–380

Rowland-Bamford AJ, Nordenbrock C, Baker JT, Bowes G, Allen LH Jr (1990) Changes in stomatal density in rice grown under various CO_2 regimes with natural solar irradiance. Environ Exp Bot 30: 175–180

Rowland-Bamford AJ, Allen LH Jr, Baker JT, Bowes G (1991) Acclimation of rice to changing atmospheric carbon dioxide concentration. Plant Cell Environ 14: 577–583

Roy CR, Gies HP, Elliott G (1989) The ARL solar ultraviolet radiation measurement programme. Trans Menzies Found 15: 71–76

Rudolph J (1994) Anomalous methane. Nautre 368: 19–20

Rundel RD (1983) Action spectra and estimation of biologically effective UV radiation. Physiol Plant 58: 360–366

Rundel RD (1986) Computation of spectral distribution and intensity of solar UV-B radiation. In: Worrest RC, Caldwell MM (eds) Stratospheric ozone reduction, solar ultraviolet radiation and plant life. NATO ASI Ser. Springer, Berlin Heidelberg New York, pp 49–62

Runeckles VC, Krupa SV (1994) The impact of UV-B radiation and ozone on terrestrial vegetation. Environ Pollut 83: 191–213

Rupert CS (1986) Cellular repair and assessment of UV-B radiation damage. In: Worrest RC, Caldwell MM (eds) Stratospheric ozone reduction, solar ultraviolet radiation and plant life. NATO ASI Ser. Springer, Berlin Heidelberg New York, pp 127–137

Ryan MG (1991) Effects of climate change on plant respiration. Ecol Appl 1: 157–167

Sage RF, Sharkey TD, Seemann JR (1989) Acclimation of photosynthesis to elevated CO_2 in five C_3 species. Plant Physiol 89: 590–596

Saleh AFM, Lantican MA, Bhuiyan SI, Agua MM (1993) Increasing productivity of droughtprone rainfed lowlands in Northern Luzon, Philippines. Second Annual Technical Meeting of the Rainfed Lowland Rice Research Consortium, Semarang, Indonesia

Salvas PL, Taylor BF (1980) Blockage of methanogenesis in marine sediments by the nitrification inhibitor 2-chloro-6-(trichloromethyl) pyridine (Nitrapin or N-serve) Curr Microbiol 4: 305

Samson MI, Buresh RJ, De Datta SK (1990) Evolution and soil entrapment of nitrogen gases formed by denitrification in flooded soil. Soil Sci Plant Nutr 36: 299–307

SAS (1987) SAS guide for personal computer, version 6. SAS Institute, Cary, NC

Sasek TW, DeLucia EH, Strain BR (1985) Reversibility of photosynthetic inhibition in cotton after long-term exposure to elevated CO_2 concentrations. Plant Physiol 78: 619–622

Sass RL, Fisher FM, Harcombe PA, Turner FT (1990) Methane production and emission in a Texas rice field. Global Biogeochem Cycles 4: 47–68

Sass RL, Fisher FM, Harcombe PA, Turner FT (1991a) Mitigation of methane emission from rice fields: possible adverse effects of incorporated rice straw. Global Biogeochem Cycles 5: 275–287

Sass RL, Fisher FM, Turner FT, Jund MF (1991b) Methane emission from ricefields as influenced by solar radiation, temperature, and straw incorporation. Global Biogeochem Cycles 5: 335–350

Sass RL, Fischer FM, Harcombe PA, Turner FT (1991c) Methane production and emission in a Texas agriculture wetland. Global Biogeochem Cycles 4: 47–68

Sass RL, Fisher FM, Wang YB, Turner FT, Jund MF (1992a) Methane emission from ricefields: the effect of flood water management. Global Biogeochem Cycles 6: 249–262

Sass R, Denmead OT, Conrad R, Freney JR, Klug M, Minami K, Mosier AR, Neue H-U, Rennenberg H, Su WH, Wang L-M (1992b) Exchange of methane and other trace gases in rice cultivation. Ecol Bull 42: 199–206

Sass RL, Fisher FM, Wang YB Turner FT, Jund MF (1993) Methane emission from rice fields: the effect of floodedwater management. Global Biogeochem Cycles 6(3): 249–262

Sass RL, Fisher FM, Lewis ST, Jund MF, Turner FT (1994) Methane emissions from rice fields: effect of soil properties. Global Biogeochem Cycles 8: 135–140

Satake T, Yoshida S (1978) High temperature-induced sterility in Indica rice at flowering. Jpn J Crop Sci 47: 6–17

Sato T, Kumagai T (1993) Cultivar differences in resistance to inhibitory effects of near-UV radiation among Asian type and Japanese lowland and upland cultivars of rice. Jpn J Breed 43: 61–68

Sato T, Kang HS, Kumagai T (1994) Genetic study of resistance to inhibitory effects of UV radiation in rice (*Oryza sativa*). Physiol Plant 91: 234–238

Schelling TC (1991) Economic responses to global warming: prospects for cooperative approaches. In: Dornbusch R, Poterba JM (eds) Global warming: economic policy responses. The MIT Press, Cambridge, MA, pp 197–221

Schlesinger WH (1992) Biogeochemistry: an analysis of global change. Academic Press, San Diego, CA

Schneider SA (1993) Scenarios of global warming. In: Kareiva PM, Kingsolver JG, Huey RB (eds) Biotic interactions and global change. Sinauer, Sunderland, MA, pp 9–23

Schütz H, Seiler W (1989) Methane flux measurements: methods and results. In: Andreae MO, Schimel DS (eds) Exchange of trace gases between terrestrial ecosystems and the atmosphere. John Wiley, Chichester, pp 209–228

Schütz H, Seiler W, Conrad R (1989a) Processes involved in formation and emission in rice paddies. Biogeochemistry 7: 33–53

Schütz H, Holzapfel-Pschorn A, Conrad R, Rennenberg H, Seiler W (1989b) A three year continuous record on the influence of daytime, season, and fertilizer treatment on methane emission rates from an Italian rice paddy field. J Geophys Res 94: 16405–16416

Schütz H, Seiler W, Rennenberg H (1990) Soil and land use related sources and sinks of methane (CH_4) in the context of the global methane budget. In: Bouwman AF (ed) Soils and the greenhouse effect. John Wiley, Chichester, pp 268–285

Schütz H, Schroder P, Rennenberg H (1991) Role of plants in regulating the methane flux to the atmosphere. In: Trace gas emission by plants. Academic Press, New York, pp 29–61

Scotto J, Cotton G, Urbach R, Berger D, Fears T (1988) Biologically effective ultraviolet radiation: surface measurements in the United States, 1974–1985. Science 239: 762–764

Seiler W (1984) Contribution of biological processes to the global budget of CH_4 in the atmosphere. In: Klug MJ, Reddy CA (eds) Current perspectives in microbial ecology. American Society for Microbiology, Washington, pp 468–477

Seiler W, Holzapfel-Pschorn A, Conrad R, Scharffe D (1984) Emission of methane from rice paddies. J Atmos Chem 1: 241–268

Seino H, Shiotsuki Y, Oya S, Hirao Y (1987) Prediction of long-distance migration of rice planthoppers to northern Kyushu considering low-level jet stream. J Agric Meteorol 43: 203–208

Semenov MA, Porter JR, Delecolle R (1993) Simulation of the effects of climate change on growth and development of wheat in the UK and France. In: Kenny GJ, Harrison PA, Parry ML (eds) The effect of climate change on agricultural and horticultural potential in Europe. Res Rep No 2, Environmental Change Unit, University of Oxford, Oxford, UK

Shangguan XJ (1993) Processes of methane production and emission from Chinese rice fields. PhD Thesis, Institute of Atmospheric Physics, Academia Sinica, Beijing, China
Shangguan XJ, Wang MX, Shen RX, Wassmann R, Toelg M, Papen H, Rennenberg H (1993a) The influence of temperature on diurnal and seasonal variation of methane emission. J Grad School Acad Sin (in press)
Shangguan XJ, Wassmann R, Wang MX, Xie XL, Toelg M, Chen DZ, Wang WD, Shen RX, Wang YS, Papen H, Rennenberg H, Seiler W (1993b) Methane emission from rice fields in Hunan Province (PR China): influence of mineral versus organic fertilizers. Biogeochemistry (in press)
Shine KP, Derwent RG, Wuebbles DJ, Moncrette JJ (1990) Radiative forcing of climate. In: Houghton JT, Callandar BA, Varney SK (eds) Climate change, the IPCC scientific assessment. Cambridge University Press, Cambridge, pp 47–68
Sinclair TR, Rawlins SL (1993) Inter-seasonal variation in soybean and maize yield under global environmental change. Agron J 85: 406–409
Sinclair TR, Diaye ON, Biggs RH (1990) Growth and yield of field-grown soybean in response to enhanced exposure to ultraviolet-B radiation. J Environ Qual 19: 478–481
Singh U, Ritchie JT (1993) Simulating the impact of climate change on crop growth and nutrient dynamics using the CERES-Rice model. J Agric Meteorol 48: 819–822
Smith CJ, Brandon M, Patrick WH Jr (1982) Nitrous oxide emission following urea-N fertilization of wetland rice. Soil Sci Plant Nutr 28: 161–171
Smith JB, Tirpak DA (eds) (1989) The potential effects of global climate change on the United States. Rep to Congr. EPA-230-05-89-050. United States Environmental Protection Agency, Washington
Smith KA, Scott A, Galle B, Klemedtsson L (1993) Use of a long-path infra-red monitor for measurement of nitrous oxide flux from soil. J Geophys Res (in press)
Smith RC, Prezelin BB, Baker KS, Bidigare RR, Boucher NP, Coley T, Karentz D, MacIntyre S, Matlick HA, Menzies D, Ondrusek M, Wan Z, Waters KJ (1992) Ozone depletion: ultraviolet radiation and phytoplankton biology in antarctic waters. Science 225: 952–959
Soil Survey Staff (1992) Key to soil taxonomy, 5th edn. SMSS Technical Monograph Number 19. Pocahontas Press, Blacksburg, VA
Sombroek WG (1990) Geographic quantification of soils and changes in their properties. In: Bouwman AF (ed) Soils and the greenhouse effect. John Wiley, New York, pp 225–244
Song YH, Heong KL (1993) Use of geographical information systems in analyzing large area distribution and dispersal of rice insects in South Korea. Korean J Appl Entomol 32: 307–316
Steele LP, Dlugokencky EJ, Lang PM, Tans PP, Martin RC, Masarie KA (1992) Slowing down of the global accumulation of atmospheric methane during the 1980's. Nature 358: 313–316
Steffen W, Walker B, Ingram J, Koch G (1992) Global change and terrestrial ecosystems: the operational plan. IGBP Report No 21. Royal Swedish Academy of Sciences, Stokholm
Steinmetz V, Wellmann E (1986) The role of solar UV-B in growth regulation of cress (*Lepidium sativum* L.) seedlings. Photochem Photobiol 43: 189–193
Steinmuller D, Tevini M (1986) UV-B-induced effects upon cuticular waxes of cucumber, bean, and barley leaves. In: Worrest RC, Caldwell MM (eds) Stratospheric ozone reduction, solar ultraviolet radiation and plant life. Springer, Berlin Heidelberg New York, pp 261–269
Stephanon G, Alahiotis SN (1983) Non-Mendelian inheritance of heat-sensitivity in *Drosophila melanogaster*. Genetics 103: 93–107
Stolarski RS (1988) The antarctic zone hole. Sci Am 258: 189–193
Stone P (1992) Forecast cloudy: the limits of global warming models. Technol Rev (February/March): 32–40
Strain BR (1987) Direct effects of increasing atmospheric CO_2 on plants and ecosystems. Tree 2: 18–21

Strain BR, Cure JD (eds) (1985) Direct effects of increasing carbon dioxide on vegetation. DOE/ER-0238, United States Department of Energy, Office of Energy Research, Washington
Strid A, Porra RJ (1992) Alteration in pigment content in leaves of *Pisum sativum* after exposure to supplemental UV-B. Plant Cell Physiol 33: 1015–1023
Suge H, Sato T, Kumagai T (1991) UV-B injury in rice plants: a genetic study. Jpn J Genet 66: 347–361
Tabbal D, Lampayan R, Bhuiyan SI (1992) Water saving irrigation technique for rice. In: Murty VVN, Koga K (eds) Proc of the international workshop on soil and water engineering for paddy field management. Asian Institute of Technology, Bangkok, Thailand, pp 146–159
Tajima K (1966) Raising rice seedlings by CO_2 enrichment. Nogyo-Gijyutsu 21: 282–284 (in Japanese)
Takai Y (1970) The mechanism of methane fermentation in flooded soils. Soil Sci Plant Nutr 16: 238–244
Takai Y, Wada E (1990) Methane formation in waterlogged paddy soils and its controlling factors. In: Scharpenseel HW, Schomaker M, Ayoub A (eds) Soils in a warmer earth. Elsevier, Amsterdam, pp 101–107
Takai Y, Koyama T, Kamura T (1956) Microbial metabolism in reduction process of paddy soil. Part I. Soil Plant Food 2(2): 63–66
Takeuchi Y, Akizuki M, Shimizu H, Kondo N, Sugahara K (1989) Effect of UV-B (290–320 nm) irradiation on growth and metabolism of cucumber cotyledons. Physiol Plant 76: 425–430
Takeuchi Y, Ikeda S, Kasahara H (1993) Dependence on wavelength and temperature of growth inhibition induced by UV-B irradiation. Plant Cell Physiol 34: 913–917
Tanaka A (1984) Introduction. In: Tanaka A (ed) Acid soils and their agricultural use, Hakuyu-shya, Tokyo, pp 7–20 (in Japanese)
Tashiro T, Wardlaw IF (1989) A comparison of the effect of high temperature on grain development in wheat and rice. Ann Bot 64: 59–65
Taylor J (1993) The mutable carbon sink. Nature 366: 515–516
Teramura AH (1983) Effects of ultraviolet-B radiation on the growth and yield of crop plants. Physiol Plant 58: 415–427
Teramura AH, Murali NS (1986) Intraspecific differences in growth and yield of soybean exposed to ultraviolet-B radiation under greenhouse and field conditions. Environ Exp Bot 26: 89–95
Teramura AH, Biggs RH, Kossuth S (1980) Effects of ultraviolet-B irradiances on soybean. II. Interaction between ultraviolet-B and photosynthetically active radiation on net photosynthesis, dark respiration, and transpiration. Plant Physiol 65: 483–488
Teramura AH, Sullivan JH, Lydon J (1990) Effects of UV-B radiation on soybean yield and seed quality: a 6-year field study. Physiol Plant 80: 5–11
Teramura AH, Ziska LH, Sztein AE (1991) Changes in growth and photosynthetic capacity of rice with increased UV-B radiation. Physiol Plant 83: 373–380
Tevini M, Teramura AH (1989) UV-B effects on terrestrial plants. Photochem Photobiol 50: 479–487
Thimijan RW, Carns HR, Campbell LE (1978) Radiation sources and related environmental control for biological and climatic effects UV research (BACER). Final Rep, United States Environmental Protection Agency, Washington
Thomas RB, Strain BR (1991) Source-sink balance as a factor in photosynthetic acclimation of cotton seedlings grown in elevated carbon dioxide. Plant Physiol 96: 627–634
Thompson AM, Cicerone RJ (1986) Possible perturbations to atmospheric CO, CH_4 and OH. J Geophys Res 9(D): 10858–10864
Tietje O, Tapkenhinrichs M (1993) Evaluation of pedotransfer functions. Soil Sci Soc Am J 57: 1088–1095
Toelg M, Shangguan XJ, Chen DZ, Wang MX, Wassmann R, Xie XL, Wang WD, Shen RX, Wang YS, Papen H, Rennenberg H, Seiler W (1994) Methane emission from rice fields in Hunan Province (PR China): temporal variation of methane emission. Biogeochemistry (in press)

Topp E (1993) Effects of selected agrochemicals on methane oxidation by an organic agricultural soil. Can J Soil Sci 73: 287–291
Trabalka JR, Reichle DE (eds) (1986) The changing carbon cycle: a global analysis. Springer, Berlin Heidelberg, New York
Trabalka JR, Edmonds JA, Reilly JM, Gardner RH, Voorhees LD (1985) Atmospheric carbon dioxide and the global carbon cycle. DOE/ER-0239, National Technical Information Service, United States Department of the Environment, Washington
Trabalka JR, Edmonds JA, Reilly JM, Gardner RH, Reichle DE (1986) Atmospheric CO_2 projections with globally averaged carbon cycle models. In: Trabalka JR, Reichle DE (eds) The changing carbon cycle: a global analysis. Springer, Berlin Heidelberg New York, pp 534–560
Travis J, Futuyma DJ (1993) Global change: lessons from and for evolutionary biology. In: Kareiva PM, Kingsolver JG, Huey RB (eds) Biotic interactions and global change. Sinauer, Sunderland, MA, pp 251–263
Tyrrell RM (1986) Repair of genetic damage induced by UV-B (290–320 nm) radiation. In: Worrest RC, Caldwell MM (eds) Stratospheric ozone reduction, solar ultraviolet radiation and plant life. Springer, Berlin Heidelberg New York, pp 139–149
Uchijima T (1976) Some aspects of relation between low air temperature and sterile spikeletes in rice plants. J Agric Meteorol 31: 199–202
US DOE–United States Department of Energy (1989) Carbon dioxide and climate: summaries of research in fiscal year 1989. DOE/ER-0425, Washington
Valent S, Chumley FG (1991) Molecular genetic analysis of the rice blast fungus. Annu Rev Phytopathol 29: 443–467
Van Breemen N, Feijtel TCJ (1990) Soil processes and properties involved in the production of greenhouse gases, with special relevance to soil taxonomic systems. In: Bouwman AF (ed) Soils and the greenhouse effect. John Wiley, New York, pp 195–223
Van Cleemput O, Ramon H, Vermoessen A (1991) Emission of C_1-C_3 hydrocarbons from soils. In: Proc of EUROTRAC Symposium 1990. The Hague, Netherlands, pp 189–191
Van de Gon HAC, Neue H-U, Lantin RS, Wassmann R, Alberto MCR, Aduna JB, Tan MJP (1992) Controlling factors of methane emission from rice fields. In: Batjes NH, Bridges EM (eds) World inventory of soil emission potentials. WISE Report 2, International Soil Reference and Information Centre (ISRIC), Wageningen, Netherlands, pp 81–92
Van Diepen CA, Wolf J, van Keulen H, Rappoldt C (1989) WOFOST: a simulation model of crop production. Soil Use Manage 5: 16–24
Van Engelen VWP, Wen T-T (1993) Global and national soil and terrain digital databases (SOTER): procedures manual. FAO-ISSS-UNEP-ISRIC, International Soil Reference and Information Centre (ISRIC), Wageningen, Netherlands
Van Waveren EJ, Bos AB (1988) ISRIC soil information system: user manual and technical manual (revised edition). Technical Paper 15, International Soil Reference and Information Centre (ISRIC), Wageningen, Netherlands
Vereecken H, Diels J, van Orshoven J, Feyen J, Bouma J (1992) Functional evaluation of pedotransfer functions for the estimation of soil hydraulic properties. Soil Sci Soc Am J 56: 1371–1378
Vergara BS (1980) Rice plant growth and development. In: Luh BS (ed.) Rice production and utilization. Avi Publishing, Westport, CT, pp 75–86
Vermoesen A, Ramon H, van Cleemput O (1991) Composition of the soil gas phase. Permanent gases and hydrocarbons. Pedology 41: 119–132
Virmani SS, Khush G, Pingali P (1993) Hybrid rice for the tropics: potential, research priorities and policy issues. Regional Expert Consultation on Hybrid Seed Production, Development and Security of Major Cereal Crops, Bangkok, Thailand
Vogels GD, Keltjens JT, Van Der Drift C (1988) Biochemistry of methane production. In: Zehnder AJB (ed) Biology of anaerobic microorganisms. John Wiley, New York, pp 707–770
Von Caemmerer SM, Farquhar GD (1981) Some relationship between the biochemistry of photosynthesis and the gas exchange of leaves. Planta 153: 367–387

Wagatsuma T, Jujo K, Tawaraya K, Sato T, Ueki A (1992) Decrease of methane concentration and increase of nitrogen gas concentration in the rhizosphere by hygrophytes. Soil Sci Plant Nutr 38: 467–476

Walker BH (1994) Landscape to regional-scale: responses of terrestrial ecosystems to global change. AMBIO 24: 67–73

Wang MX, Khalil MAK, Rasmussen RA (1987) Flux measurement of CH_4 emission from rice fields and biogas generator leakages. Chin Sci Bull 21: 1646–1649

Wang MX, Dai AG, Shen RX, Schütz H, Renennberg H, Seiler W, Wu HB (1990) Methane emission from a Chinese rice paddy field. Acta Meteorol Sin 4: 265–275

Wang MX, Dai AG, Huang J, Ren LX, Shen RX, Schütz H, Seiler W, Rasmussen RA, Khalil MAK (1993a) Methane source from China. Sci Atmos Sin 17(2): 52–64

Wang MX, Shangguan XJ, Shen RX, Wassmann R, Seiler W (1993b) Methane production, emission and possible control measures in the rice agriculture. Adv Atmos Sci 10(3): 307–314

Wang Z (1986) Rice-based systems in subtropical China. In: Juo ASR, Lowe JA (eds) Wetlands and rice in Subsaharan Africa. International Institute of Tropical Agriculture (IITA), Ibadan, Nigeria, pp 195–206

Wang Z, Delaune RD, Lindau CW, Patrick WH Jr (1992) Methane production from anaerobic soil amended with rice straw and nitrogen fertilizers. Fert Res 33: 115–121

Wang Z, Lindau CW, Delaune RD, Patrick WH Jr (1993a) Methane emission and entrapment in flooded rice soils as affected by soil properties. Biol Fert Soils 16: 163–168

Wang Z, Delaune RD, Masscheleyn PH, Patrick WH Jr (1993b) Soil redox and pH effects on methane production in a flooded rice soil. Soil Sci Soc Am J 57: 382–385

Wang Z, Crozier CR, Patrick WH Jr (1995) Methane emission in a flooded rice soil with and without algae. In: Lal R et al. (eds) Soil and global change Lewis Publishers, Chelsea, MI (In press)

Warner CW, Caldwell MM (1983) Influence of photon flux density in the 400–700 nm waveband on inhibition of photosynthesis by ultraviolet-B (280–320 nm) irradiation in soybean leaves: separation of indirect and immediate effects. Photochem Photobiol 38: 341–346

Washington WM, Meehl GA (1984) Seasonal cycle experiment on the climatic sensitivity due to a doubling of CO_2 with an atmospheric general circulation model coupled to a simple mixed-layer ocean model. J Geophys Res 89: 9475–9503

Wassmann R, Schütz H, Papen H, Rennenberg H, Seiler W, Dai AG, Shen RX, Shangguan XJ, Wang MX (1993a) Quantification of methane emissions from Chinese rice fields (Zhejiang Province) as influenced by fertilizer treatment. In: Ghazi S (ed) Contribution to global change: biosphere-atmosphere interactions. Commission of the European Communities, Luxemburg, pp 71–75

Wassmann R, Papen H, Rennenberg H, Seiler W, Wang MX, Shangguan XJ (1993b) Adequate fertilizer application as an essential element in intercomparative studies on methane emission from rice fields. ECE/PRC Workshop on Contribution to Global Change, Beijing, China

Wassmann R, Wang MX, Shangguan XJ, Xie XL, Shen RX, Wang YS, Papen H, Rennenberg H, Seiler W (1993c) First records of a field experiment on fertilizer effects on methane emission from rice fields in Hunan Province (PR China). Geophys Res Lett 20: 2071–2074

Wassmann R, Papen H, Rennenberg H (1993d) Methane emission from rice paddies and possible mitigation strategies. Chemosphere 26(1–4): 201–217

Wassmann R, Neue H-U, Lantin RS, Aduna JB, Alberto MCR, Flores MJF, Tan MJP, Denier Van der Gon HAC, Hoffmann H, Papen H, Rennenberg H, Seiler W (1994) Temporal patterns of methane emissions from wetland rice fields treated by different modes of N-application. J Geophys Res 99: 16457–16462

Wassmann R, Shangguan XJ, Toelg M, Chen DZ, Wang MX, Papen H, Rennenberg H, Seilelr W (1995) The effect of split and blend application of mineral and organic

fertilizers on methane emission from Chinese rice fields. Biology and Fertility of Soils (Submitted)
Watanabe A, Kimura M (1995) Methane production and its fate in ricefields. VIII. Seasonal variation of methane retained in soil. Soil Sci Plant Nutr 41: 225–233
Watanabe T, Seino H, Kitamura C, Hirai Y (1988) Computer program for forecasting the overseas immigration of long-distance migratory rice planthoppers. Jpn J Appl Entomol Zool 32: 52–55 (in Japanese with English summary)
Watanabe A, Katoh K, Kimura M (1993) Effect of rice straw application on CH_4 emission from rice fields. I. Effect of weathering of rice straw in the field during off-crop season. Soil Sci Plant Nutr 39: 701–706
Watanabe A, Murase J, Katoh K, Kimura M (1994a) Methane production and its fate in rice fields. V. Fate of methane remaining in rice soil at harvesting stage. Soil Sci Plant Nutr 40: 221–230
Watanabe A, Katoh K, Kimura M (1994b) Effect of rice straw application on CH_4 emission from rice fields. III. Effect of incorporation site of rice straw on CH_4 emission rates and their variation among shoots of a rice plant. Soil Sci Plant Nutr 40: 497–504
Watanabe A, Kimura M, Kasuya M, Kotake M, Katoh K (1994c) Methane in groundwater used for Japanese agriculture: its relationship to other physico-chemical properties and possible tropospheric source strength. Geophys Res Lett. 21: 41–44
Watson RT, Rohde H, Oeschger H, Siegenthaler U (1990) Greenhouse gases and aerosols. In: Houghton JT, Callandar BA, Varney SK (eds) Climate change: the IPCC scientific assessment. Cambridge University Press, Cambridge, pp 7–40
Watson RT, Meiro Fillho LG, Sanhueza E, Janetos A, Fraser PJ, Fung IJ, Khalil MAK, Manning M, McCulloch A, Mitra AP, Moore B, Rohde H, Schimel D, Siegenthaler U, Skole D, Stolarski RS (1992) Greenhouse gases: sources and sinks. In: Houghton JT, Callender BA, Varney SK (eds) Climate change 1992: the supplementary report to the IPCC scientific assessment. Cambridge University Press, Cambridge, pp 1–46
Webb EK, Pearman GI, Leuning R (1980) Correction of flux measurements for density effects due to heat and water vapour transfer. QJR Meteorol Soc 106: 85–100
Webb RS, Rosenzweig CE (1993) Specifying land surface characteristics in general circulation models: soil profile data set and derived water-holding capacities. Global Biogeochem Cycles 7: 97–108
Weier KL, Doran JW, Power JF, Walters DT (1993) Denitrification and the dinitrogen/nitrous oxide ratio as affected by soil water, available carbon, and nitrate. Soil Sci Soc Am J 57: 66–72
Weir AH, Bragg PL, Porter JR, Rayner JH (1984) A winter wheat crop simulation model without water or nutrient limitations. J Agric Sci 102: 371–382
Wessman CA (1992) Spatial scales and global change: bridging the gap from plots to GCM grid cells. Annu Rev Ecol Syst 23: 175–200
Wester U (1992) UV-B monitoring activities in Sweden. In: UV-B monitoring workshop: a review of the science and status of measuring and monitoring programs. Science and Policy Associates, Washington, pp C279–C296
Westerman (1993) Temperature regulation of methanogenesis in wetlands. Chemosphere 26: 321–328
Whalen SC, Reeburg WS (1990) Consumption of methane by tundra soils. Nature 346: 160–162
White EB, DeBach P, Garbe MJ (1970) Artificial selection for genetic adaptation to temperature extremes in *Aphytis lingnanensis* (Hymenoptera: Aphelinidae). Hilgardia 40: 161–192
Wilkerson GG, Jones JW, Boote KJ, Ingram KT (1983) Modeling soybean growth for management. Trans ASAE 26: 63–73
Williams JR, Jones CA, Dyke PT (1984) A modeling approach to determining the relationship between erosion and soil productivity. Trans ASAE 27: 129–144
Wilson CA, Mitchell JFB (1987) A doubled CO_2 climate sensitivity experiment with a global climate model including a simple ocean. J Geophys Res 92: 13315–13343
Wilson JD, Shum WKN (1991) A re-examination of the integrated horizontal flux method for estimating volatilisation from circular plots. Agric For Meteorol 57: 281–295

Wilson JD, Thurtell GW, Kidd GE, Beauchamp EG (1982) Estimation of the rate of gaseous mass transfer from a surface plot to the atmosphere. Atmos Environ 16: 1861–1867

Wolf J (1993) Effects of climate change on wheat and maize production potential in the EC. In: Kenny GJ, Harrison PA, Parry ML (eds) The effect of climate change on agricultural and horticultural potential in Europe. Res Rep No 2, Environmental Change Unit, University of Oxford, Oxford, UK

Wong SC (1979) Elevated atmospheric partial pressures of CO_2 and plant growth. I. Interactions of nitrogen nutrition and photosynthetic capacity in C_3 and C_4 plants. Oecologia 44: 68–74

Woodrow IE, Berry JA (1988) Enzymatic regulation of photosynthetic CO_2 fixation in C_3 plants. Annu Rev Plant Physiol Plant Mol Biol 39: 533–594

Wulff RD, Alexander HM (1985) Intraspecific variation in the response to CO_2 enrichment in seeds and seedlings of *Plantago lanceolata* L. Oecologia 66: 458–460

Wulff RD, Strain BR (1983) Effects of CO_2 enrichment on growth and photosynthesis in *Desdomonium paniculatum*. Can J Bot 60: 1084–1091

Wyngaard JC (1991) On the maintenance and measurement of scalar fluxes. In: Schmugge TJ, Andre JC (eds) Land surface evaporation, measurement and parameterization. Springer, Berlin Heidelberg New York, pp 289–301

Xiong Yi, Li Qingkui (1987) Chinese soils (2nd edn). Scientific Publishing House, Beijing

Xu Q (1981) Cropping system in relation to fertility of paddy soils in China. In: Proc of symposium on paddy soil. Institure of Soil Science, Chinese Academy of Sciences. Science Press, Beijing and Springer, Berlin Heidelberg New York, pp 220–230

Yagi K, Minami K (1990a) Effect of organic matter application on methane emission from some Japanese rice fields. Soil Sci Plant Nutr 36: 599–610

Yagi K, Minami K (1990b) Effects of organic matter applications on methane emission from Japanese paddy fields. In: Bouwman AF (ed) Soils and the greenhouse effect. John Wiley, New York, pp 467–473

Yagi K, Minami K (1991) Emission and production of methane in the paddy fields of Japan. JARQ 25: 165–171

Yagi K, Minami K, Ogawa Y (1990) Effects of water percolation on methane emission from paddy fields, Res. Rep Div Environ Planning 6: 105–112

Yagi K, Chairoj P, Tsuruta H, Cholitkul W, Minami K (1994a) Methane emission from rice fields in the central plain of Thailand. Soil Sci Plant Nutr 40: 29–37

Yagi K, Tsuruta H, Minami K, Chairoj P, Cholitkul W (1994b) Methane emission from Japanese and Thai paddy fields In: Minami K, Mosier A, Sass R (eds) CH_4 and N_2O. Global emissions and controls from rice fields and other agricultural and industrial sources. Yoken-do, Tokyo, pp 41–53

Yelle S, Beeson RC, Trudel MJ, Gosselin A (1989a) Acclimation of two tomato species to high atmospheric CO_2. I. Sugar and starch concentrations. Plant Physiol 90: 1465–1472

Yelle S, Beeson RC, Trudel MJ, Gosselin A (1989b) Acclimation of two tomato species to high atmospheric CO_2. II. Ribulose 1–5, bisphosphate carboxylase/oxygenase and phosphoenolpyruvate carboxylase. Plant Physiol 90: 1473–1477

Yoshida S (1976) Carbon dioxide and yield of rice. In: Climate and rice. International Rice Research Institute, Los Baños, Philippines, pp 211–222

Yoshida S (1977) Rice. In: Alvim PT, Kozlowski TT (eds) Ecophysiology of tropical crops. Academic Press, London, pp 57–87

Yoshida S (1981) Fundamentals of rice crop science. International Rice Research Institute, Los Baños, Philippines

Yoshida S, Parao FT (1976) Climatic influence on yield and yield components of lowland rice in the tropics. In: Climate and rice. International Rice Research Institute, Los Baños, Philippines, pp 187–210

Yoshida S, Satake T, Mackill DJ (1981) High-temperature stress in rice. IRRI Research Paper Series No. 67. International Rice Research Institute, Los Baños, Philippines

Yoshino MM, Horie T, Seino H, Tsujii H, Uchijima T, Uchijima Z (1988) The effects of climate variations on agriculture in Japan. In: Parry ML, Carter TR, Konijn NT (eds)

The impact of climate variations on agriculture. Vol 1: Assessments in cool temperate and cold regions. Kluwer, Dordrecht, pp 725–865
Yu W, Teramura AH, Sullivan JH (1991) Model YMT UV-B modulation system: manual of operation. Final report submitted to the United States Environmental Protection Agency, Corvallis, OR
Zheng KL, Mackill DJ (1982) Effect of high temperature on anther dehiscence and pollination in rice. SABRAO J 14: 61–66
Zhou Yixing (1991) Potential influences of the greenhouse effect on grain production in China. In: Geng S, Cady CS (eds) Climate variations and change: implications for agriculture in the Pacific rim. The Public Service Research and Dissemination Program. University of California, Davis, CA, pp 147–158
Ziska LH, Bunce JA (1993) Inhibition of whole plant respiration by elevated CO_2 as modified by growth temperature. Physiol Plant 87: 459–466
Ziska LH, Drake BG, Chamberlain S (1990) Long-term photosynthetic response in single leaves of a C_3 and C_4 salt marsh species grown at elevated atmospheric CO_2 in situ. Oecologia 83: 469–472
Ziska LH, Hogan KP, Smith AP, Drake BG (1991) Growth and photosynthetic response of nine tropical species with long-term exposure to elevated carbon dioxide. Oecologia 86: 383–389
Ziska LH, Teramura AH, Sullivan JH (1992) Physiological sensitivity of plants along an elevational gradient to UV-B radiation. Am J Bot 79: 863–871
Ziska LH, Teramura AH, Sullivan JH, McCoy A (1993) Influence of ultraviolet-B (UV-B) radiation on photosynthetic and growth characteristics in field-grown cassava (*Manihot esculentum* Crantz). Plant Cell Environ 16: 73–79
Zobler L (1986) A world soil file for global climate. NASA Technical Memorandum 87802, National Aeronautics and Space Administration, Washington

Subject Index

Springer-Verlag and the Environment

We at Springer-Verlag firmly believe that an international science publisher has a special obligation to the environment, and our corporate policies consistently reflect this conviction.

We also expect our business partners – paper mills, printers, packaging manufacturers, etc. – to commit themselves to using environmentally friendly materials and production processes.

The paper in this book is made from low- or no-chlorine pulp and is acid free, in conformance with international standards for paper permanency.

MIX
Papier aus verantwortungsvollen Quellen
Paper from responsible sources
FSC® C105338

If you have any concerns about our products,
you can contact us on
ProductSafety@springernature.com

In case Publisher is established outside the EU,
the EU authorized representative is:
Springer Nature Customer Service Center GmbH
Europaplatz 3, 69115 Heidelberg, Germany

Printed by Libri Plureos GmbH
in Hamburg, Germany